Flora's Fieldworkers

Flora's Fieldworkers

WOMEN AND BOTANY IN NINETEENTH-CENTURY CANADA

EDITED BY

ANN SHTEIR

McGill-Queen's University Press
Montreal & Kingston • London • Chicago

ISBN 978-0-2280-1112-5 (cloth)
ISBN 978-0-2280-1346-4 (ePDF)

Legal deposit second quarter 2022
Bibliothèque nationale du Québec

Printed in Canada on acid-free paper that is 100% ancient forest free (100% post-consumer recycled), processed chlorine free

This book has been published with the help of a grant from the Canadian Federation for the Humanities and Social Sciences, through the Awards to Scholarly Publications Program, using funds provided by the Social Sciences and Humanities Research Council of Canada. Funding in support of publication has also been received from the Ottawa Research and Development Centre of Agriculture and Agri-Food Canada.

Funded by the Government of Canada | Financé par le gouvernement du Canada | Canada | Canada Council for the Arts | Conseil des arts du Canada

We acknowledge the support of the Canada Council for the Arts.

Nous remercions le Conseil des arts du Canada de son soutien.

Library and Archives Canada Cataloguing in Publication

Title: Flora's fieldworkers : women and botany in nineteenth-century Canada / edited by Ann Shteir.
Names: Shteir, Ann B., 1941- editor.
Description: Includes bibliographical references and index.
Identifiers: Canadiana (print) 20220133379 | Canadiana (ebook) 2022013345X | ISBN 9780228011125 (hardcover) | ISBN 9780228013464 (PDF)
Subjects: LCSH: Botany—Canada—History—19th century. | LCSH: Women in botany—Canada—History—19th century. | LCSH: Women botanists—Canada—History—19th century. | LCSH: Women botanists—Canada—Biography. | LCSH: Botanists—Canada—History—19th century. | LCSH: Botanists—Canada—Biography. | LCGFT: Biographies.
Classification: LCC QK21.C2 F56 2022 | DDC 580.820971—dc23

This book was designed and typeset by studio oneonone in Minion 11/14

Contents

Acknowledgments

Flora's Fieldworkers: Women and Botany in Nineteenth-Century Canada developed from a workshop, "Women, Men, and Plants in 19th-Century Canada: New Resources and New Perspectives," that took place at York University in October 2017. The workshop was organized by York's Centre for Feminist Research and funded by the Social Sciences and Humanities Research Council of Canada under the Connection Grant Program. Further financial support for the workshop came from the Centre for Canadian Historical Horticultural Studies, Royal Botanical Gardens (Canada), and, at York University, from the Faculty of Liberal Arts & Professional Studies, the School of Gender, Sexuality and Women's Studies, the vice-president research innovation, and the vice-president academic and provost. As principal investigator on the grant and director of the workshop, I thank the funders and also express my gratitude to York colleagues Alison Crosby, Janet Friskney, and Julia Pyreskina for their key contributions to this event.

A grant from the Aid to Scholarly Publications Program of the Federation for the Humanities and Social Sciences provided publishing assistance for the volume. The Ottawa Research and Development Centre, Agriculture and Agri-Food Canada, awarded funds toward colour reproduction of images, and special thanks are due to Christian Malouin, Stephanie Warner, and Erin Picard and especially to Jacques Cayouette for their work on this.

Flora's Fieldworkers

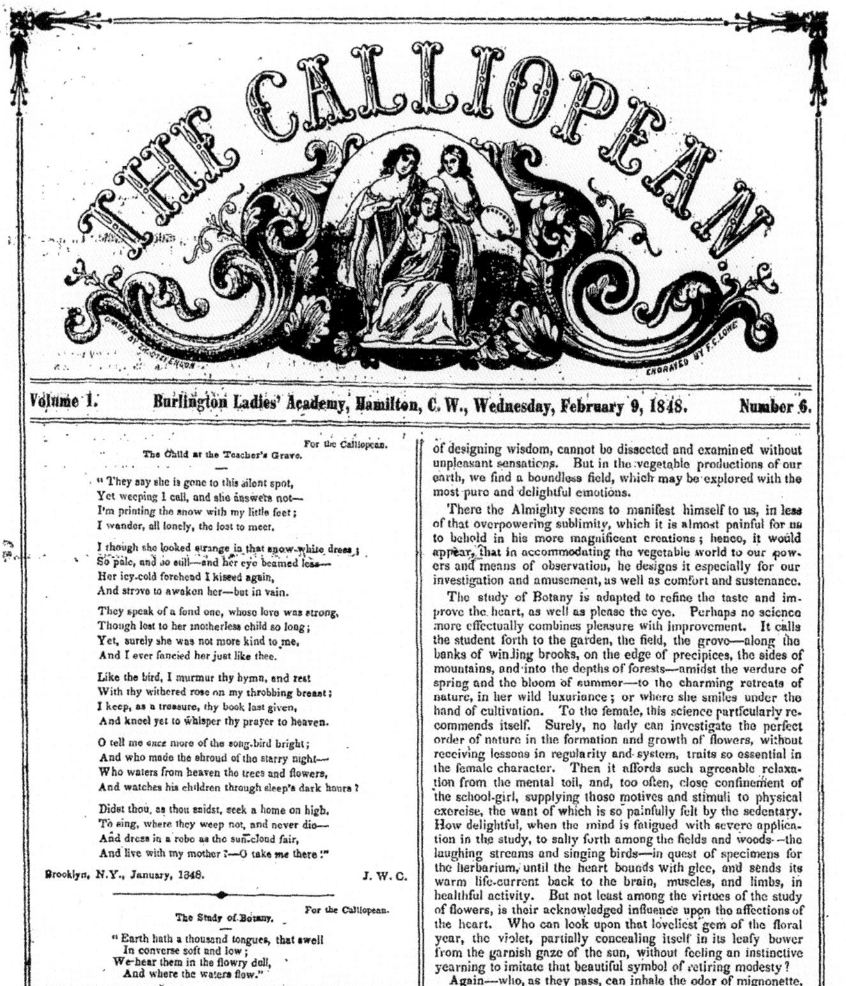

THE CALLIOPEAN

Volume 1. Burlington Ladies' Academy, Hamilton, C. W., Wednesday, February 9, 1848. Number 6.

For the Calliopean.

The Child at the Teacher's Grave.

"They say she is gone to this silent spot,
Yet weeping I call, and she answers not—
I'm printing the snow with my little feet;
I wander, all lonely, the lost to meet.

I though she looked strange in that snow-white dress;
So pale, and so still—and her eye beamed less—
Her icy-cold forehead I kissed again,
And strove to awaken her—but in vain.

They speak of a fond one, whose love was strong,
Though lost to her motherless child so long;
Yet, surely she was not more kind to me,
And I ever fancied her just like thee.

Like the bird, I murmur thy hymn, and rest
With thy withered rose on my throbbing breast;
I keep, as a treasure, thy book last given,
And kneel yet to whisper thy prayer to heaven.

O tell me *once* more of the song-bird bright;
And who made the shroud of the starry night—
Who waters from heaven the trees and flowers,
And watches his children through sleep's dark hours?

Didst thou, as thou saidst, seek a home on high,
To sing, where they weep not, and never die—
And dress in a robe as the sun-cloud fair,
And live with my mother?—O take *me* there!"

Brooklyn, N.Y., January, 1848. J. W. C.

For the Calliopean.

The Study of Botany.

"Earth hath a thousand tongues, that swell
In converse soft and low;
We hear them in the flowry dell,
And where the waters flow."

The study of nature, in any of her multitudinous forms, is highly interesting and instructive. We may soar in imagination to the vast orbs which compose the universe, and hold converse with the bright intelligences which inhabit them; but even imagination droops beneath the mighty conception of their distance and immensity.

The animal creation, though affording the most striking marks of designing wisdom, cannot be dissected and examined without unpleasant sensations. But in the vegetable productions of our earth, we find a boundless field, which may be explored with the most pure and delightful emotions.

There the Almighty seems to manifest himself to us, in less of that overpowering sublimity, which it is almost painful for us to behold in his more magnificent creations; hence, it would appear, that in accommodating the vegetable world to our powers and means of observation, he designs it especially for our investigation and amusement, as well as comfort and sustenance.

The study of Botany is adapted to refine the taste and improve the heart, as well as please the eye. Perhaps no science more effectually combines pleasure with improvement. It calls the student forth to the garden, the field, the grove—along the banks of winding brooks, on the edge of precipices, the sides of mountains, and into the depths of forests—amidst the verdure of spring and the bloom of summer—to the charming retreats of nature, in her wild luxuriance; or where she smiles under the hand of cultivation. To the female, this science particularly recommends itself. Surely, no lady can investigate the perfect order of nature in the formation and growth of flowers, without receiving lessons in regularity and system, traits so essential in the female character. Then it affords such agreeable relaxation from the mental toil, and, too often, close confinement of the school-girl, supplying those motives and stimuli to physical exercise, the want of which is so painfully felt by the sedentary. How delightful, when the mind is fatigued with severe application in the study, to sally forth among the fields and woods—the laughing streams and singing birds—in quest of specimens for the herbarium, until the heart bounds with glee, and sends its warm life-current back to the brain, muscles, and limbs, in healthful activity. But not least among the virtues of the study of flowers, is their acknowledged influence upon the affections of the heart. Who can look upon that loveliest gem of the floral year, the violet, partially concealing itself in its leafy bower from the garnish gaze of the sun, without feeling an instinctive yearning to imitate that beautiful symbol of retiring modesty?

Again—who, as they pass, can inhale the odor of mignonette, without staying to enquire whence such grateful fragrance? To his surprise, he finds it issues from yon perianth of plain and unassuming garb, neither touched with the hues, nor dressed in the ostentatious robe of the hydrangea hortensis; but so rich in precious odors, gentleness, and worth, that amid a world of gayer companions true wisdom would woo her first.

Again—we find some plants common in the tropical regions,

I.1

The Calliopean 1, no. 6 (9 February 1848), 1. The article "The Study of Botany," written by a "Young Lady" at the Burlington Ladies' Academy, Hamilton, Canada West, promotes the benefits of botany to women: "To the female, this science particularly recommends itself."

Introduction

Women and Plant Practices in Nineteenth-Century Canada beyond "the Usual Records"

ANN SHTEIR

In late 1847, a group of students at the Burlington Ladies' Academy in Hamilton, Canada West, set out to raise funds for the school library by putting together a literary periodical they named *The Calliopean*. Their overarching mandate was "to contribute … toward the intellectual and social improvement of our sex … elevate the standard of *female education* in Canada, and thus … promote domestic happiness and social virtue."[1] Two issues a month were filled with poems, essays, fiction, excerpts from writings by notable women, and biographical pieces about "Eminent Literary Ladies." Much of the content reflected ideas and debates of the day about women and their place in the world and hence about education, social roles, and gender norms. An article on "The Study of Botany," for example, written by a "Young Lady" of the school, uses language found widely in British and colonial periodicals that recommended "appropriate" activities for girls: "The study of Botany is adapted to refine the taste and improve the heart, as well as please the eye. Perhaps no science more effectually combines pleasure with improvement … To the female, this science particularly recommends itself. Surely, no lady can investigate the perfect order of nature in the formation and growth of flowers, without receiving lessons in regularity and system, traits so essential in the female character."[2] The magazine's title invokes Calliope, a goddess of eloquence and writing and one of the nine Muses in classical Greek mythology, who is figured on the masthead in the company of other female embodiments of art, music, and culture.

Plants and flowers were areas of nature study associated with girls and women in the polite culture of eighteenth- and nineteenth-century Britain and its colonial and imperial domains.[3] They brought together "pleasure" with "improvement" and both with art and gardens, as well as beliefs about moral and religious order and "the female character." *The Calliopean* article shows the imprint of these ideas in practices of natural history and science by women and for women during Canada's Victorian era.[4] But associations between women and plants are also part of a much wider frame of reference that links women to nature in the literary, visual, and philosophical traditions of many cultures.[5]

Nature in its floral aspects was embodied in Roman mythology, for example, in the personified figure of Flora as the goddess of flowers. When the poet Ovid wrote about springtime festivals in the Roman religious year, he celebrated Flora as the "mother of flowers," a joyous and sexualized figure of abundance who "scattered new seed across countless nations" through her transformative powers of love.[6] Images of women as figures of nature, fertility, and renewal developed out of Ovid's account of Flora and informed works of art across the centuries, including Sandro Botticelli's late fifteenth-century painting *Primavera.* Flora also became an icon in books about floristic, horticultural, medicinal, and botanical aspects of plants. She appears on the engraved frontispiece of the seventeenth-century gardening treatise *Flora, seu, de florum cultura* by horticulturalist John Rea as a sturdy monarch, an icon of power, who rules over all the plants in her domains, attended by Ceres, the goddess of grain, and Pomona, the goddess of fruit. Images of Flora as Queen of Flowers continued in popular books about plants, flowers, and botany well into the nineteenth century, but "Flora" also took on a new shape during the eighteenth century as the name of a genre of scientific writing about plants; William Hudson's *Flora Anglica* (1762), the standard book of British botany for decades, is one example of this.[7] As the term is used now, a Flora is a botany book in an inventory tradition that lists and describes plants specific to an area and identifies their botanical characteristics.

Five decades after *The Calliopean* wrote about botanical education for girls in relation to beliefs about "the female character," David Penhallow, professor of botany at McGill University, published an article about "the history of botanical progress" in the *Transactions of the Royal Society of Canada* in 1897 that was similarly shaped by beliefs about who should study this field and why. He reflects in his "Review of Canadian Botany from 1800 to 1895" on changing botanical orientations across the century, notably, the shift from "mere collecting" to work in plant physiology, paleobotany, and geographical distribution. Penhallow cited nearly 500 "landmark" monographs and journal articles

I.2
"Flora, Ceres, and Pomona," frontispiece, John Rea, *Flora, seu, de florum cultura*. London, 1665.

on Canadian botanical topics published in English and French and celebrated "authorities" who, in his words, "advanced knowledge" "from consciousness of a duty nobly conceived."[8]

Penhallow's vocabulary from within his late nineteenth-century Canadian scientific community recalls John Lindley in early nineteenth-century England whose teaching and writing played a key part in shaping botany as a "modern" field of study. Lindley advocated for botany as a discipline that studied plant structure and systematics based in Continental theories, and firmly rejected ideas of botanist Carl Linnaeus about plant identification and classification that had been widely adopted for popular botanical study from the mid-eighteenth century onward. Lindley, the first professor of botany at the newly formed London University, spoke about botany in his inaugural lecture in 1829 as "an occupation for the serious thoughts of man," worthy of the attention of "men of enlightened mind." He demarcated the "science" of botany from genteel forms of plant practices that he tagged as "an amusement for ladies."[9] Seeking to transform his field, he worked to create his own ideal community of botanical practice and practitioners. Lindley did not ignore women as an audience for his new-style topics, but his *Ladies' Botany* (1834–37), a "Familiar Introduction to the Study of the Natural System of Botany," set plant study solely as part of women's domestic and familial responsibilities in a Victorian separate sphere. This early nineteenth-century bifurcation between botany as a "science" and plant activities as "an amusement for ladies" puts gender firmly onto the agenda for histories of botany and for histories of gender and science more generally. Cultural associations between women and nature and women and plants had eased the way for some women into popular plant activities during the eighteenth and nineteenth centuries. But the imprint of women in nineteenth-century botanical culture underscored the importance for some botanists of separating out their preferred methods and emphases from literary and artistic activities that were particularly associated with women and women's lives at that time.

Gender, as Erika Milam and Robert Nye wrote in their introduction to *Scientific Masculinities* (2015), is "a powerful analytical tool with which to rethink science as a fundamentally gendered activity, whether or not women are present."[10] Milam and Nye direct their gender lens to topics about both men and masculinity and showcase scholarship about, for example, scientific labour in gendered spaces, processes of professionalization, and ideas about manliness in different areas of science. How, they ask, do "masculine" cultures in science express themselves at historical moments and in specific settings? Forthright in their diagnosis of the effects of such practices, they write that "as women eventually sought entry to scientific workplaces in positions of equality, the masculine cultures of the lab, the classroom, and the field were mobilized to screen out or segregate women from full participation … The result has been long-standing masculine traditions in science that are at once differentiated, flexible, nuanced, and incredibly persistent."[11] Gendered markings such as those

Milam and Nye identify are as evident in Penhallow's late-century narrative of "botanical progress" in Canada as they were at midcentury in *The Calliopean* and earlier still in Lindley's assessment of botanical culture in England during the 1820s. Each of these publications shows a different type of botany and a different way of presenting it as an area of knowledge. Each also shows cultural practices in action that shape access to learning and channel norms of inclusion and exclusion. Given gendered directions in science across the nineteenth century, it is not surprising that only one woman – botanical artist Maria Morris Miller – appears in Penhallow's Royal Society article and this because much-cited botanist George Lawson prepared the texts to accompany the drawings in editions of her *Wild Flowers of Nova Scotia.*[12]

Penhallow's research for his "Review of Canadian Botany from 1800 to 1895" built not only on primary publications by "authorities" but also on biographical information found in secondary sources. Penhallow was a community builder in his field,[13] and in this he resembles British botanist William Jackson Hooker who brought people and information together in a similar way for his inventory project *Flora Boreali-Americana: The Botany of the Northern Parts of British North America* (1829–40). Hooker's publication, described by Penhallow as "a complete summary of our knowledge of the Canadian flora" at that time,[14] includes the names of many naturalists, military officers, and others who collected and sent him plants. Commenting on Hooker's workforce, Penhallow wrote that "[i]t seems probable … that owing to their work being wholly confined to collecting … their names have gradually fallen into neglect, and the part they played in the advancement of Canadian botany – important if obscure – cannot now be ascertained." Furthermore, "it is more than probable that there were many others, of whom all trace has been completely lost. A careful search of the usual records fails to disclose any information concerning any of these botanists who were evidently well known to Sir William Hooker."[15]

Flora's Fieldworkers: Women and Botany in Nineteenth-Century Canada takes up the generosity in Penhallow's remarks about contributors whose names and work do not appear in the "usual" documents of record and who as a result have been largely erased from botanical history. It gives visibility – new, restored, or freshened – to women who were active in Flora's domains in Canada (and two imperial locations). Looking beyond importance to the history of "botanical science" based on the "published writings of key individuals" (to use Penhallow's phrasings), this volume foregrounds women in nineteenth-century Canada (and Australia) who were "mere" plant collectors as well as botanical correspondents, networkers, teachers, artists, craftworkers, and botanical and horticultural writers. Choosing plant-related activities for their own reasons and purposes, they contributed to numerous colonial, imperial, Canadian, and Australian projects. Their engagement enlarges understanding of communities of botanical and scientific practice altogether. The amplitude of their plant work provides insights for topics of current interest in women's and feminist history, histories of Canadian and imperial botany, natural history, and science, and the field of women, gender, and science.

In October 2017, a workshop on "Women, Men, and Plants in Nineteenth-Century Canada: New Resources, New Perspectives" convened researchers from universities across Canada and from Scotland and Australia, who were joined by botanists working in botanical gardens and for the government of Canada and archivists from the Archives of Ontario. Participants came from academic backgrounds in history, art history, landscape studies, literary studies, feminist studies, botany and biology, the history of education, and the history of science. They reported on research related to the historical study of plants and cultural contexts to plant work and brought interdisciplinary perspectives into the discussions.

Seeking fresh perspectives on material by and about women and men in nineteenth-century Canada who were involved in plant study, we queried what counted as knowledge of plants across those decades of cultural, social, and political change. Who shaped and disseminated this knowledge? How did access come about for those who took part in plant-related activities and what roles did families, networks, and emergent or existing institutions play? Toward learning more about individuals who took part in plant study, researchers were interested in unstudied materials as well as in older materials to be approached with fresh questions. What resources offer ways to find and study women who collected plants, contributed to floristic projects, wrote about plants, drew and painted plants, compiled botanical albums, sought membership in botanical societies, and included botany in school curricula, and who were active on behalf of plants in other ways both public and private? What did these numerous forms of engagement with plants mean to the individuals who took part in them? How does the work look when seen from their perspectives and also when juxtaposed to plant-related activities elsewhere? The comparative dimension brought to our workshop by a historian of science in nineteenth-century Australia led to discussion about possible differences in settings across the British Empire.

The volume assembled here presents research from the 2017 workshop, augmented by one contribution invited afterward. The material clusters in British colonial and settler locations in Canada, with one chapter on colonial Australia. The Canadian material ranges geographically from Halifax, Nova Scotia and St John's, Newfoundland westward to the city of Quebec and Montreal, across changes in name from "Upper Canada" (1791–1841) to "Canada West" and "Canada East" (1847–67) and "Ontario" from the time of Confederation, and over to Victoria, British Columbia. As in many other colonial and imperial nineteenth-century places, histories of botanical work in nineteenth-century Canada were shaped by disciplinary formations of botanical science, institutional accounts of universities and professional societies, and connections to British metropolitan and imperial science.

This volume reports on botanical, floristic, and horticultural activities by women in nineteenth-century English-speaking Canada (and Australia) and affirms their historical presence. Some of the women were well known at the time but others less so and subsequently largely erased for us. They came from British and British settler-colonial backgrounds, and their activities gave voice to languages of nature that were part of nineteenth-century vocabularies. Some of the plants featured here are botanical specimens, collected, named, pressed into albums, and shipped for botanical use by others. Some were labelled "rare" because they were unfamiliar in the eyes of newcomers, and others flourished back then but are "rare" now. Some are "native" plants, and others were cultivated as a way to bring "home" into a new landscape. Plants discussed here were represented through art work and craft work, and written about in books and articles for popular, educational, scientific, and commercial purposes.

Looking beyond activities mapped in this book, a broader canvas of plant work about women in nineteenth-century Canada would also include topics in farming, food, and medicine such as those featured in recent publications by Sarah Carter, Nancy Turner and Katherine Turner, and Kristin Burnett.[16] It would reach beyond British and settler-colonial communities of practice and include francophone women and plant study in nineteenth-century Quebec and other areas, women in Black Nova Scotia communities, nineteenth-century immigrant and ethnic communities, and diasporic communities from Africa and the Caribbean. Above all, it would bring Indigenous women into a more inclusive history of women and plant work in nineteenth-century Canada still to be written. Aware of this need, contributors in a number of chapters in this volume refer to Indigenous history and artistic traditions and to the interplay between settlers and their Indigenous neighbours, and they point out evidence found in their research of how Indigenous knowledge was supplanted and erased. A chapter from the history of material artifacts, for example, finds appropriation of Indigenous beadwork and symbolism in the embroidery style on a colonial quilt in Canada. By contrast, the chapter about colonial Australia shows settler women learning about plants from Indigenous Peoples and, through botany, shaping a notion of identity beyond imperial Britain.[17]

Key terms used in analysis of science and nineteenth-century culture that came under scrutiny during the workshop reverberate across chapters in this volume. What kinds of plant practices did "botany" denote in nineteenth-century Canada and elsewhere? Who was or was not a "botanist," by self-description or as designated by others? How and in what ways was a "botanist" different from terms such as "botanizer," "botanophile," or "natural historian" then in use? These terms, in flux at that time, generate questions about hierarchies of value and boundary-keeping that fold into the social history of science and the historiography of modern science more generally. The term "amateur" as self-description or as designated by others echoes in the same ways. Should "amateur" be set aside as historically disparaging to those not fitting normative profiles of expertise or class or gender? Or, should "amateur" be reclaimed as a way to show a broader citizenry of interest in nature and science in both past and present? Questions like these about amateurs in the history of science have been a focus in recent scholarship that, challenged by feminist scholarship and social and cultural historians, integrates work by women and other communities into a fuller picture of practices and contributions.[18] Might the emotive root meaning of "love" in "amateur" now also be reclaimed in historical research and re-appropriated in contemporary practices in light of interest in the role of emotions in science and scientists? A question posed in a recent collection about the history of science and the emotions is pertinent in this regard: "How does the historian identify emotions in a culture that deliberately and methodologically excluded emotions from its texts, narratives, practices, and practitioners, [and] separated the factual-scientific from the personal-experiential?"[19]

RECOVERING WOMEN AND BOTANY IN NINETEENTH-CENTURY CANADA

By the turn into the twenty-first century, a full generation of scholarship had identified issues of women, gender, nature, and knowledge in many areas of history, culture, and society. Historian of technology Ruby Heap wrote about this in her 2006 introduction to a special issue of the journal *Scientia Canadensis* on "Women and Gender in Canadian Science, Engineering, and Medicine."

Heap (herself a discussant at our workshop) reviewed many themes that retain their urgency in feminist scholarship on women and science, including "women's exclusion, marginalization and subordination; the lack of recognition for the work accomplished; the resilience of women and the variety of strategies they developed to persist and challenge the stereotypes about their ability to succeed; [and] the intersection of private and public activities."[20] Heap brought together scholars in feminist studies with those working on science and technology studies in Canada so as to analyze "socially defined roles of women in science and … the gendered norms within the culture of science." Amber Lloydlangston's research, for example, studied sex-specific work by women for Canada's federal Department of Agriculture during the years 1887 to 1919. She found that women initially had opportunities there as correspondents and plant collectors in unpaid work but later were offered only gender-segregated and low-waged jobs as lab technicians testing agricultural seed. This was, she writes, "repetitive, routine, and fine work, [necessitating] the patience, tolerance of boredom, docility, and manual dexterity supposedly 'natural' to women."[21] Through a close look at one government institution Lloydlangston detailed an organizational and gendered hierarchy taking shape for women.

For the most part, research on women in Canadian science has focused on work by women in professional science since the late nineteenth century and also on English Canada.[22] Marianne Gosztonyi Ainley provided foundational mapping on this topic in 1990 in the path-breaking collection *Despite the Odds: Essays on Canadian Women and Science*. Her edited volume brought historical and biographical studies together with contemporary concerns and showcased women scientists whose "hidden contributions to science and society as researchers, teachers, and editors of scientific journals have so far been minimized or neglected."[23] Ainley's own chapter traced work by women natural scientists in nineteenth-century Canadian botany and zoology from the "pre-professional period" into decades when training and careers for women developed within Canada's scientific community. Like other researchers on women and science who have catalogued and sought to demonstrate "contributions" by women to scientific knowledge,[24] Ainley was intent on showing the seriousness of women's engagement in their activities. Her publication "Science in Canada's Backwoods" (1997) made vigorous claims for re-evaluating botanical work by Catharine Parr Traill during the nineteenth century that historians and others might categorize and then dismiss as "amateur." Setting out the fieldwork, botanical study, and environmental experience that underlay

Traill's investigations and ecological concerns, Ainley argued that Traill's body of writing, when taken as a whole, "provides a postcolonial alternative to earlier western scientific texts."[25] She particularly highlighted Traill's attention to plant knowledge from Indigenous communities.

The historical lens and critical feminist perspectives in Marianne Ainley's work prepared the ground for research on other women in nineteenth-century Canadian science and nature study. Kerrie Kennedy's article "Science Culture in the Nineteenth Century" reconstructs research by two women in the early 1860s who studied the culture of silkworms and reported their findings in papers for the Botanical Society of Canada, a society newly established to promote "scientific" botany.[26] One woman, "Mrs Lawson," was interested in research that would enhance the production of silk as fabric and thereby provide employment for women; the other woman, "Miss Gildersleeve," had been conducting experiments for two years to assess which vegetation for silkworms would yield the best cocoons.[27] Lawson's paper led to a response from Gildersleeve that challenged Lawson's work, and this led in turn to Lawson's firm rejoinder. Kennedy, looking at the back-and-forth of their language, tone, and arguments against a backdrop of gendered practices and power dynamics, found "attempts to negotiate a space within the changing persona of botany as a scientific pursuit while trying to balance the social expectations of suitable behavior for respectable women." Lawson, for example, initially positioned her work on silkworms as a topic "for ladies," perhaps as the way to accommodate her own research interests in gender-appropriate ways. Later, however, putting tropes of femininity aside, she claimed scientific authority for her work on various grounds and aligned Gildersleeve with "amateur" enthusiasms instead.[28]

Kennedy had access to Lawson and Gildersleeve's research from the early 1860s because their papers were reprinted in Kingston's local newspaper, *The Daily News*. The Botanical Society of Canada, based at Queen's University, had been founded mainly to enhance the professional standing of its elite and academic male members, but it also cultivated polite learning and admitted women as "Lady Members." For this reason, publishing some Society papers as reports in the local newspaper may have been intended as outreach to general, non-academic, and female readers in Kingston. Yet, reporting on the exchange between Lawson and Gildersleeve in a local newspaper rather than in the formal *Annals* of the Botanical Society of Canada places their work at the margins of scientific culture at the very time when boundaries were being forged.

The publications by Lloydlangston, Ainley, and Kennedy about women in nineteenth-century Canadian botany are examples of work that current historians have identified as a priority for future research on women in Canada. In the recent edited volume *Reading Canadian Women's and Gender History* (2019), Nancy Janovicek and Carmen Nielson do a stock-taking of their field and identify "ongoing silences and absences" in how Canadian women's and gender history has been written. They call for "recovery history" as "an ongoing political project"[29] and list research areas that include, but are by no means limited to, Indigenous women, women activists, migrants, transgender people, and racialized women. Contributors to their volume explore gender, ethnicities, social class, racializations, and other intersecting practices that shape lives. They reflect, furthermore, on cultural meanings of gender and how these were naturalized, enforced, navigated, and resisted in specific times and places. Historians have called as well for research into pre-Confederation Canada that would explore gendered dimensions of life for women and for men in that patriarchal colonial society,[30] and Katherine McKenna's chapter "Class, Race, and Gender Roles in Early British North America" is one contribution to this project. McKenna searched a wide variety of resources for evidence of the gendered range of opportunities that were available to elite women, farm women, and Indigenous and African Canadian women in the English-speaking colonies up to 1850 and assessed the opportunities that "bush masculinity" and "gentry masculinity" offered to men as well. "English Canada," she wrote "became a place where white middle-class men could exercise their independence … [but] for women, the roles they could play and the spaces they could occupy became increasingly restricted."[31]

Janovicek and Nielson's historiographical survey of Canadian women's and gender history addresses silences and absences and also leads to questions about source materials and how to use them. What counts as a "credible source" for new scholarship that can enlarge the history of specific groups and historical understanding more generally? Why are there hierarchies of knowledge within history as a discipline such that public history, for one, is ranked as inferior to academic work? How can feminist historians bring in voices from diverse communities and in locations such as museums that are outside the academy?[32] These concerns, echoing of course well beyond Canada, can inform research on women in nineteenth-century plant study as well.

In *Flora's Fieldworkers*, archival sources, newly available manuscripts, herbaria specimens, data bases, and artifacts from material culture are among resources from the nineteenth century through which to address silences and absences about women who pursued botanical interests and a multitude of plant practices in Canada. Botanical art, albums of wildflower drawings, scrapbooks, and ephemera appear here as well, along with catalogues from plant nurseries, popular educational and pedagogical materials, newspapers, and other print and visual materials that have been unavailable, unexplored, or less explored thus far, including materials more easily accessible now through digital means. Correspondence has been an invaluable resource for tracing women's activities in nineteenth-century plant work in Canada, as elsewhere.[33] A particularly satisfying example of this is a letter written by Harriet Sheppard, one of four women who contributed to Hooker's *Flora Boreali-Americana*, about collecting plants with her children on the shores of the St Lawrence in the summer of 1829. Harriet Sheppard's account is available to historical study because it was attached to a letter sent to Hooker by her husband and was archived as a result in Hooker's *Director's Correspondence* in the Royal Botanic Gardens, Kew.[34] Documentation like this illustrates the value but also the challenge of using institutional holdings in official archives that are patterned on male authority and textual records and that replicate the historical marginalization of women's experiences.[35]

OVERVIEW

The chronological arc of research in this volume extends from botanical fieldwork by women in early nineteenth-century Nova Scotia, Quebec, and Newfoundland (Reid; Cayouette and Khoo; Shteir) to work as writers and teachers in Canada West and Canada East at mid-century (Peterman; Stanworth), and as collectors, artists, organizers, and horticulturalists in later nineteenth-century Australia, British Columbia, and Ontario (Maroske; Huneault; Pringle; Bazely and McPherson). Some women also found access to knowledge of plants through university degree programs at the turn into the twentieth century, when, as Ainley put it in a study of women and science at English Canadian universities, the institutionalization and professionalization of science in Canada "opened new avenues for women," at least initially.[36] However, it is apparent that, increasingly over the second half of the century, the botanical activities that women were involved in were placed outside "the usual records" of what counted as "scientific."

Most of the women were collectors, learning on their own or developing skills through direct tutelage or written instructions from an experienced hand. Some collected by invitation for British imperial projects led by botanists in Glasgow, Melbourne, and London (Reid; Vandenberg; Shteir; Maroske).

Some collected plants for their own interest, prepared plant lists, compiled herbaria, gathered data about botanical phenomena, and made their results known in public ways (Reid; Cayouette and Khoo; Pringle; Peterman; Stanworth; Huneault; Galbraith; Bazely and McPherson). They collected in Quebec's Eastern Townships, the Muskoka area of Ontario, along the coasts of Newfoundland, and in hot and dusty Australian landscapes. The rigours of this work should not be discounted.

Members of this cohort came to plant knowledge and activities in nineteenth-century Canada within British traditions of interest in plants and gardens. Whether they collected and cultivated plants as elite women in colonial Quebec, Nova Scotia, and Newfoundland (Reid; Vandenberg; Shteir) or as settler-colonists in Ontario (Peterman; Nicholas), cultural associations between plants and "home" were a backdrop for floral, botanical, and horticultural work. Formal and informal education were also pathways to botany as an expression of British moral and gendered values (Shteir; Maroske; Stanworth), and British floral practices in arts and crafts gave women ready access to visual work with plants (Maroske; Peterman; Huneault; Nicholas).

Access to botanical work came through ethnicity and through family. Early in the century, elite women in the British colonies in Canada were recruited through transatlantic social networks and some could pursue their interests farther out into empire as well (Reid; Vandenberg; Shteir). British beliefs in natural theology informed the approach to nature that Catharine Parr Traill learned from her natural history mentor Gilbert White (Peterman), and the Scots Presbyterian orientation of intellectual life shaped teaching and learning in 1860s Montreal (Stanworth). Botanical interests shared among family members featured in the development of floristic work by a young woman in rural Ontario that also led her to promote native plants (Pringle). On the other side of the British Empire, settler women collecting for a flora of Australia enlisted family members in botany as a family business (Maroske).

Social and institutional developments in nineteenth-century Canada (and Australia) also widened opportunities for plant work by women. Early learned associations such as the Literary and Historical Society of Quebec and the Society for the Encouragement of Sciences and the Arts gave a few women affiliation to scientific societies in early decades,[37] and the categories of "Lady Member" and "associate member" gave some women provision for botanical connection later on (Pringle; Stanworth; Galbraith). Just as women in late eighteenth-century England had turned to botanical writing as a source of income, so women in nineteenth-century Canada combined their knowledge of plants and literary skills into publications in multiple genres and for several different audiences (Peterman; Bazely and McPherson). Women profiled in this volume were employed as teachers (Pringle; Stanworth) and as organizers working in public organizations (Pringle; Bazely and McPherson). They had access to formal training in visual arts, were schooled in drawing, painting, and printing techniques, and enjoyed significant careers in botanical book illustration (Maroske; Peterman; Huneault; Bazely and McPherson). They also had access to professional training in horticultural work and landscape architecture (Bazely and McPherson).

INTRODUCING THE CHAPTERS

The interdisciplinary research in this collection casts new and more voluminous light on plant-related activities by women in nineteenth-century English Canada and Australia and provides resources for further work on women, nature, and knowledge in historical perspectives. Chapters are grouped into five thematic parts that examine floristic history and its uses, cultures of plant collecting, the history of natural history, botanical art and craftwork, and public plant practices. Tracking from early to mid- and late-century topics, they show that botanical activities enlarged over time but also that gendered attitudes shaped what individual women were able to do. The book's many illustrations highlight botanical culture as a visual culture that was based in looking closely, observing and describing carefully, and depicting what one sees in diverse textual, artistic, and material forms.

Chapters in part 1, "Approaching Lady Dalhousie: New Resources, New Perspectives," introduce an influential elite woman who was active in early nineteenth-century botanical and horticultural circles in Nova Scotia, Quebec, and India. Christian Broun Ramsay, Countess of Dalhousie, arrived in Halifax from Scotland in 1816 when her husband George Ramsay, the 9th Earl of Dalhousie, was appointed as a colonial administrator. Her profile resembles that of other women across the British Empire whose lives were shaped by the imperial careers of husbands and fathers.[38] Class and social status that assigned duties for "gentle women of the Empire" to perform as a public face for Britain[39] also gave Lady Dalhousie access to intellectual and cultural activities, along with resources for pursuing her interests. In "A Botanical Journey of Discovery: Lady Dalhousie in British North America," landscape historian Deborah Reid traces how a privileged young woman with a private interest in plants developed into a zealous plant collector, networker, and promoter of botanical and horticultural projects that bridged the early nineteenth-century Atlantic world. Based on access to largely unstudied letters, journals, and diaries held in private family materials and archives in Scotland, Reid brings methodological attention and interpretive skills into searching for the personal plant fervour inside the public figure of a woman who received scant recognition for botanical work in her time.

In contrast to Reid's biographical approach, the chapter by Jacques Cayouette and Faye-Yin Khoo examines Lady Dalhousie's plant collecting as a resource for both historical study and contemporary botanical application. "Lady Dalhousie's Orchids and Other Rare Plants in Lower Canada, 1820–28: Resources for Historical Study" is a data-driven analysis of her work. Research botanist and botanical historian Jacques Cayouette studied documents from Lady Dalhousie's extensive activity as a collector of native orchids and plants, including sources not studied before. A manuscript acquired by the Nova Scotia Museum in Halifax in 1985, for example, contains nearly 300 entries for plants that Lady Dalhousie collected in Quebec. Cayouette and Khoo used nine different sources to assemble a list of these plants and cross-referenced the data to recent Quebec government publications about rare and endangered plants in the province. Their table "Rare plant species collected in Quebec by Lady Dalhousie" is appended to the chapter. Based on that material, they identified thirty-six species that are considered at risk today because of changes to habitats and other environmental and social factors. This research demonstrates the scientific value of Lady Dalhousie's work and the utility of historical data like hers for current practices and policies.

Virginia Vandenberg's chapter, "Gender, Botany, and Imperial Networks: Reflections on a Letter," travels with Lady Dalhousie to India and situates her plant work inside global networks and practices of imperial science culture. En route for Lord Dalhousie to serve as commander-in-chief of the British Army stationed there, Lady Dalhousie wrote to Anne Mary Perceval, one of her botanical friends in Quebec, about botanical gardens and unfamiliar vegetation they saw on their journey. Her letter, an unstudied archival resource dated December 1829 and extant only as a typescript in private family papers, is included in full as an appendix to Vandenberg's chapter. Vandenberg is interested in research methodologies that can enlarge the frame of traditional histories of science. Bringing together feminist and post-colonial studies with recent scholarship on epistolarity and empire, she reads Lady Dalhousie's letter as a document from global history and also as a historical artifact that itself shapes knowledge and empire.

Part 2, "Collecting and Its Contexts," introduces women who worked with plants as collectors on Newfoundland's coastal shores in the 1830s, in the Australian bush during the 1860s–90s, and in rural and sparsely settled Ontario during the 1890s. Some were recruited for projects shaped by empire and others collected at a time when cultural identifications were becoming more national than imperial. Collecting is a social and cultural activity and whether the object is artwork, ethnographic artifacts, scientific data, books, or plants the process is shaped by personal and professional agendas as well as by historical circumstances.[40]

Large floristic projects like Hooker's *Flora Boreali-Americana* were based in records of specimens found in remote areas and under perilous circumstances, but not all plant hunting went so far afield. Recent attention by historians, geographers, and others to domestic settings for science has widened the research canvas about plant collecting and brought more kinds of work, and more contributors, into view.[41] Women botanical collectors are among the beneficiaries of this reorientation. My chapter, "'I dare not say Botanical … Mine is a real love for flowers': Mary Brenton in 1830s Newfoundland," uses biographical and archival research as well as contextual scholarship to explore the route into public botanical work by the woman who was Hooker's main source of plants from the far eastern end of British America. Mary Brenton was born into the established and transatlantic elite of colonial Halifax, Nova Scotia, and would have grown up within ideals for women of her class and ethnicity that included some interest in plants. Later she was part of the circle around the colonial governor of Newfoundland, who cultivated a strong British cultural presence and likely brought her into Hooker's project through transatlantic networks. The letters she wrote to accompany shipments of plants to Hooker in Glasgow take us into her activities and the development of her knowledge. Mary Brenton was an eager and diligent researcher, but she did not make claims for her work as "botanical." Instead, she distinguished in a letter from 1836 between "knowledge of that interesting science" and her own "real love for flowers." Her phrasing points to demarcations in languages of nature and domains of practice that were sharpening at that time.

In another part of the British Empire, historian of science Sara Maroske shows that the need for plant collectors also gave women in colonial Australia opportunities to develop botanical knowledge. Her chapter, "Baron Ferdinand von Mueller's Plant Collectors: At Home with the Australian Flora," focuses on women in the workforce for the *Flora Australiensis*, one of the colonial floras produced by the Royal Botanic Gardens, Kew. Government botanist Mueller in Melbourne used personal networks, newspaper advertisements, and circulars to recruit collectors across social class and also across Australia's continental expanse. Maroske's research identified 1,800 women, men, and children who sent Mueller specimens of local and native plants. For some of

Mueller's collectors, plants were a source of income, and others routed their knowledge about Australian plants into artwork. Another group integrated what they learned from Mueller into how they educated their children; the Australian content in Mueller's *Botanic Teachings* (1877), an introductory schoolbook and early publication in nature study in Australia, was especially important in this regard. Maroske presents Mueller's collectors as "agents of cultural change," whose involvement with native plants became part of creating an "Australian" and national culture.

Botanist James Pringle's chapter adds Alice Hollingworth to the literature of Canadian botanical history as one of a small number of women who "contributed significantly to the knowledge of the Ontario flora." Based in herbaria materials and archival sources, "Alice Hollingworth, Early Botanical Explorer in Muskoka District, Ontario" documents collecting work in the 1890s by a young woman from a settler family. Alice Hollingworth discovered rare species of orchids and ferns, compiled her own herbarium (since 2016 at the University of Guelph), exchanged specimens with other collectors in Canada and internationally, and corresponded with Toronto naturalist William Brodie and Ottawa botanists James Fletcher and John Macoun. She came into fieldwork through scientific and botanical activities in her family, collecting medicinal herbs for a brother's homeopathic pharmacy, for example. Family friends involved in natural history organizations sent her books, encouraged her interests, and gave her access to late nineteenth-century amateur and professional scientific groups. In later years Alice Hollingworth worked as an organizer and speaker for the Women's Institutes of Ontario and particularly promoted the use of native plants rather than exotic ornamentals for landscaping rural homes.

In his book *Ways of Knowing: A New History of Science, Technology and Medicine*, John Pickstone identifies "natural history," "analysis," and "experimentalism" as three approaches to knowledge of nature that shaped the history of Anglo-European practices and vocabularies. "Natural history" is about "knowing the variety of the world – about describing and collecting, identifying and classifying, utilizing and displaying."[42] In Pickstone's mapping, all three approaches were in use across the centuries, but analysis and experimentalism came during the nineteenth century to supplant the natural history way of knowing. Chapters in part 3 of *Flora's Fieldworkers*, "Natural History 'Old' and 'New,'" feature two women who pursued their botanical interests during midcentury decades when ways of knowing and languages of nature were in flux. One woman was caught in the crosscurrents as modernizing methods and values of natural science challenged and largely displaced an approach to nature associated with "old naturalists." The other woman developed and pursued her work as a plant collector and progressive educator within the solidifying practices of new-style botanical knowledge and the academic and scientific institutions associated with it.

In "Catharine Parr Traill: A Natural Historian in Changing Times," Michael Peterman chronicles botanical roadblocks encountered by Traill, author of *The Backwoods of Canada* (1836) and other popular writings from settler Canada. Traill brought together literary and scientific languages of nature in a hybrid style that was widely used in publications for "Ladies" in Victorian Canada.[43] Fully aware that her way of writing posed difficulties for her and for some readers, she addressed this directly in the preface to her well-known publication *Wild Flowers of Canada* (1868). "The scientific reader," she wrote, "may possibly expect a more learned description of the Plants and may notice many defects and omissions; while others who are indifferent to the subject, may on the other hand think that there are too many botanical terms introduced. It is difficult to please two parties."[44] Peterman's chapter reflects on Traill's long career as a botanical fieldworker and nature writer who for decades collected, dried, and pressed specimens, sent plants to botanists, and submitted collections of native plants to local and provincial fairs and exhibitions. Peterman locates her work in relation to British natural theology and an "old-world way of thinking – that of 'the natural historian,'" and calls for revisiting Traill's approach to nature and to writing.

By contrast to the changing languages for nature study that limited Traill's public botanical career, Karen Stanworth examines new-style pathways for plant work through the career of educator Isabella McIntosh. In "'Botany … a Prominent Study': Isabella McIntosh's Ferns and Natural History in 1860s Montreal," Stanworth, a historian of education, introduces the principal of a private girls' school who collected ferns as a manifestation of her interest in botanical science. Stanworth situates McIntosh's work in proximity to the revitalized Natural History Society of Montreal in the mid-1850s and the integration of modern pedagogical and scientific practices at McGill College, both under the leadership of Professor John Dawson. McIntosh was a progressive educator who experimented with object lessons in the classroom, brought natural history and physical education into the curriculum, and gave pupils access to specialized lectures in several sciences. She described botany in a year-end speech to students at her school in 1865 as a "healthful and spirit stirring recreation" that results in "scientific knowledge." She collected native ferns in the Eastern Townships and prepared specimens for exhibition, and her botanical work is cited in the transactions of the Natural History Society of Montreal. Stanworth reconstructs McIntosh's activities and achievements through resources that include school advertisements, drafts of speeches, and a scrapbook with plants pressed between its pages.

The visual culture of nineteenth-century botany is on special display in part 4 of this collection, "Seeing and Making." In August 1852, the author of an article on "Female Education," published in *The Canadian Family Herald* in Toronto, Canada West, asked, "Shall it be considered more in keeping with the wants of the age that the female fingers be trained to paint a lily or a butterfly, or that the mind be prepared by a sound and judicious study of Botany and Entomology to unfold the varied mysteries of the one, or expatiate on the beauties of the other? We speak not now of accomplishments, but of sober study."[45] Whereas earlier chapters feature plants as physical specimens to be collected, exchanged, studied, and written about, here plants are objects depicted on paper and fabric. Art historian Kristina Huneault has charted a female tradition in botanical art in nineteenth-century Canada and shown a visual record of British North America that would be altogether poorer without the wildflower drawings and watercolours made, for example, by the wives of colonial officials.[46] Arts like these are also ways into enquiry about the women who created the images. What brought them to this work and what may the images have meant to them?

Kristina Huneault's chapter "Botanical Albums as Theoretical Objects: Sophie Pemberton and the Logic of Identity" locates botanical artwork within an extensive and historically specific nineteenth-century floral culture. Sophie Pemberton was twenty-six years old and home in colonial Victoria, British Columbia, from art school in England in 1895 when she drew native wildflowers and assembled them later into albums as gifts for family members. Creating albums like these was a popular, polite, and gendered activity for women, and in keeping with conventions, she included the botanical names of plants and cited passages of verse, including Tennyson's high Victorian assertion that "Men are God's trees, and women are God's flowers." Huneault, bringing contemporary perspectives to her material, uses work by feminist philosophers and philosophers of "difference" to go inside the nineteenth-century material in ways that, she argues, also echo forward to current concerns. Botanical illustration, she explains, is a genre of art that identifies plants by "sameness" among "types" of specimens, yet the "look" of Pemberton's albums is less static than generic botanical images may suggest, with some drawings of flowers following botanical uniformities and others showing specimens with idiosyncrasies and diverse features. Huneault reads these differences as signs of a young woman seeking agency for herself, "seeing a thing plainly … unfettered by external expectations and pre-existing categories," at a time when assertions about identity between flowers and "femininity" were tightly linked to social regulation.

Plant motifs on embroidered quilts made by women in nineteenth-century Ontario take art historian Vanessa Nicholas into British settler culture, craft history, and garden history. "Slips and Seeds: Botany and Horticulture in Two Nineteenth-Century Canadian Quilts" reconstructs the creation of two previously unstudied heritage quilts. Elizabeth Bell's mid-century "Fallowfield Quilt" is made up of images of wildflowers that reflect a British cultural identification with flowers. They also reflect a design tradition of early modern English "slips" that needleworkers copied from illustrated herbals and used as embroidery patterns. Nicholas finds signs on the quilt of Bell's connection to her new home in the natural world of Canada, notably in images of trilliums that were likely based on direct observation of flowers around the family's homestead. The other quilt, by Margaret McCrum, was embroidered with pots of plants such as roses, peonies, and amaranths that would have been found in the flowerbeds and doorway gardens of later nineteenth-century rural Ontario. McCrum's quilt draws on an American quadrant-style of design and also, Nicholas suggests, on beadwork traditions and floral motifs associated with Woodlands Indigenous Peoples. Nicholas reads Elizabeth Bell's and Margaret McCrum's quilts as reflecting "the scientific sensibility of their time" and calls for attention by quilt historians and others to floral motifs and their botanical and horticultural actualities in women's homecrafts.

In *Inventing Canada*, her foundational history of science in nineteenth-century Canada, Suzanne Zeller underscores the centrality of the doctrine of utilitarianism to nation building and the public good and presents botany as a central plank of this historical project.[47] Botany promised agricultural improvement, scientific understanding of plant structures and processes, and the material potential of economic resources. How was plant knowledge of such importance communicated to different users and audiences? Part 5 of this book, "Expanding Public Practices," opens with a historical overview of botanical gardens as one expression of outreach for science. David Galbraith's chapter, "Botanical Gardens in Nineteenth-Century Canada: Individuals and Institutions," is based in archival and visual materials and focuses on key men and women in this history. In earlier centuries, the term "botanical garden" often designated spaces that were exclusively for researchers, but the nineteenth-century and Canadian direction was toward developing institutions that would combine public education and science-based research. There were clear tensions, however, about how to fit that combination to the needs of different audiences. When the Botanical Society of Canada, for example, sought to establish a botanical garden at Queen's University in the early 1860s, its goal was to "bring our members and the public into scientific modes of thought and expression, [rather] than to allow our Society to yield up its scientific character to suit the popular taste." What might this apparent bifurcation between the "scientific" and the "popular" have meant for the thirty women who were listed as members of the Botanical Society of Canada? Galbraith, head of science at the Royal Botanical Gardens (Canada), foregrounds botanical work by Lucy Lawson and Kate Crooks. He also cites references to Indigenous materials found in his research for this history.

Public botanical and horticultural activities in developing institutions are also the focus of the chapter "Women, Citizen Science, and Botanical Knowledge in Ontario, 1870–1920." Biologist Dawn Bazely and social historian Kathryn McPherson have researched women who were botanical writers and illustrators, commercial landscape architects, flower breeders, and business owners, and key figures in late nineteenth and early twentieth-century horticultural and gardening organizations. They present women in this chapter as citizen scientists of their day who blurred the lines between amateurs and professionals and were validated by botanists and others for their work. Bazely and McPherson also highlight women who used the print culture of their day as "disruptive technologies." The magazine *Canadian Horticulturist*, for example, was a popular forum from the 1870s onward where women shared interests as well as expertise based, in some cases, on training in horticulture as a science-based field. Bazely and McPherson's chapter addresses concerns within the international community of STEM (science, technology, engineering, and mathematics) about increasing scientific literacy and democratizing science so as to bring more participants into scientific knowledge. Foregrounding women whose plant work has been unrecognized and also historically erased is one part of this campaign.

The afterword by Suzanne Zeller to this volume reflects on shifting outlooks and practices in nineteenth-century Canadian science from the vantages of contemporary historiography and ecology. In "Finding Meaning in the Understory" Zeller reads the chapters chronologically in relation to the outlooks and practices that John Pickstone charted in his *Ways of Knowing* as ways of knowing in nineteenth-century science. She points out that the sequential displacements in these ideas over the course of the century – from natural historical to analytical and to experimentalist – map onto orientations in how the history of nature study itself was written. These changing orientations in turn did much to generate "the usual records" that are reflected and addressed through chapters in this book. Zeller's afterword also looks beyond historiographical approaches to Canada's science history and beyond the nineteenth century. She finds glimmers of ecological views of nature in *Flora's Fieldworkers* that bring the material into current-day challenges to prevailing modes of knowing nature and writes that "recent scholarship supports increasingly vigorously the need to continue looking and listening for alternative sources of understanding in science's rich and intriguing understory." Her epigraph is from *The Overstory* (2018), the celebrated ecological and eco-feminist novel by Richard Powers: "The tree is saying things, in words before words … *If your mind were only a slightly greener thing, we'd drown you in meaning.*"

CONCLUSION: TURNING BOTANICAL

In their 2008 introduction to "Natural Science in the New World: The Descriptive Enterprise," in *Scientia Canadensis*, Victoria Dickenson and Elsbeth Heaman trace Canadian elements of this history, from observations made by explorers, early settlers, and naturalists working within French and British imperial contexts into later nineteenth-century developments that led to replacing "old naturalists" by university-trained biologists. With findings based on an interdisciplinary symposium and reflecting both older and newer approaches to knowledge of nature, Dickenson and Heaman welcome the reinvigoration of the history of natural history by researchers across the arts and sciences, and they take the position that "the new sciences are strongest where they are best informed by natural history traditions."[48] In this, they join John Pickstone's *Ways of Knowing* and other path-breaking examples of scholarship that have invigorated study of natural history and its histories and practices in recent years.[49]

New approaches to nature and natural history are also part of a "vegetative turn" in thinking about relations between human and non-human worlds. Among scholars in many areas who focus on plant agency and sentience and on plant language and communication, feminist theorist Luce Irigaray and philosopher Michael Marder, for example, explore ideas about "vegetal being" and the urgent need to care for it through alternative philosophies and social practices.[50] In an example from literary studies of nature, Natania Meeker and Antónia Szabari read across from early modern texts into contemporary feminist writing and make a satisfying case in their *Radical Botany: Plants and Speculative Fiction* for "a turn to the plant as our guide to the end of the world, moving us toward new political possibilities and responsibilities that remain unmapped."[51]

The botanical "turn to the plant" in *Flora's Fieldworkers* is less speculative than in those examples of world making, but goes back to roots in its own ways. Chapters explore nineteenth-century plant practices by women and highlight diverse resources that make these stories known. They focus for the most part on individual women and bring the social and cultural circumstances of their plant work into view. Women's botanical activities sit also within changing historical and political contexts for nineteenth-century plant study, and in this regard give material for analyzing other colonial, imperial, and national dimensions of botany in Canada and Australia. They offer material as well for enquiry into histories of the academic study of plants and the part that gender plays in practices, hierarchies of knowledge, and norms of participation. Botany had already been demarcated as a science in nineteenth-century Canada when Mary Brenton in 1830s Newfoundland distinguished between her "love of flowers" and the "botanical" work of William Hooker. That demarcation may not have deterred Mrs Lawson and Miss Gildersleeve's experiments in silkwork culture as lady members of the Botanical Society of Canada in the 1860s, but it may well have influenced how their work was received. It was clear to Catharine Parr Traill in that same decade that her way of writing about plants was caught in crosscurrents having to do with both content and style. How deliberately or how unwittingly was a gendered path for botany pursued in nineteenth-century Canada? New materials and perspectives in *Flora's Fieldworkers* implicitly challenge how field histories have been written.

Looking beyond "the usual records" and also beyond the usual record keepers, the findings about women and plant study in this book map a broad picture of women as botanical knowers and doers: as artists and illustrators, correspondents, networkers, teachers, album makers, needleworkers, public educators, horticulturalists, and organizers and as writers in nineteenth-century traditions of popularizing, introductory, and instructional books for general audiences. Gender alone does not explain the variety and extent of women's plant-related activities at that time. Instead, it is the intersection of gender and ethnicity in British colonial and imperial cultures that elucidates how and why numbers of women came into botanical activities. This intersection brings several otherwise different communities of practice into visibility in Canada and in Australia as well. Elite women in early nineteenth-century Quebec, Newfoundland, and Nova Scotia, for example, had botanical opportunities as wives, daughters, and friends of colonial and imperial British officials during their years in British North America. Botanical interests came from England with settler women or were cultivated through family and social networks. Imperial botanical projects offered opportunities to settler women in Australia, and the ongoing British cultural nexus of those decades gave Canadian women connections and aspirations for their work as teachers, artists, and writers. Opportunities went across age and life stages.

Every story is local and individual, but chapters in this volume offer abundant evidence that women who pursued botanical knowledge and activities received much in return or in the process. Friendship, sociability, intellectual engagement, and interests shared with family were part of their stories. So too were aesthetics and religious or spiritual affirmation. There are glimpses of joy in research about nature, emotional connection, solace in work of one's own choosing, energy for self-fashioning and career advancement, and the satisfaction of contributing to floristic knowledge. This book brings history to botanists and botany to historians and both to workers in the field of women, gender, and science. Enlarging understanding of the circumstances of women and men who sought knowledge of plants in nineteenth-century Canada gives Flora's fieldworkers their histories, and this matters for practices and practitioners now and in the future.

ACKNOWLEDGMENTS

I gratefully acknowledge conversations with friends and colleagues – in person, by email, Skype, and Zoom, and on the page – as this volume took shape. Thank you to Jacques Cayouette, the late Marika Ainley, David Galbraith, Sara Maroske, Bernard Lightman, and Kristina Huneault. Thank you to Stephanie Morningstar for discussing Indigenous matters with me. My deep appreciation goes as well to Jane Couchman, Joan Gibson, Frieda Forman, Cynthia Zimmerman, Claire Pizer, Cynthia Holz, Sheila Malovany-Chevallier, Bronwen Cunningham, and Sarah Shteir. Clio Windust provided exemplary technical assistance and sweet comradeship. It is a pleasure to acknowledge the good cheer and steadfastness of contributors to *Flora's Fieldworkers*. Special thanks go to the readers of the manuscript for McGill-Queen's University Press whose engaged attention strengthened the work in all its parts and especially as a book for readers in diverse fields of interest and study.

NOTES

1 *The Calliopean*, Early Canadiana Online 1 (23 November 1847): 8.

2 Ibid., 6 (9 February 1848): 41–2. On the Burlington Ladies' Academy, a "relatively early experiment in women's education in Ontario," see Sonser, "Literary Ladies and *The Calliopean*," 368. Curricular and pedagogical practices of the school included "Lectures, given formally and informally, on subjects connected with the health, manners, and appropriate duties of young ladies" and also "courses of Lectures, with experiments and illustrations … on Chemistry and Astronomy" (*The Calliopean* 1 [23 November 1847]: 16).

3 See Shteir, *Cultivating Women, Cultivating Science*. Scholarship on this topic includes George, *Botany, Sexuality and Women's Writing 1760–1830*; Kelley, *Clandestine Marriage*; Scourse, *The Victorians and their Flowers*; and Page and Smith, *Women, Literature, and the Domesticated Landscape*. See also Zytaruk, *Nature on the Page*, ch. 4; Jordan, *Picturesque Pursuits*; and Jiang, "Angel in the House, Angel in the Scientific Empire."

4 On natural history in nineteenth century Canada, see Berger, *Science, God, and Nature*.

5 See Hadot, *The Veil of Isis*, and Merchant, *The Death of Nature*.

6 Shteir, "Iconographies of Flora," 7.

7 Ibid., 14.

8 Penhallow, "Canadian botany from 1800 to 1895," 10. References to Penhallow's vocabulary appear *passim*.

9 Lindley, *Introductory Lecture*, 17, 10. On Lindley, see Shteir, "Gender and 'Modern' Botany in Victorian England" and *Cultivating Women, Cultivating Science*, ch. 6, "Defeminizing the Budding Science of Botany, 1830–1860."

10 Milam and Nye, "An Introduction to *Scientific Masculinities*," 2.

11 Ibid., 14.

12 On Maria Morris Miller and botanical art by women in nineteenth-century Canada, see Huneault, *I'm Not Myself at All*, ch. 4 and her chapter in this volume.

13 On Penhallow, see *Dictionary of Canadian Biography*. Penhallow supported professional work by women in botany at McGill and worked hard to ensure academic appointments for botanist Carrie Derick. See Gillett, "Carrie Derick (1862–1941) and the Chair of Botany at McGill," 81.

14 Penhallow, "Canadian Botany from 1800 to 1895," 10. On Hooker's *Flora Boreali-Americana*, a compilation of botanical information about approximately 5,000 plants found in locations across what is now Canada, see Pringle, "The History of the Floristic Exploration of Canada," 302–5.

15 Penhallow, "Canadian Botany from 1800 to 1895," 12.

16 Carter, *Imperial Plots*; Turner and Turner, "'Where our women used to get the food'"; Burnett, *Taking Medicine*.

17 See chapters by Nicholas and Maroske. See also chapters by Vandenberg, Peterman, and Galbraith. Pringle mentions Alice Hollingworth's interest in archaeological work on Indigenous habitation in the Muskoka District.

18 See Pandora, "Amateurs." For an overview on the amateur in the history of science, see Allen, "Amateurs and Professionals."

19 Dror et al., "*History of Science and the Emotions*."

20 Heap, "Introduction: Women and Gender in Canadian Science, Engineering and Medicine," 8.

21 Lloydlangston, "Women in Botany and the Canadian Federal Department of Agriculture, 1887–1919," 114.

22 For an overview on plants and plant study from before the British era in Canada and from eighteenth and nineteenth-century Quebec, see volumes in Asselin, Cayouette, and Mathieu, *Curieuses Histoires de Plantes du Canada*, including attention to numerous women herbalists and midwives in seventeenth-century New France. See also Cayouette, *À la Découverte du Nord.*

23 Ainley, *Despite the Odds*, 20.

24 See, for example, Creese, *Ladies in the Laboratory*, a survey of contributions by women in science in America, Britain, Western Europe, and British colonial South Africa, Australia, New Zealand, and Canada. Vols. I (1998) and II (2004) are based on publications listed in the *Catalogue* of the Royal Society. Vol. III (2010) expands criteria to include "amateur" work and contributions by women trained at universities; the section on nineteenth-century Canada includes biographical sketches of women collectors and artists across a wide geographical expanse.

25 Ainley, "Science in Canada's Backwoods," 93.

26 "Science Culture in the Nineteenth Century: Women and the Botanical Society of Canada" appeared in an issue of the journal *Resources for Feminist Research* on "The Nature of Feminist Science Studies" (2010). On the Botanical Society and its founder George Lawson, see the chapter by David Galbraith in this volume, and also Zeller, 228–39.

27 "Mrs Lawson" and "Miss Gildersleeve" are among thirty women named in the *Annals* of the Botanical Society of Canada. Their first names are not recorded. "Mrs Lawson" was Lucy Lawson, wife of the society's founder, Professor George Lawson. For possible identification of "Miss Gildersleeve," see David Galbraith's chapter in this volume, n50.

28 Kennedy, "Science Culture in the Nineteenth Century," 48–9.

29 Janovicek and Nielson, *Reading Canadian Women's and Gender History*, 3–4.

30 See McPherson, Morgan, and Forestell, *Gendered Past*, 1–11; and Carstairs and Janovicek, *Feminist History in Canada*, 3–22.

31 McKenna, "Class, Race, and Gender Roles," 110.
32 Janovicek and Nielson, *Reading Canadian Women's and Gender History*, 10–11.
33 See, for example, Gianquitto, "'My Dear Dr.'"; also Evans, *Darwin and Women*.
34 Shteir and Cayouette, "Collecting with 'botanical friends'"; on Harriet Sheppard, see 13–18.
35 Antoinette Burton takes up this point in her historiographical discussion "Memory Becomes Her."
36 Ainley, *Creating Complicated Lives*, 47–8. Ainley's assessment of women and Canadian science contrasts with Margaret Rossiter's contention in the first volume of *Women Scientists in America* that the professionalization of science in the United States led to the exclusion of women. The career of botanist Carrie Derick at McGill University illustrates the opportunities and still familiar obstacles as women entered gendered academic workplaces (see Gillett, "Carrie Derick (1862–1941) and the Chair of Botany at McGill").
37 Stanworth, *Visibly Canadian*, 45–54.
38 Lambert and Lester, eds. *Colonial Lives across the British Empire*, esp. 1–31
39 See Errington, "Suitable Diversions"
40 For cultural perspectives on collecting, see Baudrillard, "The System of Collecting" and McCormick, "Scottish Cultures of Collecting." For perspectives from within the history of science see Endersby, *Imperial Nature*, 54–83, and Strasser, "Collecting Nature," 312–23.
41 See Opitz et al., eds. *Domesticity in the Making of Modern Science*, and Livingstone and Withers, *Geographies of Nineteenth-Century Science*.
42 Pickstone, *Ways of Knowing*, 60. See his chapter "Natural History," 60–82.
43 For discussion of a similar combination of literary and scientific languages of nature found in writing by women in early nineteenth-century Canada, see Matthews, "Women Writers."
44 Traill, "Preface," *Canadian Wild Flowers*, 8.
45 *The Canadian Family Herald* 1 (1851): 295.
46 See Huneault, *I'm Not Myself at All*, ch 4, esp. 152–73. See also Burant, *Drawing on the Land*.
47 Zeller, *Inventing Canada*, part III.
48 Dickenson and Heaman, "Introduction: Natural Science in the New World," 9.
49 See especially Curry et al., eds, *Worlds of Natural History*, and N. Jardine et al., eds, *Cultures of Natural History*.
50 Irigaray and Marder, *Through Vegetal Being*. See also reflections by Irigaray, Marder, and others in Gagliano, Ryan, and Vieira, eds., *The Language of Plants*.
51 Meeker and Szabarai, *Radical Botany*, 201.

BIBLIOGRAPHY

Archives

Hooker, William Jackson. *Director's Correspondence.* Archives, London: Royal Botanic Gardens, Kew.

Books and Articles

Ainley, Marianne Gosztonyi. *Creating Complicated Lives: Women and Science at English-Canadian Universities, 1880–1980.* Edited by Marelene Rayner-Canham and Geoff Rayner-Canham. Montreal and Kingston: McGill-Queen's University Press, 2012.

– ed. *Despite the Odds: Essays on Canadian Women and Science.* Montreal: Véhicule Press, 1990.

– ed. "Last in the Field? Canadian Women Natural Scientists, 1815–1965." In *Despite the Odds: Essays on Canadian Women and Science*, 25–62. Montreal: Véhicule Press, 1990.

– "Science in Canada's Backwoods: Catharine Parr Traill." In *Natural Eloquence: Women Reinscribe Science*, edited by Barbara T. Gates and Ann B. Shteir, 79–97. Madison: University of Wisconsin Press, 1997.

Allen, David E. "Amateurs and Professionals." In *The Cambridge History of Science*, vol. 6, edited by Peter J. Bowler and John V. Pickstone, 15–33. Cambridge: Cambridge University Press, 2009.

Asselin, Alain, Jacques Cayouette, and Jacques Mathieu. *Curieuses Histoires de Plantes du Canada*, tome 1: 1000–1670. Quebec: Éditions du Septentrion, 2014.

– *Curieuses Histoires de Plantes du Canada*, tome 2: 1670–1770. Quebec: Éditions du Septentrion, 2015.

– *Curieuses Histoires de Plantes du Canada*, tome 3: 1760–1867. Quebec: Éditions du Septentrion, 2017.

– *Curieuses Histoires de Plantes du Canada*, tome 4: 1867–1935. Quebec: Éditions du Septentrion, 2019.

Baudrillard, Jean. "The System of Collecting." In *The Cultures of Collecting*, edited by John Elsner and Roger Cardinal, 7–24. Melbourne: Melbourne University Press, 1994.

Berger, Carl. *Science, God, and Nature in Victorian Canada.* Toronto: University of Toronto Press, 1983.

Burant, James, ed. *Drawing on the Land: The New World Travel Diaries and Watercolours of Millicent Mary Chaplin, 1838–1842.* Manotick, ON: Penumbra Press, 2004.

Burnett, Kristin. *Taking Medicine: Women's Healing Work and Colonial Contact in Southern Alberta, 1880–1930.* Vancouver: UBC Press, 2010.

Burton, Antoinette. "Memory Becomes Her: Women, Feminist History, and the Archive." In *Dwelling in the Archive: Women Writing House, Home, and History in Late Colonial India*, 3–30. Oxford: Oxford University Press, 2003.

Carstairs, Catherine, and Nancy Janovicek, eds. *Feminist History in Canada: New Essays on Women, Gender, Work, and Nation.* Vancouver: UBC Press, 2016.

Carter, Sarah. *Imperial Plots: Women, Land, and the Spadework of British Colonialism on the Canadian Prairies.* Winnipeg: University of Manitoba Press, 2016.

Cayouette, Jacques. *À la Découverte du Nord: Deux siècles et demi d'exploration de la flore nordique du Québec et du Labrador.* Quebec: Éditions MultiMondes, 2014.

Creese, Mary R.S., with contributions by Thomas M. Creese. *Ladies in the Laboratory? American and British Women in Science, 1800–1900: A Survey of Their Contributions to Research.* Lanham, MD: Scarecrow Press 1998.

– *Ladies in the Laboratory II: West European Women in Science, 1800–1900: A Survey of Their Contributions to Research.* Lanham, MD: Scarecrow Press, 2004.

– *Ladies in the Laboratory III: South African, Australian, New Zealand, and Canadian Women in Science: Nineteenth and Early Twentieth Centuries: A Survey of Their Contributions.* Lanham, MD: Scarecrow Press, 2010.

Curry, H.A., N. Jardine, J.A. Secord, and E.C. Spary, eds. *Worlds of Natural History.* Cambridge: Cambridge University Press, 2018.

Dickenson, Victoria, and Elsbeth Heaman. "Introduction: Natural Science in the New World: The Descriptive Enterprise." *Scientia Canadensis* 31, 1–2 (2008): 1–11.

Dror, Otniel E., Bettina Hitzer, Anja Laukötter, and Pilar León-Sanz. "An Introduction to the *History of Science and the Emotions.*" *Osiris* 31, 1 (2016): 1–18.

Endersby, Jim. *Imperial Nature: Joseph Hooker and the Practices of Victorian Science.* Chicago: University of Chicago Press, 2008.

Errington, Elizabeth Jane. "Suitable Diversions: Women, Gentility and Entertainment in an Imperial Outpost." *Ontario History* 102, 2 (2010): 175–96.

Evans, Samantha. *Darwin and Women: A Selection of Letters.* Cambridge: Cambridge University Press, 2017.

Gagliano, Monica, John C. Ryan, and Patrícia Vieira, eds. *The Language of Plants: Science, Philosophy, Literature.* Minneapolis: University of Minnesota Press, 2019.

Gates, Barbara T. *Kindred Nature: Victorian and Edwardian Women Embrace the Living World.* Chicago: University of Chicago Press, 1998.

– ed. *In Nature's Name: Anthology of Women's Writing and Illustration, 1780–1930.* Chicago: University of Chicago Press, 2002.

Gates, Barbara T., and Ann B. Shteir, eds. *Natural Eloquence: Women Reinscribe Science.* Madison: The University of Wisconsin Press, 1997.

George, Sam. *Botany, Sexuality and Women's Writing 1760–1830: From Modest Shoot to Forward Plant.* Manchester: Manchester University Press, 2007.

Gianquitto, Tina. "'My Dear Dr.': American Women and Nineteenth-Century Scientific Correspondence." In *The Edinburgh Companion to Nineteenth-Century American Letters and Letter-Writing,* edited by Celeste-Marie Bernier, Judie Newman, and Matthew Pethers, 435–49. Edinburgh: Edinburgh University Press, 2016.

Gillett, Margaret, "Carrie Derick (1862–1941) and the Chair of Botany at McGill." In *Despite the Odds: Essays on Canadian Women and Science*, edited by Marianne Gosztonyi Ainley, 74–87. Montreal: Véhicule Press, 1990.

Hadot, Pierre. *The Veil of Isis: An Essay on the History of the Idea of Nature*. Cambridge, MA: Harvard University Press, 2006.

Heap, Ruby. "Introduction: Women and Gender in Canadian Science, Engineering and Medicine." *Scientia Canadensis* 29, 2 (2006): 3–15.

Hooker, William Jackson. *Flora Boreali-Americana: The Botany of the Northern Parts of British North America*. London: Henry G. Bohn, 1829–40. 2 vols.

Huneault, Kristina. *I'm Not Myself at All: Women, Art, and Subjectivity in Canada*. Montreal and Kingston: McGill-Queen's University Press, 2018.

Irigaray, Luce, and Michael Marder. *Through Vegetal Being [Two Philosophical Perspectives]*. New York: Columbia University Press, 2016.

Janovicek, Nancy, and Carmen Nielson, eds. "Introduction: Feminist Conversations." In *Reading Canadian Women's and Gender History*, 3–11. Toronto: University of Toronto Press, 2019

– eds. *Reading Canadian Women's and Gender History*. Toronto: University of Toronto Press, 2019.

Jardine, N., J.A. Secord, and E.C. Spary, eds. *Cultures of Natural History*. Cambridge: Cambridge University Press, 1996.

Jiang Hong. "Angel in the House, Angel in the Scientific Empire: Women and Colonial Botany during the Eighteenth and Nineteenth Centuries." *Notes and Records of the Royal Society*, 24. doi:10.1098/rsnr.2020.0046.

Jordan, Caroline. *Picturesque Pursuits: Colonial Women Artists and the Amateur Tradition*. Carlton: Melbourne University Press, 2005.

Kelley, Theresa M. *Clandestine Marriage: Botany and Romantic Culture*. Baltimore: The Johns Hopkins University Press, 2012.

Kennedy, Kerrie. "Science Culture in the Nineteenth Century: Women and the Botanical Society of Canada." *RFR/DRF: The* Nature *of Feminist Science Studies* 33, no. 3–4 (2010): 47–70.

Lambert, David, and Alan Lester, eds. *Colonial Lives across the British Empire: Imperial Careering in the Long Nineteenth Century*. Cambridge: Cambridge University Press, 2006.

Lindley, John. *Introductory Lecture Delivered in the University of London on Thursday, April 30, 1829*. London: John Taylor, 1829.

– *Ladies' Botany, or A Familiar Introduction to the Study of the Natural System of Botany*. London: James Ridgeway & Sons, 1834-37.

Livingstone, David N., and Charles W.J. Withers, eds. *Geographies of 19th-Century Science*. Chicago: University of Chicago Press, 2011.

Lloydlangston, Amber. "Women in Botany and the Canadian Federal Department of Agriculture, 1887–1919." *Scientia Canadensis* 29, 2 (2006): 99–130.

McCormick, Kaitlin. "Scottish Cultures of Collecting on the Northwest Coast of North America, 1820s to 1860s." *Visual Culture in Britain* 17, 2 (2016): 163–80.

McKenna, Katherine. "Class, Race, and Gender Roles in Early British North America." In *Reading Canadian Women's and Gender History*, edited by Nancy Janovicek and Carmen Nielson, 110–19. Toronto: University of Toronto Press, 2019.

McPherson, Kathryn, Cecilia Morgan, and Nancy M. Forestell, "Introduction: Conceptualizing Canada's Gendered Pasts." In *Gendered Pasts: Historical Essays in Femininity and Masculinity in Canada* [1999], edited by Kathryn McPherson, Cecilia Morgan, and Nancy M. Forestell, 1–11. Toronto: University of Toronto Press, 2003.

Matthews, Charity Christine. "Women Writers and the Study of Natural History in Nineteenth-Century Canada." Electronic Theses and Dissertations (ETDS) 2008+. T, University of British Columbia. 2013. Doi: http://dx.doi.org/10.14288/1.0071932.

Meeker, Natania, and Antónia Szabari. *Radical Botany: Plants and Speculative Fiction*. New York: Fordham University Press, 2020.

Merchant, Carolyn. *The Death of Nature: Women, Ecology, and the Scientific Revolution*. San Francisco: Harper & Row, 1980.

Milam, Erika Lorraine, and Robert A. Nye. "An Introduction to *Scientific Masculinities*." *Osiris* 30, ser. 2 (2015): 1–14.

Morris Miller, Maria. *Wild Flowers of Nova Scotia*. 3rd ser. London: John Snow, 1866; 4th ser. London: Reeve and Co., 1867.

Opitz, Donald L., Staffan Bergwik, and Brigitte Van Tiggelen, "Introduction: Domesticity and the Historiography of Science." In *Domesticity in the Making of Modern Science*, edited by Donald L. Opitz, Staffan Bergwik, and Brigitte Van Tiggelen, 1–15. New York: Palgrave Macmillan, 2016.

Page, Judith W., and Elise L. Smith. "Introduction." In *Women, Literature, and the Domesticated Landscape: England's disciples of Flora, 1780–1870*, edited by Judith W. Page and Elise L. Smith, 1–14. Cambridge: Cambridge University Press, 2011.

Pandora, Katherine. "Amateurs." In *A Companion to the History of Science*, edited by Bernard Lightman, 139–42. Chichester: Wiley Blackwell, 2016.

Penhallow, D.P. "A Review of Canadian Botany from 1800 to 1895." *Proceedings and Transactions of the Royal Society of Canada* 3(IV), ser. II (1897): 3–56.

Pickstone, John V. *Ways of Knowing: A New History of Science, Technology and Medicine*. Chicago: University of Chicago Press, 2001.

Powers, Richard. *The Overstory*. New York: W.W. Norton, 2018.

Pringle, James S. "The history of the floristic exploration of Canada." *The Canadian Field-Naturalist* 109 (1996) ["1995"]: 291–356.

Rea, John. *Flora: seu, De Florum Cultura. Or, A Complete Florilege*. 3 vols. London 1665.

Rossiter, Margaret W. *Women Scientists in America* [Vol. I]: *Struggles and Strategies to 1940*. Baltimore: The Johns Hopkins University Press, 1982.

Scourse, Nicolette. *The Victorians and their Flowers*. London: Croom Helm, 1983.

Shteir, Ann B. *Cultivating Women, Cultivating Science: Flora's Daughters and Botany in England 1760 to 1860*. Baltimore and London: The Johns Hopkins University Press, 1996.

– "Gender and 'Modern' Botany in Victorian England." *Osiris* 12, ser. 2 (1997): 29–38.

– "Iconographies of Flora: The Goddess of Flowers in the Cultural History of Botany." In *Figuring It Out: Science, Gender, and Visual Culture*, edited by Ann B. Shteir and Bernard Lightman, 3–27. Hanover, NH: Dartmouth College Press/University Press of New England, 2006.

Shteir, Ann B., and Jacques Cayouette. "Collecting with 'botanical friends': Four Women in Colonial Quebec and Newfoundland." *Scientia Canadensis* 41, 1 (2019): 1–30.

Sonser, Anna. "Literary Ladies and *The Calliopean*: English Studies at the Burlington Ladies' Academy (1846–1851)." *University of Toronto Quarterly* 64, 3 (1995): 368–80.

Stanworth, Karen. *Visibly Canadian: Imaging Collective Identities in the Canadas, 1820–1910*. Montreal and Kingston: McGill-Queen's University Press, 2014.

Strasser, Bruno J. "Collecting Nature: Practices, Styles, Narratives." In *Clio Meets Science: The Challenges of History*, edited by Robert E. Kohler and Kathryn M. Olesko, 303–40. *Osiris* 30, 2012.

Traill, Catharine Parr. *Canadian Wild Flowers, Painted and Lithographed by Agnes Fitzgibbon, with Botanical Descriptions by C.P. Traill*. Montreal: John Lovell, 1868.

Turner, Nancy J., and Katherine L. Turner. "'Where our women used to get the food': cumulative effects and loss of ethnobotanical knowledge and practice: case study from coastal British Columbia." *Botany* 86, 2 (2008): 103–15.

Zeller, Suzanne. *Inventing Canada: Early Victorian Science and the Idea of a Transcontinental Nation* (1987). Montreal and Kingston: McGill-Queen's University Press, 2009.

Zytaruk, Maria. *Nature on the Page: The Print and Manuscript Culture of Victorian Natural History*. Toronto: University of Toronto Library, 2019.

PART ONE

Approaching Lady Dalhousie

New Resources, New Perspectives

1

A Botanical Journey of Discovery
Lady Dalhousie in British North America

DEBORAH REID

During her time in Halifax, Nova Scotia, and Quebec in the early part of the nineteenth century, Christian Ramsay, the Countess of Dalhousie (1786–1839), amassed an extraordinary collection of North American plants, thought to comprise some of the oldest specimens known to be in any Canadian herbarium today. Sir William Jackson Hooker used a selection of these specimens in the preparation of his celebrated botanical work on North American flora, *Flora Boreali-Americana* (1829–40). New and rare exotic species she introduced from Canada played an important horticultural role in changing the appearance of British gardens, and her contributions to horticultural history increased still more when she collected plants in India during the 1830s. Despite these achievements, Lady Dalhousie's contributions have received scant recognition within traditional narratives of garden history. Drawing on extensive personal correspondence, handwritten journals, plant lists, and memorabilia contained within archives belonging to the Broun-Lindsay and Dalhousie families, this chapter critically evaluates the use of archival sources, provides an analysis of the biographical evidence found within them, and reveals the extent to which Lady Dalhousie overcame her lack of formal botanical training and juggled the demands of wife and mother to become one of Scotland's most successful female plant collectors. The chapter also demonstrates that over the course of her lifetime, Lady Dalhousie converted an early horticultural interest in the propagation and cultivation of plants within her own private gardens into an increasingly scientific interest in the collection and identification of both rare and unknown botanical specimens in British North America, Penang, and India, many of which are now housed in herbaria in Britain, Canada, and France.[1]

Archival rediscovery of the Countess of Dalhousie starts with the handwritten diaries of her husband George Ramsay, 9th Earl of Dalhousie (1770–1838), housed at the National Records of Scotland.[2] They provide a fascinating insight into Lord Dalhousie's life and work in Canada from 1816 until 1828 and, very occasionally, tantalizing reference is made to his wife, Lady Dalhousie. Much has subsequently been written about Lord Dalhousie, but the story of Lady Dalhousie and that of her botanical journey of discovery in Canada has never been told.[3] And yet, within the same archive and among other private family sources, a rich seam of information in the form of letters and journals written by Lady Dalhousie herself was discovered, which revealed her transition from a young woman with an interest in plants to an accomplished botanist. While Lady Dalhousie's manuscripts are scattered throughout a number of archive collections, including within the correspondence of William Jackson Hooker held at the Royal Botanic Gardens in Kew, the majority can be found in her husband's family archive under the name of Maule and in papers belonging to her paternal ancestors, the Broun family of Coalstoun.[4]

The archive itself is a challenging yet invigorating environment within which to work.[5] Even the identification process itself can be a painstaking one, especially when evidencing female histories, as women are rarely catalogued under their own name. As an added complication, Christian is variously referred to as Christian Ramsay, Christian Broun Dalhousie (or CBD, which often appears on herbarium specimens collected by her), the Countess of Dalhousie, Lady Dalhousie, or simply "Lady D" as she was affectionately known and referred to by Lord Dalhousie in his diaries. In the nineteenth century, women drawn from the upper classes appear to have had a greater propensity to document their lives through the writing of letters and the keeping of journals, and, as an aristocratic woman, Lady Dalhousie was no exception. The Broun family archive, for example, contains almost 500 manuscripts written by Lady Dalhousie or relating to her and many more can be found within the Maule papers, providing a rich and diverse source of biographical material especially when triangulated with information found within correspondence from friends and family, in particular the letters and journals written by her husband.[6] However, archives "only ever yield partial understandings" according to Michelle King.[7] Evidence found was usually fragmentary and had to be pieced together in the same way that a jigsaw is assembled, except that the picture was rarely complete. Illegible and damaged manuscripts were commonplace and had, therefore, to be treated with the utmost caution since missing information can significantly affect our understanding of a given

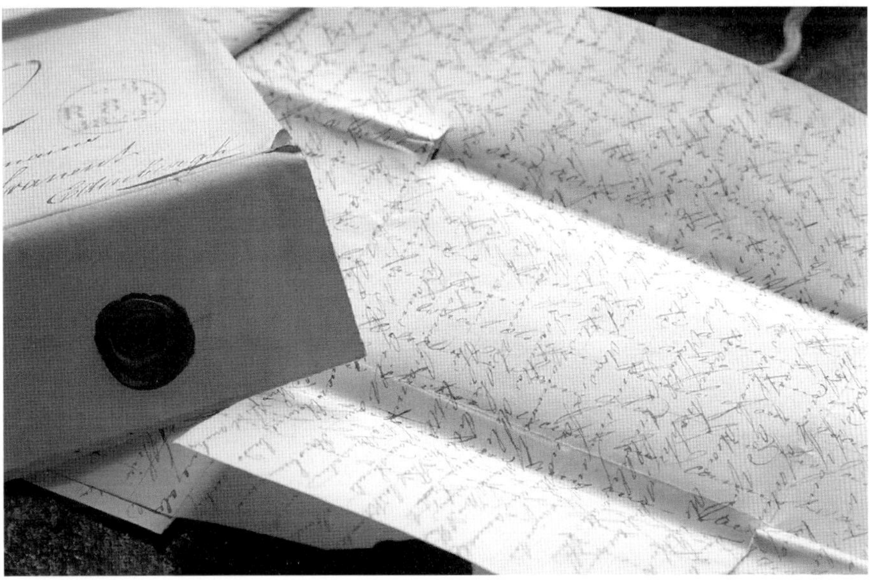

1.1
A typical letter written by Christian Ramsay in 1818 while in Halifax, Nova Scotia.

communication. The slanted copperplate handwriting adopted by Lady Dalhousie in her letters was often difficult and sometimes impossible to read, made worse by her tendency to write to the very end of the page before turning the document ninety degrees and writing sideways across her script, so that on first acquaintance the text in her letters resembled squared graph paper.[8] The process employed for making sense of found information usually took place outside the archive, using data from other primary and secondary sources to help decipher meaning and provide corroboratory evidence.

Content found within the correspondence and memoirs written by Lady Dalhousie required an epistemological approach. Her letters were usually sent to family members and friends when she was travelling and living with her husband in Nova Scotia, Quebec, and India, and while her desire to describe the people and places she met has resulted in a rich source of information on her plant-hunting expeditions, her narrative was clearly shaped by her social status and desire to create a good impression of the countries she visited for the benefit of her relatives and acquaintances at home. This was made apparent when cross-referencing her diary entries with those of her husband, which in some cases presented an alternative view of a particular encounter or experience. In general terms, Lady Dalhousie's diaries often reveal her deep appreciation of the beauty of her surroundings and an optimistic view of life, whereas Lord Dalhousie was more self-reflective in his journals and tended to display a more pessimistic outlook. With limitations, such as illegibility as an issue, and interpretive challenges notwithstanding, the journals and correspondence left behind by Lady Dalhousie in private and public archives have helped provide a richer understanding and appreciation of her life, in particular her work as a plant collector in the early part of the nineteenth century, and it is fortunate that family members preserved her personal documentation, even though at the time they may not have been aware of its value to contemporary social historians. As Alistair Thomson remarks, "History is the story we make of the stories we find. Our first duty as historians is to hear those found stories and understand what they meant in their own time."[9] The stories uncovered within the archives have made possible a fuller profile than heretofore of Lady Dalhousie and her contribution to the floristic history of Canada.

Heiress to her father's fortune, Christian Broun was born into wealth and privilege in Scotland on 28 February 1786 and became a member of the Scottish aristocracy when she married George Ramsay, 9th Earl of Dalhousie, in 1805. She also appears to have inherited a love of the natural world from her father and his family. Her father Charles Broun, the 22nd Laird of Coalstoun, was an advocate and "an authority on agriculture and the planting of hedges," according to his descendant Edith Broun Lindsay.[10] These horticultural interests were also shared by Christian's great-grandfather, Charles Broun of Gleghornie, who played an active role in managing and improving the Coalstoun estate and is credited with the planting of "some fine larch trees, brought to Scotland in 1725, a magnificent pair of copper beeches, and the beech avenue lying below Colstoun Mains, which he planted to commemorate the Union of the parliaments."[11] Her mother, Christian McDouall, died only two days after giving birth to her daughter and Christian was brought up by her father at the family seat of Coalstoun, located in the parish of Haddington in East Lothian, which at the time encompassed more than 2,000 acres of agricultural pasture and parkland. Although little is known about her early formative years, it is possible that her father shared his interest in horticulture with his daughter, at least until she reached the age of thirteen, when she was sent to England for schooling. Daughters of the Scottish gentry and richer lairds and merchants generally were educated at home by governesses in preparation for marriage. The absence of a mother may well account for her father's decision to send Christian to a school run by the Misses Carver in Doncaster. A letter that survives from one of the Misses Carver to Charles Broun provides a summary of Christian's education, which consisted of "French, Music, drawing and the writing of figures, English reading and Geography of the globe."[12] When not engaged in these pursuits her time was spent "taking exercise and amusement."[13] Her education was typically narrow, confined to those subjects deemed acceptable feminine accomplishments, and intended only as preparation for a good marriage, but Miss Carver's letter also reveals a glimpse of Christian's personality – her "charming" and "most amiable disposition" and her application with the "utmost attention" to her work, while at the same time being "full of animation and spirits," characteristics that she was to display throughout her life.[14]

An early awareness of nature, especially trees, is also evident in her journal entries. In the summer of 1802 at the age of sixteen, Christian undertook a tour of England and Wales with the Misses Carver and described the orangery she visited at Margam near Penrice in South Wales. "Behind the orangerie," she wrote, "is a fine winding gravel walk planted on each side with beautiful arbutis trees. We then turned into another walk which led to a lovely grove of orange and lemon trees … Behind these, are uncommonly fine tulip trees, one 100 feet high. I never saw a sweeter place than Margam, the grounds are beautiful in themselves and behind is a mountain clothed with trees to the very top."[15] A further account of her visit to Bonvilston, a Welsh village set in the Vale of Glamorgan, illuminates her interest in and growing knowledge of plants, as she wrote how

> we drove thro' a beautiful little village called Bonvilston which perfectly realises all that has been written about the romantic cottages in

> Wales. They are all neatly thatched and whitewashed, with neat little flower gardens and they are all literally covered either with Roses, Woodbine, Ivy or Vines trained over them, they have all casement windows and indeed altogether give one more the idea of what I have read of Italian cottages rather than British ones. The hedges are filled with wild Roses, the most beautiful I ever saw, Woodbines and Fox Glove in the greatest perfection.[16]

On a further tour of England and Scotland seven years later, Christian, now the twenty-three-year-old and married Lady Dalhousie, displayed a more confident and assured vocabulary when describing a garden she had visited close to Exeter in Devon. "The flower garden," she explained, "is perfect in its stile [style]. The flowers and shrubs grow in the greater luxuriance and are disposed in careless groups interspersed with lawns green and smooth … the magnolias grow to the size of laurels and the beautiful verdure of their leaves was contrasted by the smooth whiteness of their tulip shaped flowers."[17]

Despite Lady Dalhousie's obvious fascination with the natural world, and in particular her enthusiasm for trees and flowering plants, there was little opportunity for her to convert this interest into any formal horticultural or botanical qualification. In the eighteenth century, myths and literature linked women and gardens with beliefs about modesty and innocence as "feminine" virtues. Following the introduction of new exotic flowers from all over the world, botanizing became a fashionable domestic activity among aristocratic women from the 1760s onwards. This phenomenon can be attributed in part to Queen Charlotte who was a keen amateur botanist at her gardens in Kew. By the first half of the nineteenth century, women "read botany books, attended public lectures about plants, corresponded with naturalists, collected native ferns, mosses, and marine plants, drew plants, developed herbaria for further study, and used microscopes."[18] However, while men could develop their interest in plants through the formal academic study of botany at university, usually as part of a medical degree, such university training was denied to women in Britain until the late 1860s.[19] In addition, women's role within the world of botany became increasingly marginalized. As Catherine Horwood explains, in the late eighteenth and early nineteenth centuries two competing systems of plant identification, one developed by the Swedish botanist Carl Linnaeus (1707–1778) and the other by the Swiss botanist Augustin Pyramus de Candolle (1778–1841), had led to a "battle for scientific fame" during which women "found themselves increasingly excluded from involvement in the male-dominated world of academic botany."[20] Attempts by male proponents in Britain, principally John Lindley (1799–1865), the first professor of botany at London University, to differentiate scientific botany from drawing room botany led to the marginalization of women from the 1830s onwards.[21] Gendered distinctions between science on the one hand and polite accomplishments on the other personified the botanist as "male and masculine and the botanophile [as] usually female and feminine."[22] Women's role within the world of botany diminished, and by the 1850s, botany had been shaped very much as a science for men. As a result, British women who travelled and collected plants in the opening decades of the nineteenth century were as rare as the specimens they found.[23]

Although there is no evidence to suggest that Lady Dalhousie sought any formal advancement of her botanical knowledge and nothing within her manuscripts to indicate that she attended talks on horticulture or botany, she was an active gardener. As a young woman, she cultivated her early interest in flowers and trees at Coalstoun, often making reference to having spent a morning or afternoon working in the garden there.[24] At the back of one of her diaries she notes planting the middle of the shrubbery in spring 1806 and the flower garden in 1813 but unfortunately for researchers gives no specific details about the species she planted: written accounts of the garden at Coalstoun or visual references to it have yet to be found.[25] It may, therefore, be possible to link Lady Dalhousie's eventual turn toward botany to her growing knowledge of plants and her predilection for gardening and horticulture.

Her marriage to George Ramsay in May 1805 put Lady Dalhousie in the privileged position of being able to garden not only at Coalstoun but also at Dalhousie Castle, the estate of the Ramsay family, where she recorded "planting hollies and American plants in 1808" and "roses and hardy shrubs" in the spring of 1809.[26] George Ramsay had joined the army at the age of eighteen in 1788 and served in Gibraltar, the West Indies, Ireland, and Egypt before taking up the duties of brigadier general on the staff in Scotland in 1803.[27] According to Whitelaw it was a good match, for the Ramsay family was not rich and the marriage, despite the sixteen-year age difference, appears to have been a happy one, grounded it would seem by their shared interest in plants.[28] A letter written by Lord Dalhousie in 1812 to his young wife from Villafranca in Spain, where he was serving in the army of Lord Wellington, reveals his desire to share with her the horticultural detail of his surroundings, describing the mountains as being covered with "thick evergreen Oak, and in the bottoms and gullies, very magnificent woods of Walnut and Chestnut which are generally pollarded."[29] When at home, Lord Dalhousie took an active role in the development of his land and the castle in which he had been born. He was

1.2
Dalhousie Castle and parkland ca. 1824 after James Nasmyth.

especially interested in trees and committed to the large-scale planting of woodland throughout his estate; a memorandum made by his gardener, Joseph Archibald, in 1820, details 59,500 trees in the Dalhousie nurseries which were fit to plant out, including 26,000 oaks, 22,300 Scotch fir and 5,700 ash trees.[30] In 1824, the noted landscape gardener and horticultural writer John Claudius Loudon commented on the success of Lord Dalhousie's efforts, describing the parkland at Dalhousie Castle as being "surrounded by extensive and romantic pleasure-grounds. The river Esk washes the base of the castle; and its lofty wooded banks afford delightful summer walks."[31]

Lady Dalhousie reciprocated her husband's interest in agriculture and horticulture, and their marriage appears to have been a partnership based on a mutual interest in the cultivation of plants. Early on, they embarked on a series of projects to remodel the castle and gardens. They employed Scottish landscape gardener Walter Nicol (1769–1811) in around 1806 to create a principal approach from the north and in the same year they contracted garden architect John Hay of Edinburgh (1758–1836) to design the garden and its hot houses.[32] Writing in *The Gardener's Magazine* in 1826, Archibald provided a description of the four-acre site, which housed a range of glazed structures 203 feet in length, including a greenhouse, two vineries and two peach houses; borders and plots for American plants; banks of rhododendrons and shrubbery borders.[33] Rather than enclose the gardens with straight walls, a series of squares and parallelograms were created using curved and winding walls, which according to Archibald had been "much admired by every person of taste who has visited it."[34] When the garden writer Jane Loudon visited Dalhousie Castle in 1841, she noted the "very fine American thorns and Acers in the shrubberies" which Lord Dalhousie had sent home from Canada and remarked that "the late earl and his lady, who was an excellent botanist, were both very fond of this place; and the whole of the pleasure-ground bears evident marks of their taste in laying it out."[35] Despite giving birth to three sons in the first seven years of her marriage, Lady Dalhousie was also involved in the management of the estate and gardens, particularly during her husband's long periods of absence, as evidenced in letters written by Lord Dalhousie to his wife. Writing once again from active duty in Spain, he sanctioned her to "by all means plant the rough corner in the sheep park, and enclose it at once with hedge and ditch [and] go on with that levelling at Goody's Hole."[36] In a subsequent letter to his wife, he indicated that the newly created garden at Dalhousie Castle was under her jurisdiction. "I have ordered a paper of directions about the grass parks at D. [Dalhousie] Castle, also about the planting," he wrote, and "I leave to you to give what other directions may be occasionally wanted, and the garden of course is under your especial care and management."[37] This division of labour and expertise continued throughout their married life, with Lord Dalhousie having overall responsibility for the utilitarian aspects of managing the farms and estates, while Lady Dalhousie took control of the ornamental flower gardens.

The opening of the nineteenth century heralded the decline of the English landscape garden style in favour of a renewed focus on the formal display of plants in Regency gardens.[38] Lady Dalhousie made frequent visits to London where she patronized many of the plant nurseries of the day, including James Lee and John Kennedy's The Vineyard at Hammersmith, renowned for its ability to supply and cultivate newly introduced exotic plants.[39] John Kennedy, the partner of the firm, and his son Lewis also produced planting designs and at the height of his success in the 1820s Lewis specialized in the design of geometric gardens and park landscapes.[40] Frustratingly, Lady Dalhousie's diary entries do not reveal the plants she bought or the inspiration that she may have sought on her nursery expeditions, but they do highlight her exposure to the horticultural fashions of the day, namely, the thirst for new plant introductions and the move away from the informality of the English landscape garden to more formal garden designs, both of which had been adopted by

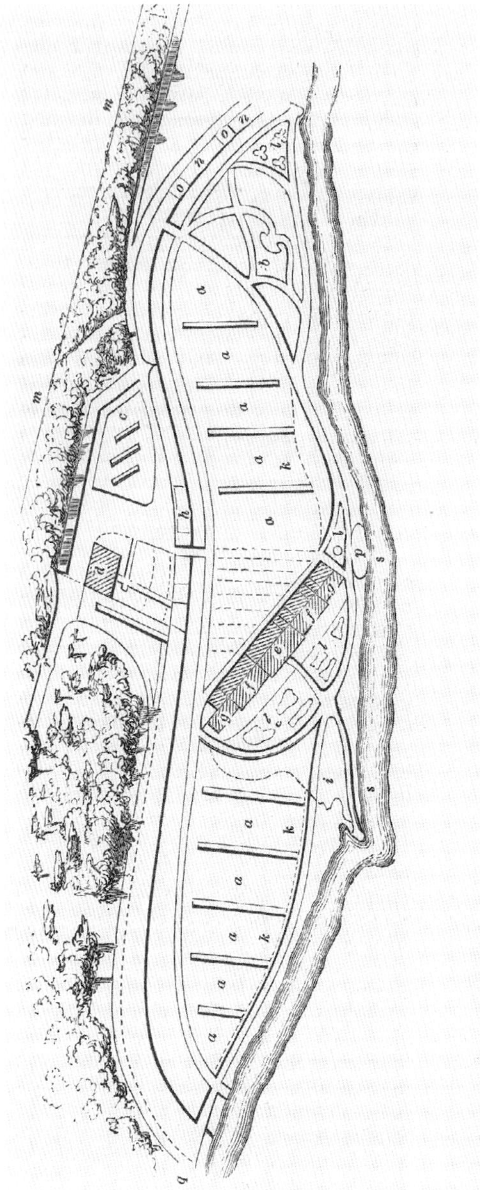

1.3
Plan of the garden at Dalhousie Castle, designed by John Hay in 1806.

Key:
a vegetables and small fruit
b border for American plants
c melon ground
d gardener's house
e greenhouse
f vineries
g peach-houses
h open shed
i bank of rhododendrons
k line of variegated hollies
l flower-beds on grass
m sunk fence
n flued wall
o stoke holes
p shrubbery borders
q walk towards castle
s South Esk river

Lady Dalhousie and her husband at Dalhousie Castle. They also indicate her growing interest in plants, with many of the pages of her journals interlaced with dried specimens of ferns, flowers, and foliage.

The opportunity for Lady Dalhousie to apply her practical knowledge of plants beyond the confines of the family home came with her husband's imperial appointment as lieutenant governor of Nova Scotia in 1816. Many of Wellington's peninsular officers embarked on careers as colonial administra-

1.4
Lady Dalhousie with her third son, the Hon. James Ramsay, by William Douglas, 1816.

tors after the wars and Lord Dalhousie was no exception.[41] Leaving behind their two elder sons in the care of Dr Pearson who kept a school at East Sheen in the London borough of Richmond, Lord and Lady Dalhousie, together with their youngest son James, set sail for Halifax in Nova Scotia on 11 September 1816, with Lady Dalhousie wryly noting down her reading matter on the long voyage: "Read *Shipwreck*!"[42] She proved a "capital sailor" according to her husband who wrote, "when most of us were objects of pity, she felt it not in the least."[43] Following their first few months in Halifax, Lady Dalhousie confided in a letter to her cousin Miss Christian Dalrymple that "it was a very severe trial for me to leave the children, and every person, thing and place, that for 30 years I had been accustomed to love," but despite this separation she found Halifax more agreeable than she expected, and her delight in the native flora was some recompense.[44] "I expect to see many beautiful plants and flowers," she wrote. "Heath or Whins were never heard of or seen in America, instead of them the Common round the town is covered with the Kalmia latifolia, Rhodora Canadensis, Ledums, Fern-leaved Gale etc etc all plants which we cultivate with infinite care in our flower gardens!"[45] It is likely that her mention of "Heath" would have referred to the heather, *Calluna vulgaris*, and "Whins" was a colloquial term for *Ulex europaeus* or the common gorse, both of which originated from Western Europe. What is evident from her comment is not only the joy at seeing native plants growing in their natural habitats but also her determination to discover more of the local flora.

This desire to seek out native plants was often at odds with her official role, however. Within weeks of arriving in Halifax, Lady Dalhousie was immersed in the duties incumbent on her position as the governor's wife. She hosted her first of many balls in November 1816, held regular "drawing rooms" for the ladies, and accompanied her husband at official functions. To enliven the society in Nova Scotia, Lord and Lady Dalhousie set up an amateur theatre company and Christian produced caricatures of the Halifax elite, examples of which can be seen in the Nova Scotia Museum in Halifax.[46] Being the wife of the governor was a guarantee of social acceptance, but Lady Dalhousie did not always enjoy the role she was expected to play. On occasion, she felt uncomfortable in the company of the ladies, confiding in a letter to her cousin that "the young ladies never speak at all, and the elder ones seem to think it necessary to be always of my opinion; then they always flatter me – some well and some ill – and if I was not well aware of, and on my guard against this, I should certainly run much risk of fancying myself very near perfection."[47] It is possible, although this is not stated explicitly in her journals, that her interest in plants gave Lady Dalhousie an outlet to escape from the obsequious social circle within which she moved. She appears to have found time within her busy schedule not only to notice and collect plants but also to initiate what would become a regular exchange of seeds and rare native plant specimens between herself and her gardener Archibald at Dalhousie Castle during her time in Canada. In a letter sent by her eldest son to Lord Dalhousie in 1817, evidence of Lady Dalhousie's early botanical transactions is conveyed. Lord Ramsay asks his father to "Pray tell Mama that the Saracenia [*Sarracenia*] purpurea is doing very well but that Archibald has not got a Rhodora [*Rhododendron*] canadense."[48] In larger British landscapes, the American garden was a common feature in the early nineteenth century (usually planted to display the floral wealth of North America, it justified its existence as a piece of "botanical geography") and this perhaps explains why Archibald was so keen to procure this particular specimen of rhododendron to add to the growing collection of rare native plants from North America at Dalhousie Castle.[49]

Tragedy and anxiety also played a role in Lady Dalhousie's botanical journey of discovery. Her involvement with plants became a source of comfort to her when, in the summer of 1817, a packet boat brought news from England that Charles, the Dalhousies' ten-year-old middle son, had died in a measles epidemic and that their eldest son Lord Ramsay had also contracted the disease.[50] They had to wait a further two weeks before news came that Lord Ramsay had recovered, and in November of the same year Lord and Lady Dalhousie made plans for him to join them in Halifax. Lady Dalhousie was consumed by grief for Charles and it is poignant to note that the only blank pages in her journals coincide with the loss of her children and husband.[51] Evidence from her journals and letters indicates that the collection, identification, and preparation of plant specimens were a source of comfort to Lady Dalhousie throughout her time in Canada and India. Physician and botanist Robert Graham (1786–1845), who held the position of regius keeper at the Royal Botanic Garden in Edinburgh from 1820 to 1845, recognized this in an obituary following her death. He wrote that Lady Dalhousie "first began the study of botany as a solace in circumstances of severe domestic affliction, and she clung to it ever after with a devotion which marked her consciousness of the blessings it had conferred."[52]

By the spring of 1818 references in Lord Dalhousie's journal indicate that Lady Dalhousie had started to collect once more. He describes her "driving out daily in an open Barouche as in summer," not only to be seen socially but also to collect plants, and the following summer Lady Dalhousie made an excursion to Mahone Bay, situated eighty-five kilometres to the west of Halifax, where she collected plants and minerals.[53] Lord Dalhousie's journals also indicate her growing interest in and knowledge of the flora and fauna in and around Halifax, which she was happy to share with her husband. "This evening," he recalls, "the ground is covered with a brushwood mixed of Kalmia, Rhodora and another shrub, bright red, which Lady D tells me is called 'Candleberry Myrtle' – in contrast to this red are the higher woods, Beech, Birch, Poplar & Maple, all shades of delicate red & yellow leaves, dropping as by every puff of wind."[54]

Lady Dalhousie's "study of botany" seems to have been largely self-taught. In addition to the practical experience gained from working within the gardens at Coalstoun and Dalhousie, much of her plant knowledge appears to have been derived from books. An avaricious reader, she either noted with increasing frequency in her journal entries what she had been reading or included a list of books read at the end of each journal. Having analyzed the types of books she was reading and the years in which they were read, it is apparent that what started as a general interest in science and the natural world seems to have developed into the specific study of plants. From 1813, her reading matter included classics such as Shakespeare's *Macbeth* and Milton's *Paradise Lost*, but by the time she was living in Halifax, her interest had widened to include scientific work such as Robert Bakewell's *Introduction to Geology*, as well as *Conversations on Chemistry*, first published anonymously in 1805 by Jane Haldimand Marcet (1769–1858), who wrote introductory books on science and natural philosophy, principally aimed at a female readership to counteract the notion that science was unsuitable for women.[55] On 28 January 1823, she refers to reading "Botany" which, when cross referenced with the inventory of books in the library at Dalhousie Castle, would appear to refer to *English Botany*, a work by the naturalist James Sowerby and botanist Sir James Edward Smith, published in eight volumes between 1790 and 1814.[56] This is the earliest evidence of Lady Dalhousie reading a work specifically on botany and it coincides with her stay in Halifax.

Dissection of her reading matter reveals that during her time in Canada, Lady Dalhousie was in the early stages of developing, largely through self-study and practical application, both her scientific and botanical knowledge. By the time she was collecting in India in the 1830s, much of her plant knowledge appears to have been derived from reading botanical work and it seems reasonable to speculate that Lady Dalhousie consulted books on botany and specific flora in order to help in the identification of plants.[57] She was also keen to procure books as soon as they were published. This is corroborated by a letter found in the archive from her friend Isabella Houstoun, dated 22 February 1830, which evidences Lady Dalhousie's acquisition of botanical publications. "I shall begin 'faute de mieux' with Botany," wrote Isabella. "Wallich's first number is come out and is splendid. Dr. G. [Dr Robert Graham] is anxious to know what you would wish done with yours; whether they are to be sent to India or to D. castle, the latter I think … Dr. G. likewise bids me tell you that there is a supplement to English Botany published in monthly numbers at 2/6 a number, that you must get by different hands and very well executed, several of the numbers are already out."[58]

Plant-hunting expeditions were another feature of Lady Dalhousie's botanical life in Nova Scotia. In July 1819, she accompanied her husband to the city of Quebec where they undertook a number of excursions, including a visit to the falls of Chaudière on the river Chaudière at Charny, close to its mouth with the St Lawrence. Despite having to walk the final two and a half miles to reach the falls in oppressive heat, she spent time there drawing and noting the plant life.[59] Watercolours of botanical specimens believed to have been painted by Lady Dalhousie around this time can still be seen today in the archive collection at the Nova Scotia Museum.[60] Despite Lord Dalhousie's general acceptance of his wife's botanizing excursions, evidence found within his journals indicates a degree of gendered disapproval of some of her activities, which he felt were appropriate for men but not for women to engage in. "Yesterday," he wrote in January 1820, "Lady D. with a party of 13 sleighs again went to the Dartmouth Lakes … a distance of 30 miles. An unlucky accident however happened by which Mrs. Col. Goodriche has had her arm broken by the unsetting of her sleigh. I am not surprised at this, having dreaded some accident every time they have gone out. It is a feat of hardihood & danger for Ladies, tho' nothing for Gentlemen to perform. I have all along discouraged this romping fashion & my apprehensions have proved but too well founded."[61] There is further evidence that Lord Dalhousie did, on occasion and albeit with the best of intentions, restrict his wife's desire to travel, forbidding her to accompany him on a tour of inspection of the military settlements of Richmond and Perth in August of the same year and writing in his journal that "she would willingly go with us on the tour, but I fear the difficulties of the new settled country we are to pass, & dare not risk her attempt of it."[62] It is possible, although not overtly stated in either journal entries or correspondence, that her husband's seemingly paternalistic attitude and desire to protect his wife may have stemmed from the sixteen-year age difference between them. What is certainly evident is that Lady Dalhousie's adventurous zeal and readiness to travel and botanize were, at times, tempered by the cultural restrictions placed on her as a wife and mother.

Happily for her, however, a change in circumstances of her husband's career presented a further opportunity for Lady Dalhousie to explore the flora of Canada and consolidate her plant knowledge. On 12 April 1820, Lord Dalhousie was promoted to captain general and governor-in-chief of Upper and Lower Canada following the unexpected death of the Duke of Richmond, and the family moved to the city of Quebec a few months later. On her arrival, communication between Lady Dalhousie and her gardener Archibald indicates the determination of both to fully exploit the botanical riches of British North America. In a letter to Lord Dalhousie, dated 2 June 1820, Archibald urged Lady Dalhousie to "make a good collection of plants" for him in Canada and suggested that "a good many plants of a kind ought to be sent as some of

1.5
Watercolour of pink flowered plant by Lady Dalhousie ca. 1820.

them are very likely to die and small plants of such would pack into little room."[63] This letter is interesting because it reveals that Archibald may have been influential in advising Lady Dalhousie on the size and quantity of plants to send from Canada to Scotland. The archive also yielded one other piece of information which holds the key to answering how she was able to successfully send home the "trees and grafts of apples, and also an additional collection of American plants" promised to Archibald.[64] An invoice issued four months later, dated 23 October 1820, lists "450 plants and seedlings, 98 species of seed and 10 apple trees" bound for Dalhousie Castle, which were sent by the nurseryman Robert Cleghorn of Blinkbonny Garden in Montreal.[65] A note accompanies the invoice from Cleghorn, explaining the "great pains" that "have been taken to pack them in such a manner as they will, with common care on the passage, reach Dalhousie Castle in good order."[66] It is evident, therefore, that Lady Dalhousie was collecting and transporting plants with the help of her gardener Archibald and the nurseryman Robert Cleghorn who was to become a close friend of the Dalhousies.[67] This botanical exchange, initiated in Nova Scotia, continued for the rest of the Dalhousie's time in Canada and the botanic garden in Edinburgh also benefited from the process. An entry in the Royal Botanic Garden Edinburgh's donations book of plants and seeds reveals that Lady Dalhousie actively collected the Canadian seeds that she gifted in 1822, including seed from the thistle *Cnicus bicolor* which she records as having "gathered from a plant that reached my eye as I sat on Cherub [a horse] 15 hands high."[68] She also subsequently donated Canadian plant specimens to the Royal Botanic Garden Edinburgh in 1826 and again in May 1831 when she sent seed from East India.[69]

Plants were again a salvation to Lady Dalhousie as life in the city of Quebec fell short of expectations. Their official residence, the Château Saint-Louis, was ill equipped and in need of refurbishment, and Quebec seemed to them a more constrained and provincial society.[70] Whenever possible, they escaped to their official summer residence which was "prettily situated on the bank of the Richelieu about a mile from Sorel on [the] St Lawrence" and provided the perfect base for Christian's plant-hunting activities.[71] Admitting in a letter to her son James that she was "wearied of Quebec and its gaieties," Lady Dalhousie expressed the sole desire when in Sorel "to drive my phaeton and to gather flowers."[72] She was, however, constantly aware of the inadequacy of her botanical knowledge and described in another letter from Sorel how she was "overwhelmed with the number of lovely wild flowers that are springing around us … and as we find no-one here who can tell us the name of one flower, we are driven by necessity to find them all out from books."[73] Lord Dalhousie also provides a vivid account of his wife's botanical pursuits with his visiting niece, writing that "Lady D. & Georgina are daily out in search, & daily also return with some new treasure found – then fly to the pressing Board, & to Botanical Books to ascertain the Plant found – their occupation is very interesting."[74] There is little doubt that these early years spent collecting in Sorel provided Lady Dalhousie with both plant knowledge, mostly gained through a program of discovery and self-improvement, and the experience of collecting, labelling, and drying specimens, which were to prove so valuable to her during her later plant-hunting expeditions in India.

The Quebec years were punctuated by a visit home to Scotland in 1824 during which Lady Dalhousie made contact with two botanical men who were to provide her with encouragement and support in her future plant-hunting

1.6
The official summer residence of Lord and Lady Dalhousie at Sorel by John Frederick Fitzgerald De Roos, 1826.

expeditions. Lord Dalhousie's governorship of Upper and Lower Canada had been beset by problems, and he began to experience first signs of the bad health that was to plague him for the rest of his life. He described being laid low by an inflammation of the eyes and "headaches & fullness which weaken & derange me very much."[75] He also received news from Scotland that his estates had been mismanaged, and as a result, he applied for an extended leave of absence to return home in June 1824. Within days of arriving back at Dalhousie Castle, Lady Dalhousie made the first of several trips to the Royal Botanic Garden in Edinburgh where she made the acquaintance of Regius Keeper Dr Robert Graham and also to the Botanic Gardens in Glasgow, which had been created by the Royal Botanical Institution of Glasgow in 1817.[76] On her return to Canada a year later, she began to correspond with William Jackson Hooker, the professor of botany at Glasgow University, promising in one letter to send him a "very small and imperfect collection of plants" that she hoped would include some "new, or at least rarer species" of orchid.[77] She ended her correspondence with this diffident offer of help: "I shall endeavour to send your paper of directions into various quarters of the provinces – & to procure plants – but over that I do not feel very sanguine of success – zeal on my own part, shall not, depend upon it – be wanting, & it will give me pleasure to hear in what way I can be useful."[78]

Lady Dalhousie's social class was clearly an advantage, as it allowed her to meet and, more importantly, to engage with men of considerable influence in the world of Scottish botany, but her language was often self-deprecating when referring to her botanical work and she appeared to lack confidence in her own ability. This was not a view shared by her gardener, Archibald. He listed more than twenty rare natives of North America collected by her, which had flowered in great perfection at Dalhousie Castle.[79] One of these, *Habenaria blephariglottis*, the white-fringed orchid, had made its debut in Hooker's *Exotic Flora* in 1825, and it is described by Hooker as a charming plant which "does not appear to have been known in our collections, till it was introduced into the garden of Dalhousie Castle by the Right Honourable the Countess of Dalhousie, who sent it from Canada."[80] Lady Dalhousie's contribution was, therefore, clearly significant in terms of introducing new and rare Canadian plants to Britain.

Her correspondence with Hooker reveals that Lady Dalhousie was an important source of plants from Lower Canada and his gratitude was expressed by enrolling her as an honorary member of the Royal Botanical Institution of Glasgow in 1825.[81] In the course of the next three years she continued to send specimens to Hooker in response to his request for help in the preparation of his *Flora Boreali-Americana*, designed to contain information on plants found across the northern parts of British America. During this time, Lady Dalhousie found herself part of a triad of female collectors within colonial Quebec, each contributing to Hooker's project.[82] Diary entries chart the growing friendship between Lady Dalhousie and fellow collector Anne Mary Perceval (1790–1876), wife of the director of customs for the port of Quebec, who according to Shteir and Cayouette became Hooker's "linchpin" for botanical contributions from Quebec province.[83] Following the formation by her husband of the Literary and Historical Society of Quebec in 1824, Lady Dalhousie also made the acquaintance of Harriet Sheppard (1786–1858), a keen collector who contributed many plant specimens to *Flora Boreali-Americana*. Although the extent of Lady Dalhousie's influence within this group is ambiguous, she was credited by Sheppard's husband with "imbuing her lady friends with a love of botany; some of whom made marked advances in this branch of natural history."[84]

Lady Dalhousie determined to maximize her remaining time in Canada from a botanical perspective. Together with her husband, she had taken an active role in around 1823 in the development of a botanic garden to be called the King's Gardens on the Île Ste-Hélène, an island in the St Lawrence River near Montreal and used by the British government as a fort.[85] The garden, according to McFarlane the gardener, was intended to exceed in number any "collection of American plants (in cultivation) on this side of the Atlantic" and he was instructed by Lady Dalhousie to not only increase the collection but also to improve its cultivation.[86] An inventory of 232 North American plants cultivated in the King's Gardens in 1824, found within the Dalhousie archive, indicates that approximately 75 per cent of them were native to Quebec.[87] Unfortunately, Lady Dalhousie's instructions to McFarlane for plants to be sent to her in Sorel were challenged by the commander of the island's fort, Captain Cardew, who "did not see how Lady Dalhousie could presume to give orders to anyone under his Command and by not attending to these orders it would be a means of giving her a check not to do so for the future."[88] It is not clear whether Cardew's stance was simply indicative of the hierarchy attached to British military command or a sign that he was not prepared for his authority to be challenged by a woman. Tensions developed between Captain Cardew and Lord Dalhousie, who supported his wife's development of the garden, and a decision was finally taken in July 1827 to remove the whole collection of plants to Sorel.[89] This arrangement suited Lady Dalhousie and she continued to actively collect both in Sorel and Quebec. At times, her work took her into rather challenging environments, as when Lord Dalhousie recorded that "Lady D. returned much gratified with St. Anne's [Sainte-Anne-de-Beaupré], but the black fly on the mountain woods of St. Fereole [Saint-Ferréol-les-Neiges] punished her face most dreadfully, shutting up both her eyes, and otherwise so disfigured that I could scarcely recognise her."[90] Some of the specimens she collected in 1827 were donated to the Literary and Historical Society of Quebec but sadly this collection is believed to have perished in fires that swept the society's headquarters in 1854 and 1860.[91]

Diary entries spanning the years of her botanical work first in Canada and later in India reveal Lady Dalhousie's sense of inadequacy about herself as a plant collector. While she recorded many detailed descriptions of the plants she collected and referred to them by their correct taxonomic names, she rarely used the term "botany" or "botanizing" to describe her activities and did not refer to herself as a "botanist." She acknowledged her involvement in "botanical researches" in a letter to her eldest son in 1822 but in most of her diary entries used the term "arranging" to refer to her work with plant specimens. In an entry dated 20 March 1829 she recorded visiting the botanic garden in Edinburgh and spending the next day "arranging my plants."[92] "Arranging" does not appear to have been a term widely associated with botany at the time, which would suggest that Lady Dalhousie used it simply to refer to her own handling of plant specimens. Yet evidence of the success of her botanical endeavours is unmistakable in the personal herbarium of almost 300 specimens, which Lady Dalhousie brought home to Dalhousie Castle from British North America in 1829. In the late 1990s, this collection was discovered in the ancestral home of Lady Dalhousie's family in Scotland and was subsequently purchased by the Royal Botanical Gardens in Canada. According to the gardens' taxonomist Dr James Pringle, the plant identifications in this collection are remarkably accurate, and he is of the opinion that Lady Dalhousie and her friends in Quebec may have identified most, if not all, of the specimens themselves, aided only by Lady Dalhousie's extensive botanical library.[93] Of national importance, her British North American collection is thought to comprise some of the oldest specimens of Canadian plants known to be in any Canadian herbarium today, and work to digitally image these specimens for ongoing taxonomic research is currently being undertaken by the Royal Botanical Gardens (Canada).[94]

Once again, travel was to play a part in Lady Dalhousie's development as a botanist. Her husband's career in British North America had taken her to Halifax and Quebec and was about to transport her to the outer reaches of the British Empire. As 1827 drew to a close, Lord Dalhousie was increasingly unhappy in his position as governor, and ministerial support for him from the British Government also appeared to be waning. The prospect of military service in India loomed. Early the following year, he was appointed commander-in-chief of the British Army in India and the Dalhousies left Canada for good on 9 September 1828. Although saddened at the thought of once more being separated from her home and family, it is clear in a letter to her son James that Lady Dalhousie knew where her responsibilities lay. "I cannot well bear the idea of being once more absent for years," she wrote, "& yet it is only what is my duty if your father goes."[95] The British government had for many years sought to exploit and dominate India in its quest to expand its empire, and according to Catherine Hall a "set of related discourses" prevailed in Britain, which "constructed India as a degraded place in need of civilisation."[96] As one of only a small but growing number of British women residing in India before direct rule was imposed in 1858, Lady Dalhousie's role as a military commander's wife would have been an important element in this "civilising" process, just as her husband's duty in India was to oversee colonial rule.

While in India during the years 1829 to 1832, botanizing for Lady Dalhousie was a welcome distraction from the duties incumbent on her role. She amassed

1.7 *Right*
Gentianella quinquefolia (L.) Small subsp. *quinquefolia* (present identification) collected by Lady Dalhousie in Sorel on 3 September 1827.

1.8 *Opposite*
Herbarium specimen of the fern *Asplenium dalhousiae*, collected by Lady Dalhousie in 1831 and named after her by William Jackson Hooker.

an extraordinary collection of more than 3,000 rare and native plant specimens there, many later donated to the Botanical Society of Edinburgh and the Royal Botanic Gardens, Kew in London. Among her finds were plants not previously known in Britain, including orchids and the ferns *Asplenium dalhousiae*, *Cheilanthes dalhousiae*, and *Pteris dalhousiae*, which were all named after her by William Jackson Hooker. Other species of plants collected by Lady Dalhousie in Simla and subsequently named in her honour include *Cynanchum dalhousiae* Wight; *Epipactis dalhousiae* Wight; *Goldfussia dalhousiana* Nees; and *Ophelia dalhousiana* Griseb. One other, *Pterisanthes dalhousiae* Planch, was collected by Lady Dalhousie in Penang. Although she did not collect species of *Dalhousiea*, the name for this genus of tropical leguminous shrubs was proposed by Dr Robert Graham to commemorate her botanical prowess

and it is a greater mark of respect that he chose to name a genus after her rather than a species.[97] The pioneering nature of her work in India was also personally acknowledged by Hooker, who wrote to her lamenting the lack of botanists who were "able & willing like your Ladyship, when in that country, to collect & preserve dried specimens with which our herbaria might be supplied."[98] Had it not been for the illness of her husband and son which resulted in their early return home to Dalhousie Castle in 1832, Lady Dalhousie might have gone on to discover many more floristic riches.

Hooker further praised the work of Lady Dalhousie when dedicating volume 60 of *Curtis's Botanical Magazine* to her in December 1833.[99] He wrote, "The Countess of Dalhousie, … notwithstanding the varied duties of her exalted station, both in the Canadas and in Bengal, has rendered essential service to botany by her extensive collections, and by the introduction of many interesting species to the gardens of this country."[100] It is clear from the sentiments expressed in his dedication that Hooker placed great value on Lady

Dalhousie's botanical work, but it is also interesting to note his recognition first and foremost of the duties expected of her "exalted station." Lady Dalhousie's position as wife of a senior imperial officer both encouraged and inhibited her success as a plant collector. On the whole, her husband supported her botanical interests and provided the circumstances for her to accompany him on tours of duty overseas, a unique opportunity even for a woman of her class in the opening decades of the nineteenth century. At times, however, Lady Dalhousie's extensive duties as a governor's wife and mother to three sons prevented her from allocating time to collecting plants and, on occasion, her capacity to explore the botanical riches of British North America was curtailed by her husband. Furthermore, the attitude of Captain Cardew forced Christian to abandon her plans for a botanic garden on the Île Ste-Hélène.

With access to formal education and training, Lady Dalhousie could have achieved so much more from a botanical perspective. It is clear, when viewed through a gendered lens, that this was not a possibility in her lifetime. Even aristocratic women like her were not privy to a university education, unlike their male counterparts, and Lady Dalhousie relied instead on deriving her botanical knowledge from books and practical experience. She was often frustrated by gaps in her botanical knowledge and felt her shortcomings keenly, admitting in a letter from India to William Jackson Hooker that she "was totally unaware that the cold – the hot – and the rainy season each furnish a flora totally distinct, and I was not a second season there, to correct my mistake."[101] She was unerringly deferential and self-effacing in her correspondence with botanists, such as Hooker, who went on to benefit from her work by incorporating it in published works of their own.

Having nursed her husband through his final illness in 1838, there was little time left for Lady Dalhousie to continue her botanical work. She died suddenly in Edinburgh on the 22 January 1839 at the age of fifty-two, having only minutes before been discussing a list of plants with her friend Edward Bannerman Ramsay, a clergyman of the Scottish Episcopal Church who was appointed dean of the diocese of Edinburgh in 1841. Dr Graham recorded in his obituary on Lady Dalhousie that "almost her last conversation turned upon the subject, and the intrusion of death, without even momentary warning, found in her hand a list of plants which she had just prepared as objects of attention during the ensuing season."[102] In a fitting tribute, Ramsay described Lady Dalhousie as being "eminently distinguished for a fund of the most varied knowledge, for a clear and powerful judgment, for acute observation, a kind heart, [and] a brilliant wit."[103] In her short lifetime, Lady Dalhousie

1.9
Portrait of Lady Dalhousie by Sir John Watson Gordon, 1837.

had overcome her lack of formal botanical education, balanced the demands placed on her as a wife and mother, and made important contributions to the floristic exploration of British North America in the early part of the nineteenth century. Her early love of plants blossomed into a welcome distraction from her formal duties and soon became a solace in times of grief. She developed an abiding interest in and knowledge of the plants she encountered on her travels. Thanks to the letters and journals she left behind, it has been possible to trace the history of Lady Dalhousie's life and work in Nova Scotia and Quebec, and it is testimony to her passion and determination as a plant collector that the importance of her botanical specimens in Canada is still being discussed today.

NOTES

1 Plant specimens collected by Lady Dalhousie are currently housed in the following herbaria: Royal Botanic Garden Edinburgh; Royal Botanic Gardens, Kew in London; the Royal Botanical Gardens (Canada) in Hamilton, Ontario; and Jardin des Plantes in Paris.

2 Edinburgh, National Records of Scotland (NRS), GD45/3: papers of the 9th Earl of Dalhousie as Lieutenant Governor of Nova Scotia and Governor-in-Chief of Canada, 1816–1828.

3 In 1978 historian Marjory Whitelaw edited the first of three volumes covering the tenure of Lord Dalhousie as governor general of Canada from 1820 to 1828; see Whitelaw, *The Dalhousie Journals.*

4 London, Royal Botanic Gardens, Kew Archives, Director's Correspondence (hereafter DC), volume 44: Miscellaneous letters to William Jackson Hooker, 1818–30; NRS, Papers of the Maule Family, Earls of Dalhousie, GD45; Haddington, East Lothian Council Archives (ELCA) John Gray Centre, Colstoun Papers, EL568. Although the family collections are privately owned, they have been made available to the general public on request.

5 For contemporary accounts of the intricacies and subjectivities of working with archival sources, see Steedman, *Dust*, and Farge, *The Allure of the Archives.*

6 ELCA, Colstoun Papers, EL568; NRS, Papers of the Maule Family, Earls of Dalhousie, GD45.

7 King, "Working With/In the Archives," 20.

8 In Scotland, a form of handwriting known as copperplate had developed by the end of the eighteenth century and was widely taught in schools until the mid-nineteenth century. See Veitch, "Scottish Handwriting." Lady Dalhousie's letters consisted of a single sheet of paper, usually written on one side, which was then folded four times to create two side flaps and one at the top and bottom. By tucking the top fold into the bottom of the letter, a self-made envelope was created and sealed using wax.

9 Thomson, "Life Stories and Historical Analysis," 101.

10 Broun Lindsay, "Colstoun: The Story of a Scots Barony," 29. Coalstoun is also written as Colstoun and the house has remained within the ownership of the Broun family. At the time of writing (2021) the present laird is Ludovic Broun-Lindsay and the house operates as a hotel and functions venue.

11 Ibid.

12 ELCA, EL568/2/491: letter from Miss Carver to Charles Broun, 30 November 1799.

13 Ibid.

14 Ibid.

15 ELCA, EL568/3/55: notes taken by CR during a tour through part of England and South Wales in the summer of 1802. The orangery at Margam, the family home of Thomas Mansel Talbot (1747–1813), was one of the most popular sights of the day and remains the longest-surviving orangery in Europe.

16 Ibid.

17 ELCA, EL568/3/67: journal of a tour of England and Scotland, 7 June to 20 October 1809.

18 Shteir, *Cultivating Women, Cultivating Science*, 3.

19 In 1869, Sophia Jex-Blake (1840–1912) and six other women, known as the Edinburgh Seven, became the first women to matriculate in medicine at a British university (the University of Edinburgh). However, the University of Edinburgh did not allow women to graduate until 1894 and the first female doctors eventually graduated in 1896. See University of Edinburgh, "Sophia Jex-Blake and the Edinburgh Seven," accessed 8 January 2018, http://www.ed.ac.uk/medicine-vet-medicine/about/history/women/sophia-jex-blake-and-the-edinburgh-seven.

20 Horwood, *Gardening Women*, 21–2.

21 Reid, "Unsung Heroines of Horticulture," 53–4.

22 Shteir, *Cultivating Women, Cultivating Science*, 32.

23 Reid, "Unsung Heroines of Horticulture," 125.

24 ELCA, EL568/3/60: CR diary, 30 September 1811 to 1 April 1819.

25 ELCA, EL568/3/62: CR diary, 1 June 1824 to 30 June 1830.

26 Ibid. A distance of seventeen miles separated the estates of Coalstoun and Dalhousie Castle.

27 Burroughs, "Ramsay, George, 9th Earl of Dalhousie," 1–12; Villeneuve, *Lord Dalhousie. Patron and Collector*.

28 Whitelaw, *The Dalhousie Journals*, 2:6.

29 ELCA, EL568/3/15: letter from GR to CR, 11 October 1812.

30 NRS GD45/19/168: memorandum made by J. Archibald, 27 January 1820. Joseph Archibald was the gardener at Dalhousie Castle from 1807 until 1825.

31 Loudon, John, *An Encyclopaedia of Gardening*, 1087.

32 Tait, *Landscape Garden in Scotland*, 140.

33 Archibald, "Dalhousie Castle & Gardens," 252.

34 Ibid.

35 Loudon, Jane, *The Ladies' Magazine of Gardening*, 318.

36 ELCA, EL568/3/volume 15: letter from GR to CR, 5 November 1812.

37 Ibid.: letter from GR to CR, 15 December 1812.

38 Laird, *Formal Garden Traditions of Art and Nature*, 141.

39 ELCA, EL568/3/60: CR diary, 30 September 1811 to 1 April 1819.

40 Jacques, *Georgian Gardens*, 196.

41 Burroughs, "Ramsay, George, 9th Earl of Dalhousie," 1.

42 ELCA, EL568/3/60: CR diary, 10 October 1816. Thought to be a reference to a poem written by William Falconer entitled *The Shipwreck*.

43 ELCA, EL568/3/volume 14: letter from GR to Maclean of Ardgour, 2 November 1816.

44 ELCA, EL568/3/volume 14: letter from CR to Christian Dalrymple (hereafter CD), 9 April 1817.

45 Ibid.

46 Halifax, Nova Scotia Museum, NSM MS Dalhousie Papers: caricatures of the Halifax elite by Lady Dalhousie.

47 ELCA, EL568/3/volume 14: letter from CR to CD, 9 April 1817.

48 ELCA, EL568/3/257: letter from Lord Ramsay to GR, 4 August 1817. *Sarracenia purpurea* (common name: huntsman's cup) is a carnivorous pitcher plant. The northern variety grows from the boreal zone to New Jersey.

49 Tait, "American Garden at Millburn Tower," 84.

50 Packet boats carried post office mail packets to and from British embassies, colonies, and outposts in the nineteenth century.

51 Christian's eldest son, Lord Ramsay, died in 1832 at the age of twenty-six and her husband's death followed six years later in 1838.

52 Graham, "Extracts," 52.

53 NRS GD45/3/541: journal of GR, 11 March 1818. Some of the minerals she collected on this trip were subsequently donated to the Edinburgh Museum, which at the time was part of the University of Edinburgh but is now known as the National Museum of Scotland.

54 NRS GD45/3/541: journal of GR, 24 October 1818.

55 Bakewell, *Introduction to Geology*; Haldimand Marcet, *Conversations on Chemistry*. It is possible that the information Lady Dalhousie gleaned from these works helped in the collection of minerals during an excursion to Mahone Bay, Nova Scotia, in 1818.

56 Sowerby and Smith, *English Botany*; ELCA, EL568/3/61: CR diary, 1 September 1819 to 31 May 1824; ELCA, EL568/3/113: catalogue of books in Dalhousie Castle library, c.1830.

57 One of Lady Dalhousie's journals written in India ends with this list of books read: Blumenbach, *A Manual of the Elements of Natural History*; Lloyd, *Botanical Terminology*; Hooker, *Flora Scotica*; Knapp, *Journal of a Naturalist*; Haldimand Marcet, *Conversations on Vegetable Physiology*; Roxburgh, *Hortus Bengalensis*; and *Flora Indica*. See Reid, "Unsung Heroines of Horticulture," 101.

58 ELCA, EL568/3/volume 16: letter from Isabella Houstoun to CR, 22 February 1830. The "number" to which Miss Houstoun refers is the first of three volumes by the surgeon and botanist Nathaniel Wallich entitled *Plantae Asiaticae Rariores*, which

no doubt would have been an invaluable reference source for Lady Dalhousie in the identification of her Indian plant specimens.

59 NRS GD45/3/542: journal of GR, 3 July 1819.

60 Halifax, Nova Scotia Museum (hereafter NSM), 85.119.29A-C: "Three botanical studies" by CR.

61 NRS GD45/3/543: journal of GR, 11 January 1820.

62 NRS GD45/3/543: journal of GR, 13 August 1820.

63 NRS GD45/14/515: letter from Joseph Archibald to GR, 2 June 1820.

64 NRS GD45/14/515: letter from Joseph Archibald to GR, 1 October 1820.

65 NRS GD45/3/439: Copy of an invoice sent to Lord Dalhousie from Robert Cleghorn, Blinkbonny Garden, Montreal, 23 October 1820.

66 Ibid.

67 Asselin, Cayouette, and Mathieu, *Curieuses Histoires*, 167–70.

68 Royal Botanic Garden Edinburgh (hereafter RBGE) Archive: *Donations Book 1823–1824*, "Canada Seeds from Lady Dalhousie, Dec 1822," f.3.

69 RBGE, *Donations Book 1826–1845*, f.1; *Donations Book 1830–38*, f.35.

70 Whitelaw, *The Dalhousie Journals*, 2:11.

71 NRS GD45/3/543: journal of GR, 14 July 1820. Sorel is situated some forty miles north east of Montreal at the confluence of the rivers Richelieu and St Lawrence.

72 NRS GD45/14/557: letter from CR to JR, 15 April 1826.

73 ELCA, EL568/3/261: letter from CR to Mrs Anderson, 20 May 1823.

74 NRS GD45/3/546: journal of GR, 8 June 1823. Georgina Hay was the thirteen-year-old daughter of Lord Dalhousie's sister, Mary Hay.

75 NRS GD45/3/545: journal of GR, 21 December 1822.

76 ELCA, EL568/3/62: CR diary for 1824. Unfortunately all correspondence for Robert Graham while regius keeper at RBGE was sold and its current whereabouts are unknown.

77 Kew Archives, DC, vol. 44: letter from CR to William Jackson Hooker (hereafter WJH), 31 October (thought to be 1825). William Jackson Hooker was regius professor of botany at Glasgow University from 1820–41 before being appointed the first full-time director of the Royal Gardens at Kew in April 1841.

78 Ibid.

79 Archibald, "Dalhousie Castle & Gardens," 255.

80 Hooker, *Exotic Flora*, t.87.

81 Her membership is corroborated by the Royal Botanical Institution of Glasgow (RBIG) Minute Book for 1825 in which Christian is listed as an honorary member alongside two other Scottish women, the Marchioness of Hastings and Lady Liston; see Glasgow, Mitchell Library Archives, D-TC11/1/1-14: minute book of the RBIG, 1825.

82 Shteir and Cayouette, "Collecting with 'botanical friends,'" 1–97.

83 Ibid, 11. Lady Dalhousie records dining with Mrs Perceval on several occasions in 1823 and outings were also arranged, such as a trip to Sillery Cove in May 1824, see ELCA, EL568/3/61: CR diary, 24 May 1824.

84 Sheppard, "Natural History Society," 55.

85 Library and Archives Canada RAC 1898, Papiers d'Etat, Bas-Canada (MG11-CO42), Q.176-1, 483 and RAC 1938, Appendice II, Inventaire des papiers de Lord Dalhousie (MG24-A12), 1816–33, V, 38; XII, p. 100. I am indebted to Louisa Blair at the Literary and Historical Society of Quebec for drawing my attention to these archives.

86 NRS GD45/3/175: letter from McFarlane to GR, 24 July 1827.

87 NRS GD45/3/439: a list of North American plants cultivated in the King's Gardens, St Helen's, 6 November 1824. My thanks to James Pringle and Jacques Cayouette, for their help in transcribing this list.

88 NRS GD45/3/175: letter from McFarlane to GR, 24 July 1827.

89 Ibid.

90 NRS GD45/3/549: journal of GR, 8 July 1827.

91 Dalhousie, "Catalogue of Canadian Plants," 255–61.

92 NRS GD45/14/546: letter from CR to Lord Ramsay, 28 September 1822; ELCA, EL568/3/62.

93 Pringle, "Canadian botanical specimens collected 1826–1828 by the Countess of Dalhousie, acquired by the Royal Botanical Gardens," 16. A full list of the specimens in Lady Dalhousie's Canadian herbarium can be found on pages 6–14.

94 Ibid, 17.

95 NRS GD45/14/557: letter from CR to JR, 16 February 1828.

96 Hall, "Of Gender and Empire," 52.

97 RBGE Archive: *Wallich's Catalogue*, no. 5339. Technically, George Bentham published the genus, as Graham had provided no description.

98 ELCA, EL568/3/255: letter from WJH to CR, 20 July (undated).

99 Curtis, *Curtis's Botanical Magazine*, frontispiece. The magazine, conducted by Samuel Curtis with plant descriptions supplied by WJH, has been published continuously since 1787 and is the longest-running botanical periodical.

100 Ibid.

101 Kew Archives, DC, vol 44: letter from CR to WJH, 4 March 1833.

102 Graham, "Extracts," 52.

103 Ramsay, *Reminiscences of Scottish Life and Character*, 165.

BIBLIOGRAPHY

Archives

Herbarium, Royal Botanic Garden Edinburgh, Edinburgh

National Records of Scotland, Edinburgh: GD45 Papers of the Maule Family, Earls of Dalhousie

Mitchell Library Archives, Glasgow

John Gray Centre, East Lothian Council Archives, Haddington: EL568 Colstoun Papers

Nova Scotia Museum Archives, Halifax: NSM MS Dalhousie Papers

Royal Botanical Gardens (Canada), Hamilton, Canada

Archives, Royal Botanic Gardens, Kew: Director's Correspondence, volume 44

Library and Archives Canada, Ottawa

Books and Articles

Archibald, Joseph. "Dalhousie Castle & Gardens; the Botany of the Neighbourhood, and various Remarks." *The Gardener's Magazine* 1, no. 3 (1826): 251–8.

Asselin, Alain, Jacques Cayouette, and Jacques Mathieu. *Curieuses Histoires de Plantes du Canada 1760–1867*, vol. 3. Quebec: Septentrion, 2017.

Bakewell, Robert. *Introduction to Geology*. London, 1813.

Blumenbach, Johann Friedrich. *A Manual of the Elements of Natural History.* London: 1825.

Broun Lindsay, Edith C. "Colstoun: The Story of a Scots Barony." *Transactions of the East Lothian Antiquarian and Field Naturalists' Society* 4 (1948): 19–33.

Burroughs, Peter. "Ramsay, George, 9th Earl of Dalhousie." Dictionary of Canadian Biography Online. Accessed 25 October 2011. http://www.biographi.ca/009004-119.01.

Curtis, Samuel. *Curtis's Botanical Magazine; or Flower Garden Displayed* 60, new ser. 7 (1833).

Dalhousie, The Countess of. "Catalogue of Canadian Plants: collected in 1827 and presented to the Literary & Historical Society." *Transactions of the Literary and Historical Society of Quebec* 1, ser. 1 (1829): 255–61.

Falconer, William. *The Shipwreck: A poem. In three cantos*. London, 1762.

Farge, Arlette. *The Allure of the Archives.* New Haven and London: Yale University Press, 2013.

Graham, Robert. "Extracts from a Report on the Progress and State of Botany in Britain, from March 1838 to February 1839 [...], 14 March 1839." *Third Annual Report and Proceedings of the Botanical Society* (1838–39).

Haldimand Marcet, Jane. *Conversations on Chemistry, Intended More Especially for the Female Sex.* London, 1813.

– *Conversations on Vegetable Physiology.* London: Longman, Rees, Orme, Brown & Green, 1829.

Hall, Catherine. "Of Gender and Empire: Reflections on the Nineteenth Century." In *Gender and Empire*, edited by Philippa Levine, 46–76. Oxford: Oxford University Press 2004.

Hooker, William Jackson. *Exotic Flora* 2. Edinburgh: February 1825.

– *Flora Boreali-Americana; or the Botany of the northern parts of British America.* London: Henry G. Bohn, 1829–40.

– *Flora Scotica.* London: 1821.

Horwood, Catherine. *Gardening Women: Their Stories from 1600 to the Present.* London: Virago, 2010.

Jacques, David. *Georgian Gardens: The Reign of Nature.* London: Batsford, 1983.

King, Michelle T. "Working With/In the Archives." In *Research Methods for History*, edited by Simon Gunn and Lucy Faire, 13–29. Edinburgh: Edinburgh University Press, 2012.

Knapp, John L. *The Journal of a Naturalist.* London, 1829.

Laird, Mark. *The Formal Garden Traditions of Art and Nature.* London: Thames and Hudson, 1992.

Lloyd, G.N. *Botanical Terminology.* Edinburgh, 1826.

Loudon, Jane. *The Ladies' Magazine of Gardening.* London: William Smith, 1842.

Loudon, John C. *An Encyclopaedia of Gardening; comprising the theory and practice of horticulture, floriculture, arboriculture and landscape gardening.* London: Longman, Hurst, Rees, Orme, Brown and Green, 1824.

McIntosh, Charles. *The Book of the Garden.* Edinburgh and London: Blackwood, 1853.

Pringle, James S. "Canadian botanical specimens collected 1826–1828 by the Countess of Dalhousie, acquired by the Royal Botanical Gardens." *Canadian Horticultural History/Histoire de l'horticulture au Canada* 3, no. 1 (1995): 1–21.

Ramsay, E.B. *Reminiscences of Scottish Life and Character.* London: Edmonston and Douglas, 1859.

Reid, Deborah. "Unsung Heroines of Horticulture. Scottish Gardening Women, 1800 to 1930," unpublished doctoral thesis, University of Edinburgh, 2015.

Roxburgh, Dr William. *Flora Indica; or Descriptions of Indian Plants.* Serampore: Mission Press, 1820–24.

– *Hortus Bengalensis.* Serampore, 1814.

Sheppard, William. "Natural History Society. Annual Conversazione. Hon. Mr. Sheppard's address." *Canadian Naturalist and Geologist* 1, (1864): 53–7.

Shteir, Ann B. *Cultivating Women, Cultivating Science: Flora's Daughters and Botany in England 1760 to 1860.* Baltimore and London: The Johns Hopkins University Press, 1996.

Shteir, Ann, and Jacques Cayouette. "Collecting with 'botanical friends': Four Women in Colonial Quebec and Newfoundland." *Scientia Canadensis* 41, 1 (2019): 1–97.

Sowerby, James, and James Edward Smith. *English Botany; or, coloured figures of British plants, with their essential characters, synonyms, and places of growth*. London, 1790–1814.

Steedman, Carolyn. *Dust*. Manchester: Manchester University Press, 2001.

Tait, A.A. "The American Garden at Millburn Tower." In *British and American Gardens in the Eighteenth Century*, edited by Robert P. Maccubin and Peter Martin, 84–91. Williamsburg: The Colonial Williamsburg Foundation, 1986.

– *The Landscape Garden in Scotland 1735–1835*. Edinburgh: Edinburgh University Press, 1980.

Thomson, Alistair. "Life Stories and Historical Analysis." In *Research Methods for History*, edited by Simon Gunn and Lucy Faire, 101–17. Edinburgh: Edinburgh University Press, 2012.

Veitch, Kenneth. "Scottish Handwriting in the Eighteenth and Nineteenth Centuries. A Concise Guide." Sources in Local History Online. Accessed 5 May 2018. http://www.dumfriesandgalloway.hss.ed.ac.uk/wp-content/uploads/2016/08/Scottish-Handwriting-A-Concise-Guide.pdf.

Villeneuve, René. *Lord Dalhousie. Patron and Collector.* Ottawa: National Gallery of Canada, 2008.

Wallich, Nathaniel. *Plantae Asiaticae Rariores; or, Descriptions and figures of a select number of unpublished East Indian plants* 3. London, 1830–32.

– *Wallich's Catalogue. A Numerical List of Dried Specimens of Plants in the East India Company's Museum, collected under the superintendence of Dr Nathaniel Wallich, of the Company's Botanic Garden at Calcutta*. One folio volume. London, 1828.

Whitelaw, Marjory, ed. *The Dalhousie Journals*. Canada: Oberon Press, 1978–82.

2

Lady Dalhousie's Orchids and Other Rare Plants in Lower Canada, 1820–1828

Resources for Historical Study

JACQUES CAYOUETTE AND FAYE-YIN KHOO

Historical plant collections are important for establishing knowledge about the flora of a specific area at a specific time, and they make it possible to compare earlier and current flora in light of environmental changes. The best plant collections from Lower Canada in the 1820s were gathered by the Countess of Dalhousie. Her love for orchids and other rare plants, along with her botanic knowledge, are responsible for plant specimens and information that are still significant today. The 1820s predated major transformations of plant habitats by industrialization, urbanization, and farming, all developments that contributed to the decrease or even the local extinction of certain species in Quebec. Many of these rare species are native orchids that are threatened by the loss of their unique habitats, especially in bogs, wet locations, and habitats rich in limestone, as well as by the interest of gardeners who transplant specimens from their wild habitats.[1] This chapter examines nine sources[2] that contain plant specimens or information that Lady Dalhousie collected in Lower Canada between 1820 and 1828 and that reveal the presence of thirty-six species that are considered at risk in Quebec, according to the Quebec government. These nine sources include public documents such as catalogues, archives, citations from floras and other botanical publications, garden surveys, plant albums, and herbaria containing her collections. Some of these sources have never been examined and can now be linked with the other sources, as shown in table 2.1 discussed below. When taken together, these resources demonstrate the value of historical plant information for contemporary practices. They also highlight the importance of Lady Dalhousie's work as a collector to the historical study of plants in Canada.

Christian Broun Ramsay (1786–1839), better known as the Countess of Dalhousie, was very fond of plants.[3] She became a student of botany during the British Enlightenment, and her marriage to George Ramsay (1770–1838), the 9th Earl of Dalhousie, gave her access to British possessions where she could fulfill her passion for plants. She was among a small number of elite women in British North America who botanized for themselves or for foreign botanists. Her friend Anne Mary Perceval (1790–1876) established contacts with English botanist William Jackson Hooker (1785–1865) and sent him an important number of British North American vascular plants for his major project of *Flora Boreali-Americana*. Her collections are, unfortunately, dispersed in many North American herbaria in low numbers.[4] Another friend, Harriet Sheppard (1786–1858), submitted her plant collections in Lower Canada to Hooker and to other botanists as well,[5] but the unfortunate loss of her personal herbarium by a fire in 1842[6] makes her material alone unsuitable for cross-temporal flora comparison studies. By contrast, the plant collections, reports in catalogues, and other documents by Christian Broun Ramsay represent, by far, the best resources for establishing flora comparisons between the past and present.

Lady Dalhousie had a particular interest in "new" and "rare" plants, especially orchids, and expressed to Hooker the hope that "among the various Orchideae there may be some new or at least rare species."[7] What, however, did the designations "new" and "rare" mean for her? Many live orchids that she sent to her garden in Scotland were new to Hooker, who also received separate specimens at Glasgow. Given that some flowered in the British Isles for the first time, they were often unknown to British gardeners. That, however, does not mean that the specimens in question have never been previously discovered. The inscription "perhaps a new species" in her plant album of 1823 may refer to a species she believed was actually new to science and "rare" may refer to a plant that she, herself, discovered for the first time. Since she was already an expert by the time that she made comments regarding "new" and "rare" plants, it is likely that species she identified to be rare were indeed rare. Many of her plants were collected in the Montérégie and Montreal regions of southwestern Quebec, where the highest amount of rare Quebec plants is found. It is, therefore, highly possible that her discoveries correspond with what is still considered "rare" according to the Quebec government's criteria. Plants are often regarded as "at risk," versus "rare," based on standards set by the Quebec government. For the purpose of this chapter, the words "rare" and "at risk" will be used almost interchangeably.

Six documents on rare and endangered plants in Quebec published between 1983 and 2016 were consulted to establish the rarity status of the plant species collected or reported by Lady Dalhousie. The earliest of these documents was published in 1983, in the *Syllogeus* series issued by the Canadian National Museums, which was intended to document rare plants of all the Canadian provinces and territories.[8] A document issued by the Canadian Museum of Nature in 1990 lists Quebec's rare species recorded in the review of rare vascular plants of Canada.[9] The other four publications were produced by the Quebec government and its Ministry of Environment, which had several different names from 1992 to 2016.[10] These four documents support the Québec Act Respecting Threatened or Vulnerable Species (E-12.01) that was created in 1989 to cover at-risk native vascular plant species. Criteria used to designate plants as "rare" varied and evolved over time from one publication to another as a result of updates based on inventories of herbarium specimens and numerous certified field surveys. The most recent compilation more or less follows the lists by the Committee on the Status of Endangered Wildlife in Canada (COSEWIC)[11] and the international organization NatureServe.[12]

According to these documents, thirty-six of the plant species collected or documented by Lady Dalhousie in the nine sources referenced are considered at risk in Quebec. These findings are summarized in table 2.1. Column 3 indicates which of these six documents were used to determine their rarity status. Their currently accepted names are listed in alphabetical order in the first column and the names identified by Lady Dalhousie are in the second column. The nine sources of plant collections and botanical reports by Lady Dalhousie are presented in chronological order across the top, along with the years they were collected in the following row. The table is marked to indicate which rare plant is found in each of the sources. The localities are provided if they are known. At the bottom of the table, there is a total count of rare plant species that is in each source. It can be observed that one of the sources contains as many as twenty-one rare plant records, while others only have as few as two. These thirty-six at-risk plants were collected from early spring to September. They were collected in Lower Canada from west to east, from Montreal to the Mingan Islands.

Nine sources, containing plant specimens or information that Lady Dalhousie collected during the 1820s, were selected. Botanic tools include floras, like *Exotic Flora* (source 2) and *Flora Boreali-Americana* (source 9), herbarium specimens such as the ones at the Canadian Museum of Nature (source 4), the Darlington Herbarium of West Chester (source 5), the Edinburgh Herbarium (source 6), and the Hamilton Herbarium (source 7), and references from

Table 2.1
Rare plant species collected in Quebec by Lady Dalhousie

Modern names	Names used by Lady Dalhousie	Source used to determine rarity (see notes)	Sources 1 – Archibald, 1826	2 – Hooker, *Exotic Flora* coll. [1820–23]	3 – Dalhousie Manuscript, 1823 coll. 1823	4 – Dalhousie Herbarium, 1823 coll. 1823	5 – West Chester Album coll. 1823	6 – Edinburgh Herbarium coll. early 1824	7 – Hamilton Herbarium coll. 1826–28	8 – Dalhousie Catalogue, 1829 coll. 1827	9 – Hooker, *FBA* coll. [1825–28]
Adlumia fungosa	*Corydalis fungosa*	i, iii–vi							X	X	
Andersonglossum boreale	*Cynoglossum amplexicaule*	i, vi			X					X	
Anemone quinquefolia	*Anemone nemorosa*	i			X	Sorel		Banks of the Etchemin			X
Arethusa bulbosa	***Arethusa bulbosa***	**i, iii–v**	**X**	**X**	**X**	**Sorel**	**Sorel**		**Sorel**	**X**	**X**
Asplenium rhizophyllum	*Asplenium rhizophyllym*	i, iii–vi								X	
Calypso bulbosa* var. *americana	***Calypso borealis***	**iv–vi**			**X**	**Montreal**				**X**	
Claytonia virginica	*Claytonia virginica*	i, iii–vi			X	Île de Grâce (Archipel du Lac Saint-Pierre)				X	

Table 2.1 *continued*

Modern names	Names used by Lady Dalhousie	Source used to determine rarity (see notes)	Sources 1 – Archibald, 1826	2 – Hooker, *Exotic Flora* coll. [1820–23]	3 – Dalhousie Manuscript, 1823 coll. 1823	4 – Dalhousie Herbarium, 1823 coll. 1823	5 – West Chester Album coll. 1823	6 – Edinburgh Herbarium coll. early 1824	7 – Hamilton Herbarium coll. 1826–28	8 – Dalhousie Catalogue, 1829 coll. 1827	9 – Hooker, *FBA* coll. [1825–28]
Cypripedium arietinum	***Cypripedium arietinum***	**i, iii–vi**	**X**						**Sorel**	**X**	
Cypripedium reginae	***Cypripedium spectabile***	**i, iii–vi**	**X**		**X**	**Sorel**			**Sorel**	**X**	
Decodon verticillatus	*Lythrum verticillatum*	i			X	Sorel				X	
Galearis rotundifolia	***Orchis discolor* ??; *Platanthera rotundifolia***	**iv–vi**							**[Mingan Islands]**		**X**
Galearis spectabilis	***Orchis grandiflora*; *Orchis spectabilis***	**i, iii–vi**							**X**	**X**	**X**

Gentianella quinquefolia subsp. *quinquefolia*	*Gentiana amarelloides; Gentiana lutea*	–		X	Sorel	Sorel		Sorel	X	
Gentianopsis crinita	*Gentiana crinita*	i, iii–vi							X	
Gentianopsis virgata subsp. *victorinii*	*Gentiana fimbriata*	i–vi					[city of Quebec area]			
Goodyera pubescens	***Goodyera pubescens; Neottia pubescens***	**i, iii–vi**	**X**						**X**	
Hypericum ascyron subsp. *pyramidatum*	*Hypericum ascyroides; Hypericum Prolificum*	i, v, vi							X	"About Montreal"
Ionactis linariifolia	*Aster rigida?*	i–vi		X	Sorel	X		Sorel	X	
Justicia americana	*Justicia pedunculosa*	i–vi		X	Île Ste-Hélène					
Lonicera dioica s.l.	*Lonicera* sp. *Xylosteum ciliatum*	i						X	X	

Table 2.1 *continued*

Modern names	Names used by Lady Dalhousie	Source used to determine rarity (see notes)	Sources 1 – Archibald, 1826	2 – Hooker, *Exctic Flora* coll. [1820–23]	3 – Dalhousie Manuscript, 1823 coll. 1823	4 – Dalhousie Herbarium, 1823 coll. 1823	5 – West Chester Album coll. 1823	6 – Edinburgh Herbarium coll. early 1824	7 – Hamilton Herbarium coll. 1826–28	8 – Dalhousie Catalogue, 1829 coll. 1827	9 – Hooker, *FBA* coll. [1825–28]
Lysimachia quadrifolia	*Lysimachia quadrifolia*	i, iii–vi			X	Sorel					
Oenothera pilosella	*Oenothera*	ii–vi							X		
Panax quinquefolius	*Panax quinquefolia*	i–vi								X	
Pedicularis canadensis	*Pedicularis canadensis*	i			X	Sillery			Sillery		
Penstemon hirsutus	*Penstemon pubescens*	i, vi									X
Persicaria arifolia	*Polygonym sagittatum*	vi				Sorel					
Phegopteris hexagonoptera	*Polypodium hexagonopterum*	i–vi								X	

Physostegia virginiana subsp. *virginiana*	*Dracocephalum virginianum*	i, vi				X	Sorel	Sorel		Sorel	X
Platanthera blephariglottis	***Habenaria blephariglottis***	**i, iii–v**	**[prob. Sillery]**		**[prob. Sillery]**						**X X**
Platanthera macrophylla	***Orchis macrophylla***	**i, iii–vi**	**X**			**Sorel**					
Pterospora andromedea	*[Pterospora andromedea]*	i, iii–vi							[prob. Sillery]		
Ranunculus rhomboideus	*Ranunculus ovalis?*	iv–vi				Île Ste-Hélène			Île Ste-Hélène		
Scutellaria parvula	*Scutellaria parvula*	i			X	Île Ste-Hélène				Sorel	X
Staphylea trifolia	*Ptelea trifolia*	i, iii–vi				Sorel					
Trichostema brachyatum	*Isanthus ceruleus*	i, iii–vi			X	Île Ste-Hélène					
Verbena simplex	*Verbena angustifolia*	i, iii–vi			X	Île Ste-Hélène					
Total			6	2	16	20	4	2	15	21	7

There are thirty-six rare species in the nine sources examined, according to six documents on rare plants in Quebec published between 1983 and 2016. The table is marked to indicate which rare plant is found in each of the sources. Its location is provided when known. The total number of rare plants found in each source is tallied at the bottom. Orchid rows are in bold.

The six documents on rare plants in Quebec are:

i André Bouchard et al., *Les plantes vasculaires rares du Québec/The rare vascular plants of Quebec*, Canadian National Museums, 1983.

ii George W. Argus and Kathleen M. Pryer, *Les plantes vasculaires rares du Canada; notre patrimoine naturel*, Canadian Museum of Nature, 1990.

iii Gildo Lavoie, *Plantes vasculaires susceptibles d'être désignées menacées ou vulnérables au Québec*, Quebec Ministry of Environment, 1992.

iv Jacques Labrecque and Gildo Lavoie, *Les plantes vasculaires menacées ou vulnérables du Québec*, Quebec Ministry of Environment, 2002.

v CDPNQ, *Les plantes vasculaires menacées ou vulnérables du Québec*, ed. 3., Le Centre de données sur le patrimoine naturel du Québec, 2008.

vi Bernard Tardif et al., *Les plantes vasculaires en situation précaire au Québec*, CDPNQ, 2016.

Table compiled by Jacques Cayouette and Faye-Yin Khoo.

the literature, such as the report of gardener Joseph Archibald (source 1) and the catalogue of Lady Dalhousie in 1829 (source 8). Contacts and exchange with disciplines outside botany, such as history and archives, permitted the discovery and the use of the Halifax manuscript with Lady Dalhousie's list of her plant collections in 1823 (source 3). Additional information regarding the nine sources will be given in chronological order, as presented in the table, starting with Joseph Archibald's article in *Gardener's Magazine* entitled "Dalhousie Castle and Gardens."

SOURCE 1: ARCHIBALD'S "DALHOUSIE CASTLE AND GARDENS"

The Dalhousies had an estate and garden in Scotland not far from Edinburgh, where the gardener Joseph Archibald managed about 200 acres of plantation.[13] At that time, it was a common practice for gardeners to report in the *Gardener's Magazine* on the state of extensive private or institutional gardens not only in the British Isles but elsewhere in Europe and even in North America. According to him, Lady Dalhousie began shipping live North American plants in 1817, first from Nova Scotia where Lieutenant Governor Dalhousie was on duty from 1816 to 1820, and later from Lower Canada, from 1820 to 1828, including what is now the province of Quebec. One section of the garden, labelled "borders and plots for American plants," was reserved for North American plants.[14] According to Archibald, they were all rare in America and some flowered for the first time in the British Isles. However, it is difficult to determine the significance of plant rarity according to Archibald because the plants he received were mostly orchids or were flowering for the first time in the British Isles. The plants cited by Archibald suggest that Lady Dalhousie had a preference for rare orchids. Six orchids that are rare in Quebec were reported in "Dalhousie Castle and Gardens," as observed in table 2.1.

SOURCE 2: HOOKER'S *EXOTIC FLORA*

Exchanges of desirable garden plants, such as living plants and seeds, are often made between botanical gardens to achieve collections greater in size, quality, and diversity. Private gardeners in earlier times frequently sent some of their material to botanical gardens too. Lady Dalhousie shared most of her living orchids with William Hooker at the Glasgow Botanic Garden. Hooker recognized the uniqueness of her plant specimens and included them in his *Exotic Flora* project. This three-volume work has no pagination but the plant species

2.1
Habenaria blephariglottis, *Exotic Flora*, vol. 2, plate 87, 1825.

are numbered consecutively.[15] Hooker published many books like this during his career, often including colour illustrations, elaborate plant descriptions including their uses, plant localities, and notes from the collectors.[16]

Three of the rare plants from Quebec in Hooker's *Exotic Flora* are the same as the three orchids from the garden at the Dalhousies' castle. As shown in table 2.1, these are *Platanthera blephariglottis*, *Platanthera macrophylla*, and *Arethusa bulbosa*. The comments from Hooker are particularly interesting regarding their provenance and approximate date of collection. In the first volume, under the entry for another orchid name, *Pogonia ophioglossoides*, we learn that there were also plants of the rare *Platanthera blephariglottis* accompanying the roots that Lady Dalhousie sent from Canada.[17] The arrival of these two orchids in Scotland thus predates the publication of Hooker's first volume in 1823. One of the flowers of this *Platanthera* drawn and published in volume 2 comes from Lady Dalhousie's material. It was revealed in the second volume that the Scottish botanist John Goldie (1793–1886), who collected plants in Canada from 1816 to 1819, also brought roots of *Platanthera blephariglottis* for his garden "from swampy grounds near Quebec."[18] This population was discovered in the city of Quebec area by William Sheppard, a close friend of Lady Dalhousie, and her live *Platanthera* plants may have also come from there.[19] William Sheppard (1784–1867), who arrived from England as a young man, became a businessman also involved in timber and played important roles in civic and intellectual life in colonial Quebec. He and his wife, Harriet Campbell Sheppard, shared interests in botany and natural history. The case of the *Platanthera macrophylla* is quite peculiar. It was collected by Lady Dalhousie "near Montreal in 1823," which probably refers to Sorel, where the Dalhousies had a second official residence. In *Exotic Flora* volume 2 (1825), it was named *Habenaria orbiculata*, but the corresponding specimen from Sorel in Lady Dalhousie's 1823 Herbarium is a *Platanthera macrophylla*. For the *Arethusa bulbosa* illustrated and described in Volume 3 (1827), Hooker indicated that he received it from Lady Dalhousie in late 1823 and it flowered in June 1824.[20] This collection may have originated from the Sorel area and makes a direct link with the manuscript entitled *Sorell 1823* and its counterpart album of plants from the Canadian Museum of Nature entitled *The Countess of Dalhousie Herbarium, 1823*.

SOURCE 3: HALIFAX MANUSCRIPT, 1823

A manuscript at the Nova Scotia Museum in Halifax, in the hand of Lady Dalhousie, is entitled *Plants / Native of Canada / Sorell 1823*. Manuscripts and plant collections discovered among the belongings of the Dalhousie descendants in 1984 were acquired by the museum in late 1985.[21] *Plants / Native of Canada / Sorell 1823* contains fifteen pages with 283 entries and 328 plant names, of which 231 bear genus and species with Latin names. Common English plant names, their habitats, abundance, whether native or cultivated, and flowering times are occasionally noted in the second column. The list mostly follows the same plant names and order as found in the flora compiled by botanist Frederick Pursh. Pursh (1774–1820) came to North America from Germany in 1799 to collect useful plants. He travelled within the United States for twelve years and was in contact with many botanists. He later prepared a North American continental flora which he published in December 1813 while he was in London. In 1816, he was planning to publish a *Flora Canadensis* when he returned to

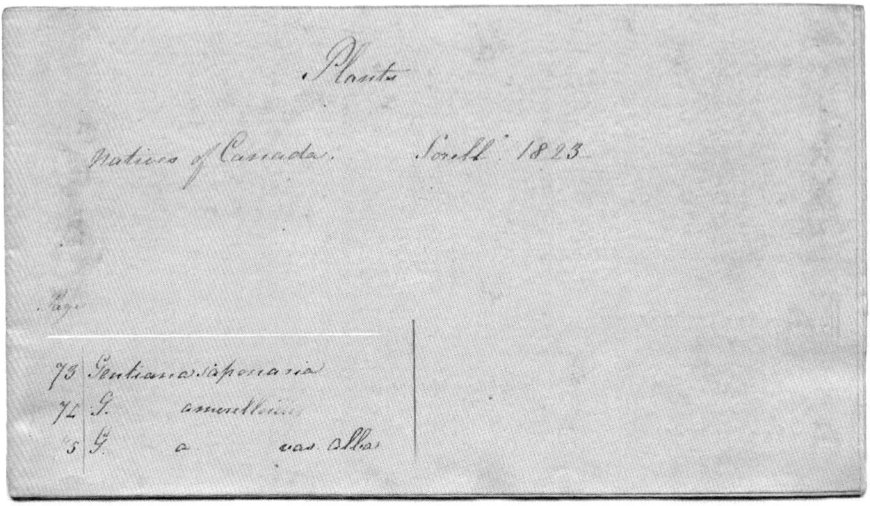

2.2
Extract from Lady Dalhousie's manuscript *Plants / Native of Canada / Sorell 1823.*

Canada, but it was never completed before his death in Montreal in 1820.[22] The Dalhousies had the 1816 edition of Pursh's Flora, and Lady Dalhousie used it extensively.[23]

SOURCE 4: *THE COUNTESS OF DALHOUSIE HERBARIUM*, 1823

Although sixteen of the plant species listed in the Dalhousie manuscript in Halifax are considered rare in Quebec, mistakes are possible due to misinterpretations of the plant names. Luckily, the discovery of a plant album entitled *The Countess of Dalhousie Herbarium* has permitted the correction of some misidentifications. This plant album was auctioned and acquired by the Canadian Museum of Nature seemingly at about the same time as the acquisition of the manuscript by the Halifax Museum (source 3). After examining the Halifax manuscript with its 283 entries and the 283 pages of the album, it became apparent that these two are closely related. Further investigation reveals that the plants in the album correspond almost completely to the plant names in the manuscript. The major differences are that *The Countess of Dalhousie Herbarium* album bears localities, dates, and more plant identifications and sometimes a different collector than the countess or identification for the same entry. There are some additional loose unidentified specimens in the album, but no rare Quebec plants are detected among them.

2.3
Cover of *The Countess of Dalhousie Herbarium* album, 1823.

A large number of twenty at-risk Quebec species was found in the album, but these are not completely the same as the sixteen at-risk species from the manuscript. In a few instances, a rare plant species is mentioned in the manuscript, but its corresponding page in the album is blank or it contains a different species (possibly due to an initial misidentification by the countess). The plants were mostly collected in Sorel from May to September 1823. Others came from locations in the Archipel du Lac Saint-Pierre, e.g., Île [du] Moine and Île de Grâce. Some were collected on Île Ste-Hélène just south of Montreal, one was from Montreal, and another was from Spencer Wood in Sillery, Quebec, collected by Anne Mary Perceval.[24] Some include comments by William Sheppard.[25] The twenty rare species from the album are peculiar because seven of these species are not found in the other eight references examined. They are *Claytonia virginica*, *Justicia americana*, *Lysimachia quadrifolia*, *Persicaria arifolia*, *Staphylea trifolia*, *Trichostema brachyatum*, and *Verbena simplex* (table 2.1).

SOURCE 5: WEST CHESTER ALBUM

Twelve plants from the Dalhousie manuscript and *The Countess of Dalhousie Herbarium* album are also part of a plant album assembled by Anne Mary Perceval that was sent to the American botanist William Darlington in West Chester, Pennsylvania. Perceval (sometimes spelled Percival), from the British regime elite, stayed in Lower Canada for about two decades and was one of Lady Dalhousie's best friends in the city of Quebec. She was a plant collector in the that area who provided material to various botanists of her homeland and in the United States and networked with botanists and plant collectors. In 1826, she prepared a bound album of 189 dried specimens collected in 1823, mostly from around the city of Quebec. It was sent to William Darlington (1782–1863), who was also a physician and a member of the American Congress.[26] The album belongs to the Special Collection Library, Darlington Herbarium in West Chester, Pennsylvania, and an electronic version is kept at DAO.[27] Most plant specimens were collected by Mrs Perceval, with twelve collected by Lady Dalhousie, in Sorel. Four out of the twelve collected by Lady Dalhousie are rare in Quebec, namely, *Arethusa bulbosa*, *Gentianella quinquefolia* subsp. *quinquefolia*, *Ionactis linariifolia*, and *Physostegia virginiana*.

Among those, as revealed by sources 3, 4, and 5, *Gentianella quinquefolia* subsp. *quinquefolia* is a special case. This rare plant represents an addition to the native flora of Quebec and no longer exists.[28] It was added to the Quebec list of at-risk species with the status of "extirpated."[29] According to Canadian botanist James S. Pringle, this rare gentian represents the northern limit of that Appalachian subspecies in North America.[30] The Sorel population found in 1823 must be noteworthy because it was reported in two flower colours in the 1823 Sorel manuscript (source 3) (see fig. 2.2),[31] and the report of a rare white colour variant of a normally blue one suggests a significant population. In the *1823 Herbarium* album, two different collections appear on pages 74 and 75, the first one in August 1823 and the second in September, with a note indicating a different collector other than the countess herself. The collection is annotated as "found by G.H. Ramsay" in the Dalhousie album and "Sorel, G.H." in the Perceval album. This refers to Georgina Hay, the thirteen-year-old niece of Lady Dalhousie. An extract from Lord Dal-

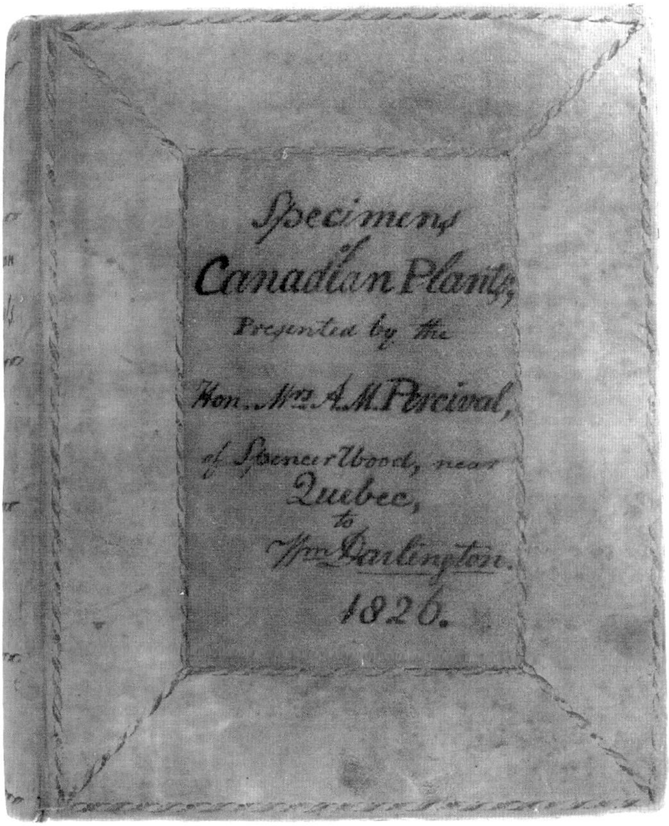

2.4
Cover of *Specimens of Canadian Plants presented by the Hon. Mrs A.M. Percival of Spencer Wood, near Quebec, to Wm. Darlington, 1826.*

housie's Journal, dated 8 June 1823, mentions that his wife and their niece regularly collected plants together.[32]

SOURCE 6: HERBARIUM, ROYAL BOTANIC GARDEN EDINBURGH

The study of this source is limited to the photographs of selected specimens that were taken and brought back to the herbarium of Agriculture Canada in Ottawa by Canadian botanist William G. Dore in 1978. The collector's index indicates that Lady Dalhousie's plant specimens come from Canada and other countries she resided in or visited.[33] Her Canadian collections are not dated in the collector's index, but the information on the labels of the photographed specimens indicates that some were collected in early 1824.[34]

2.5
Plant specimens of *Gentiana amarelloides* var. *alba* from *The Countess of Dalhousie Herbarium*, 1823, folio 75.

Two rare plants of the Ranunculaceae family, collected in Quebec, in the Edinburgh Herbarium are now identified as *Anemone quinquefolia* and *Ranunculus rhomboideus* (table 2.1). The *Ranunculus* was collected on the Île Ste-Hélène close to Montreal on 8 May [1824], where the Dalhousies planned to build a botanical garden.[35] This species is very rare in Quebec and was previously discovered by Montreal botanist Andrew F. Holmes "near Montreal" in the early 1820s, and the specimens are kept at the New York Botanical Garden.[36] This species has not been found in Quebec since Lady Dalhousie's collection and it is now considered extirpated in Quebec.[37] This rare *Ranunculus* is also present in the Dalhousie album (source 4). It was collected on the Île Ste-Hélène in July 1823 by a currently unidentified collector "Mr. S."[38] A systematic survey of Canadian collections by the Countess of Dalhousie at the Royal Botanic Garden Herbarium of Edinburgh would likely add more at-risk species from Quebec to the lists. Canadian botanists Dore and Bernard Boivin estimated that the thirty specimens belonging to the Liliaceae and Ranunculaceae families represent about 10 per cent of all Lady Dalhousie's Canadian collections at the Edinburgh Herbarium. Coincidentally, the collection size of about 300 specimens (ten times thirty specimens) is the same size as the one we noted in her herbarium in Hamilton (source 7).[39]

SOURCE 7: THE HAMILTON HERBARIUM

The Royal Botanical Gardens in Hamilton, Ontario, acquired further plant collections from Lady Dalhousie[40] about the same time that the Canadian government acquired documents from the Dalhousies in the late 1980s.[41] This collection was the subject of a publication by botanist James Pringle, along with a complete list of updated identifications. Pringle's publication is very important and informative, and the identifications by Lady Dalhousie are very accurate. The collection has 297 specimens, with 266 belonging to the countess. Most of these plant collections come from Sorel, but some are from Upper Canada. Out of the 266 specimens, there are fifteen at-risk species. The two rare species that have not been found in other sources are *Gentianopsis virgata* subsp. *victorinii* and *Oenothera pilosella*. There is also another record of the rare native *Gentianella quinquefolia* subsp. *quinquefolia* (3 September 1827),[42] which is extirpated in Quebec.

One of the most interesting pieces of data in the Hamilton Herbarium is the plant specimen "*Orchis discolor*, 21 July 1828." In 1995, Pringle revised this plant to *Amerorchis rotundifolia*, and its currently accepted name is *Galearis rotundifolia*.[43] The location is not indicated, but the Dalhousies were on a tour of the northern shore of the St Lawrence Gulf in July 1828 before returning home to Scotland. The governor needed to stop by the Mingan Islands for business reasons and Lady Dalhousie most likely collected the orchid there.[44] It was new to the countess and she pointed this out to Hooker: "The only thing which appears to me at all worth your acceptance is an Orchis which does not exactly agree with any described tho' very near to the 'discolor.'"[45] This orchid is a species that grows in environments high in limestone and is only found in Anticosti and on the Mingan Islands, on the north shore of the St Lawrence River and Gulf.[46] It is still rare in Quebec and its collection suggests that Lady Dalhousie was probably the first plant collector on the Mingan Islands.

2.6
Herbarium specimen of *Orchis discolor??*, collected by Lady Dalhousie 21 July 1828.

SOURCE 8: 1829 DALHOUSIE CATALOGUE

The Hamilton Herbarium and the 1829 Dalhousie Catalogue represent the last series of plants collected in Lower Canada by the countess, between June 1826 and August 1828. Unlike the previously mentioned sources in this chapter, these two sources include specimens of grasses, sedges, ferns, and fern-related groups. The plants in the catalogue are roughly the same as those that Pringle identifies in the Hamilton Herbarium.

The Dalhousie Catalogue was published in the first volume of the *Transactions of the Literary and Historical Society of Quebec* in 1829 when Lady Dalhousie gifted her 1827 collection to the Literary and Historical Society of

Quebec.[47] This society, founded by Lord Dalhousie in 1824, was a repository for many natural history collections. In that same year, Lady Dalhousie is thought to have given the society a collection of British North American plants. Unfortunately, this 1824 collection has never been found and was likely destroyed by fires in the 1850s and 1860s.[48] The 1829 catalogue of the 1827 collection contains 384 entries, of which 310 bear a complete identification with genus and species names. It includes the highest numbers of species that are at risk in Quebec out of all nine sources reviewed. There are twenty-one rare species, and five of these are exclusive to this source. They are *Asplenium rhizophyllum*, *Gentianopsis crinita*, *Hypericum ascyron* subsp. *pyramidatum*, *Panax quinquefolius*, and *Phegopteris hexagonoptera*. Given that there are seventy-four entries without complete identifications or have names that are difficult to attribute to a given species, the number of rare plants listed in the 1829 Dalhousie Catalogue could actually be greater than twenty-one. Unfortunately, the likely loss of that collection does not permit further analysis and interpretation.

SOURCE 9: HOOKER'S *FLORA BOREALI-AMERICANA; OR, THE BOTANY OF THE NORTHERN PARTS OF BRITISH NORTH AMERICA*

William Jackson Hooker's *Flora Boreali-Americana*, the last of the nine sources of Dalhousie plants examined here, differs from the others because it provides no information about when the Dalhousie specimens were collected. This source is composed of citations of plants that Lady Dalhousie collected in Lower Canada and sent to Hooker for his *Flora Boreali-Americana*. The countess most likely started her correspondence with Hooker in 1825,[49] and the plants sent may represent collections made sometime between 1825 and her departure from Quebec in 1828.

This 500-page major production by Hooker was published from 1829 to 1840 and includes more than 5,000 vascular plant species from British North America. Since Hooker never visited the North American continent, he had to rely on explorers, voyageurs, and residents to obtain plant collections and data. The Countess of Dalhousie was one of only four women who collected plants for Hooker; one was responsible for Newfoundland and Labrador while the other three were active mostly in Lower Canada.[50] Lady Dalhousie's acquaintances, Anne Mary Perceval and Harriet Sheppard, also contributed to collecting plants for Hooker and they frequently botanized together. Citations of the Quebec trio in the flora do not represent all the specimens that were sent to Hooker. All of the plant specimens in Hooker's personal herbarium are now part of the Herbarium at Royal Botanic Gardens, Kew,[51] and no survey of its Quebec specimens has been made. Of the forty-eight known citations in Lady Dalhousie's collection in *Flora Boreali-Americana*, seven plants are considered rare in Quebec. *Penstemon hirsutus* is the only rare plant that is not found in the eight other sources referenced in this discussion.

Christian Broun Ramsay, plant collector in early nineteenth-century Canada between 1820 and 1828, had a preference for new and rare plants, with an emphasis on orchids. The information compiled from the nine different sources, presented in table 2.1, illustrates the extent of Lady Dalhousie's botanical skills. The second column of the table shows that most of her identifications are the same as currently accepted names shown in the first column. Scientific plant names do often change, sometimes due to a redundancy (numerous names for the same species), so it is important to note that some of her identifications are names that are currently accepted as synonyms. This means that there are only a few misidentifications. These plant collections also emphasize her preference for orchids and show the diversity of her botanical interests. One-quarter of the thirty-six species listed in table 2.1 are wild orchids; these are shown in bold. For example, *Galearis rotundifolia* was new to Dalhousie and she highlighted this in a letter to Hooker.[52] The orchid *Arethusa bulbosa* is particularly important because it is reported in eight out of the nine sources that were examined for this review. It may also be present in the Edinburgh Herbarium, but we do not have complete access to this source. Aside from orchids, Lady Dalhousie also had a keen interest in ferns and fern-related groups. She made numerous collections in Asia, including many that were new to science, and some of the foreign ferns she collected were later named after her.[53] Her collections in the Hamilton Herbarium include seventeen fern and fern-related species. She reported on twenty-nine ferns in her 1829 catalogue and on six in Hooker's *Flora Boreali-Americana*.[54] The 1829 catalogue lists two rare ferns from Quebec, *Asplenium rhizophyllum* and *Phegopteris hexagonoptera*. She even wrote to Hooker about her plans to prepare a publication on the ferns that she collected during her various expeditions[55] but was unfortunately unable to accomplish this project.

Table 2.1 depicts Lady Dalhousie's significant contributions to botany in regard to the number of rare plants she reported or collected. The table therefore represents a highly important source of information for botanists and rare plant managers, from the Quebec Ministry of Environment to private conservation agencies such as Canadian Nature Conservancy. The year and locality for the first discovery of a rare specimen provide essential information

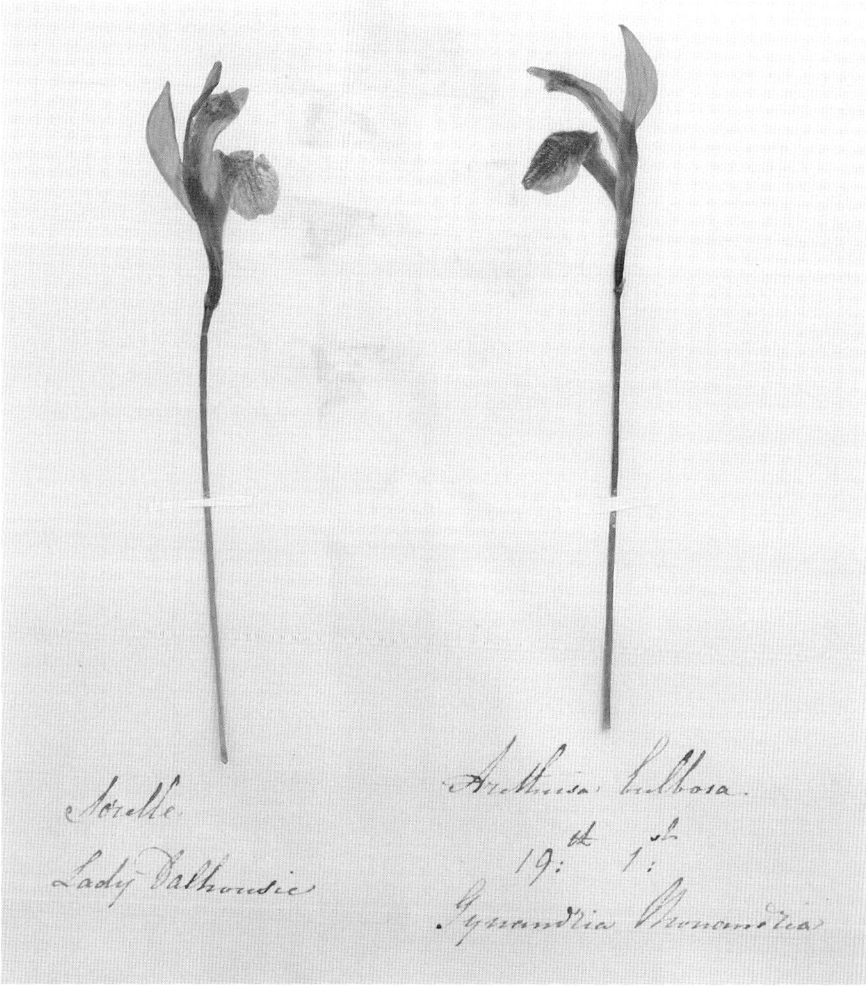

2.7
Herbarium specimens of *Arethusa bulbosa*, collected in Sorell by Lady Dalhousie, from the album Anne Mary Perceval sent to botanist William Darlington in 1826.

for understanding any change in habitats and distribution over time. Most of the habitats for Lady Dalhousie's orchids have been highly modified, if not destroyed. In the case of species that are now considered extirpated from Quebec, such as *Gentianella quinquefolia* and *Ranunculus rhomboideus*, information regarding their early occurrences would be helpful in the search for their potential rediscovery. The most recent list from the Quebec government (2016) includes twenty-seven rare taxa based on Lady Dalhousie's references for which the *Centre de Données sur le Patrimoine naturel du Québec* is in the process of updating the conservation status.[56] These twenty-seven taxa represent seven percent of the latest Quebec list. It is interesting to note that the proportion of rare plants in three of the most complete lists of her collections is around seven percent too: twenty out of 283 (7 per cent) in source 4, fifteen out of 266 (6 per cent) in source 7, and twenty-one out of 310 (7 per cent) in source 8. This is a good example of the importance of her contribution. Dalhousie's thirty-six rare taxa count (overall) could be increased with the additional examination of sources for information about plants she collected that were only partly surveyed here or not surveyed at all. There are, for example, some plant collections by Lady Dalhousie at herbaria in Florence and Geneva.[57] Moreover, only a small fraction of her collections at the Edinburgh Herbarium was examined for this review. There are also more of her plant collections at the Kew Herbarium than the forty-eight species for which Hooker cited her in his *Flora Boreali-Americana*. In addition, Hooker sent an unknown number of her specimens to the *Museum national des Plantes* in Paris in the 1840s and 1850s, and research for this chapter surveyed only a few families.[58]

Lady Dalhousie's high reputation as a plant collector was widely known and remarked upon in her time. Archibald praised her as "a lady, whose zealous and indefatigable exertions in botanical matters have seldom, I think, been surpassed; perhaps not often equalled; few having attained such proficiency as her ladyship in the science."[59] Hooker declared in *Exotic Flora* that "we are indebted to the Right Honourable the Countess of DALHOUSIE, who, with a liberality and kindness that I am proud to acknowledge, immediately upon receiving an application which had been made to her Ladyship for Canadian plants, sent to our Botanic Garden some boxes well stored with botanical rarities, especially Orchideae, from the vicinity of Montreal."[60] Hooker was most likely referring to her collections from Sorel in 1823. In 1825, he praised her further in "On the Botany of North America," a publication highlighting botanists and plant collectors who were contributing to knowledge of the Canadian flora. "In the first rank of these," he wrote, "we are proud to be able to mention the Right Honourable the Countess of Dalhousie, the lady of his Excellency the Governor, whose rank and influence, no less than her superior acquirements and great love of science, entitle us to hope for much from her in the promotion of our wishes."[61]

Plant data provided by the Countess of Dalhousie are very reliable. Collecting dates are accurate and they correspond to current flowering times.[62] The locations she identified are consistent with currently known habitats and ranges of the plant species, and in some cases can be linked directly to specific travel information provided by Governor Dalhousie in his journals.[63] Lady Dalhousie's accomplishments, particularly with respect to orchids and rare plants of Quebec, represent an invaluable legacy to botanical science. Given the physical difficulties and limitations of knowledge of her time, it is impressive that she managed to provide such remarkably extensive and significant collections which, admired by leading botanists of her time, can now contribute to the conservation of threatened plants.

ACKNOWLEDGMENTS

Many thanks to Ann Shteir for inspiration, extensive revisions, and shared information. Thank you to Alain Asselin and Jacques Mathieu, co-authors of volume 3 of *Curieuses histoires de plantes du Canada*, to Stephen J. Darbyshire, Jacques Labrecque, and Étienne Léveillé-Bourret for their suggestions on the manuscript, and to Jennifer Doubt, Kristina Huneault, Deborah Reid, and Ernie Small for their help. Thanks also to the following people and institutions for allowing the rights to reproduce the illustrations: Douglas Holland, Missouri Botanical Garden, USA; Lisa M. Bower, Nova Scotia Museum, Halifax; Christina Jenness, Canadian Museum of Nature, Quebec; Ron McColl, Darlington Herbarium, West Chester University, PA, USA; David Galbraith, Royal Botanical Gardens (Canada).

NOTES

1 On the proper use of vascular plants in gardening, see Lamoureux and Nantel, *Cultiver des plantes indigènes.*

2 The nine sources are: (1) Archibald, "Dalhousie Castle & Gardens," 251–8; (2) Hooker, *Exotic Flora*, vol. 1: 1823, 70; vol. 2, 1825, 87, 145; vol. 3, 1827, 170; (3) Nova Scotia Museum, Halifax, Mss. 85.119.34, 1823; (4) *The Countess of Dalhousie Herbarium, 1823*. Canadian Museum of Nature; (5) *Specimens of Canadian Plants, Presented by the Hon. Mrs. A.M. Percival. 1826*. J. Cayouette has compiled a complete list of plants in Anne Mary Perceval's album "Specimens of Canadian Plants." See Asselin, Cayouette, and Mathieu, *Curieuses histoires*, Appendix 8; (6) Selection of specimens of Lady Dalhousie in the *Royal Botanic Garden Herbarium* at Edinburgh; (7) Pringle, "Canadian Botanical Specimens"; (8) Dalhousie, "Catalogue of Canadian Plants collected in 1827"; (9) Hooker, *Flora Boreali-Americana.*

3 Shteir and Cayouette, "Collecting with 'botanical friends,'" 5–9; Pringle, "Canadian Botanical Specimens"; Asselin, et al., *Curieuses histoires*, 134–41; Blair, *Un Orignal en Flammes*; Browne, "Ramsay, Christian, countess of Dalhousie."

4 On Anne Mary Perceval see Pringle, "Anne Mary Perceval," 7–13; Cayouette, "Anne

Mary Perceval," 1–11; Shteir and Cayouette, "Collecting with 'botanical friends,'" 9–13; Asselin, Cayouette, and Mathieu, *Curieuses histoires*, 142–8.

5 On Harriet Sheppard see Shteir and Cayouette, "Collecting with 'botanical friends,'" 13–18.

6 LeMoine, *Album Canadien*, 96.

7 Dalhousie, *Director's Correspondence*, DC 44:56, 31 October, no date but most likely in 1825.

8 Bouchard et al., "Les plantes vasculaires rares du Québec."

9 Argus and Pryer, *Les plantes vasculaires rares du Canada*, 177–9.

10 Lavoie, *Plantes vasculaires susceptibles*; Labrecque and Lavoie, *Les plantes vasculaires menacées*; CDPNQ, *Les plantes vasculaires*; Tardif et al., *Les plantes vasculaires*.

11 COSEWIC, www.cosewic.gc.ca.

12 NatureServe: a Network Connecting Science with Conservation, www.nature reserve.org.

13 Archibald, "Dalhousie Castle & Gardens," 251.

14 Ibid., 255.

15 The title page of *Exotic Flora*, vol. 1 (1823) reads as follows: "*Exotic Flora, containing figures and descriptions of new, rare, or otherwise interesting Exotic Plants, especially of such as are deserving of being cultivated in the gardens; together with remarks upon their generic and specific characters, natural orders, history, culture, time of flowering, etc.*"

16 Stafleu and Cowen, *Taxonomic Literature*, vol. II: H-Le, ed. 2 (1979) 283–301.

17 See Hooker *Exotic Flora* I, 70, 1823. The orchids were sent by Lady Dalhousie to Dalhousie Castle between 1820 and 1823 and then to the Glasgow Botanic Gardens for Hooker. *Habenaria* (now *Platanthera*) *blephariglottis*, probably from the city of Quebec area, and *Pogonia ophioglossoides*, flowered for the first time at Dalhousie Castle. The roots of *Habenaria* (now *Platanthera*) *orbiculata* and of *Arethusa bulbosa* were sent to Hooker probably from Sorel in 1823 and flowered at the Glasgow Botanic Gardens in 1824: *Exotic Flora*, vol. 2, plate 145, 1825; vol. 3, plate 170, 1827.

18 Goldie, "On the culture of North American plants," 129–35; on Goldie, see Asselin, Cayouette, and Mathieu, *Curieuses histoires*, 90–103.

19 On William Sheppard, see Shteir and Cayouette, "Collecting with 'botanical friends'"; Asselin, Cayouette, and Mathieu, *Curieuses histoires*, 118–33; Cayouette, "Pursh explore le Haut-Canada," 6–7; Cayouette, "Pursh dans l'est du Bas-Canada, I."

20 Hooker, *Exotic Flora*, vol. 3, 170.

21 Elwood, "Discovery and Repatriation," 108–16.

22 Pursh, *Flora Americae Septentrionalis*. On Pursh, see Cayouette, "Pursh explore le Haut-Canada," 6–7; Cayouette, "Pursh dans l'est du Bas-Canada"; Cayouette, "Enquête sur la *Flora* perdue," 14–16; Cayouette, "corrigendum," 2.

23 Gallichan, "La bibliothèque personnelle," 92–3.

24 *The Countess of Dalhousie Album, 1823*, folio 200.

25 Ibid., folios 27.1 and 193.

26 On Darlington see Ewan, "Darlington, William," vol. 3, 562–3; Asselin, Cayouette, and Mathieu, *Curieuses histoires*, 143–4.

27 DAO is the acronym for National Collection of Vascular Plant Herbarium, Agriculture and Agri-Food Canada, in Ottawa.

28 Pringle, "Canadian Botanical Specimens," 18.

29 Jacques Labrecque, MDDELCC, Quebec, pers. com. to Jacques Cayouette, 16 February 2021.

30 Pringle, "*Gentianella*"; Gillett, "Revision of the North American Species of *Gentianella* Moench," 243–6.

31 Dalhousie, *Plants / Native of Canada / Sorell 1823*, Nova Scotia Museum, Halifax, Mss. 85.119.34.

32 Whitelaw, ed. *The Dalhousie Journals*, II, 144.

33 Hedge and Lamond, *Index of Collectors*, 73.

34 Letter from William G. Dore to Guy Baillargeon, 29 January 1988, private archives of Jacques Cayouette; on C.B. Dalhousie collections at the Royal Botanic Gardens Edinburgh, see Bernard Boivin, ms, undated, private archives of Jacques Cayouette.

35 Shteir and Cayouette, "Collecting with 'botanical friends,'" n23; Asselin, Cayouette, and Mathieu, *Curieuses Histoires*, 135–6, 139, 143–4.

36 An earlier collection of *Ranunculus rhomboideus* in the Montreal area was made around 1820 by botanist Andrew Fernando Holmes (1797–1860); see Cayouette, "Les découvertes botaniques," 6–7. Lady Dalhousie's collection has been added to the Quebec file of rare plant species that are now extirpated.

37 Labrecque and Lavoie, *Les plantes vasculaires menacées*, 168; CDPNQ, *Les plantes vasculaires*, 122; Tardif et al., *Les plantes vasculaires*, 321.

38 *The Countess of Dalhousie Album, 1823*, folio 162.2. "*Ranunculus ovalis*? St. Helens July 1823, Mr. S." This collector has not yet been identified.

39 See Pringle, "Canadian Botanical Specimens."

40 Ibid.

41 Elwood, "Discovery and Repatriation," 108–16.

42 Pringle, "Canadian Botanical Specimens," 9, 18.

43 Ibid., 9, 11.

44 Whitelaw, *The Dalhousie Journals*, III: 157.

45 Dalhousie, *Director's Correspondence*, DC 44: 59, 27 February 1829.

46 Tardif et al., *Les plantes vasculaires*, 253.

47 Pringle, "Canadian Botanical Specimens," 3.

48 Blair, *Une Histoire du Morrin Centre*, 197–200. Two major fires occurred on 1 February 1854 and in 1862.

49 Dalhousie, *Director's Correspondence*, DC 44:56, 31 October, no year shown but most likely 1825.

50 Hooker, "Flora of the British possessions," 92: "the botanical productions of Canada, which have been received from the Lady Dalhousie, Mrs. Percival, Mr. Sheppard …"

51 On the fate of the Hooker's private herbarium, see Stafleu and Cowen, *Taxonomic Literature*, 283.

52 See endnote 45. This orchid was obviously new to Lady Dalhousie. In Pursh's Flora, *Orchis bicolor* refers to an orchid with one basal leaf and a few flowered inflorescences. Lady Dalhousie's orchid is now known as *Galearis rotundifolia*. In July 1828, the Dalhousies were on a boat tour of the north shore of the St Lawrence Gulf because Lord D. wanted to stop on the Mingan Islands for business reasons. As *Galearis rotundifolia* is a lime-loving plant (calcicolous), it was most likely collected there. These islands are well known for their limestone beds and their special flora. See Marie-Victorin and Rolland-Germain, *Flore de l'Anticosti-Minganie*; Tardif et al., *Les plantes vasculaires*, 253.

53 See Shteir and Cayouette, "Collecting with 'botanical friends.'" *Asplenium dalhousiae* was collected by Lady Dalhousie at Simla, in the Himalayas, and described by W.J. Hooker in 1837 (Hooker, *Icones Plantarum*, plate 105). Two fern species collected in Asia by Lady Dalhousie were named after her by Hooker: *Cheilanthes dalhousiae*, also from Simla (Hooker, *Species Filicum*, 2: 80, 1852) and *Pteris dalhousiae*, from Penang, Malaysia (Hooker, *Species Filicum*, 2: 170, 1858).

54 Pringle, "Canadian Botanical Specimens," 6–14; Dalhousie, Catalogue; Hooker, *Flora Boreali-Americana*.

55 Dalhousie, *Director's Correspondence*, DC 53: 38, 4 February 1838.

56 Tardif et al., *Les plantes vasculaires*.

57 On the specimens of Lady Dalhousie in Florence, see Steinberg, "The collectors and collections." The specimens in Florence and Geneva are probably related. Botanist Augustin Pyramus de Candolle (1778–1841) acquired specimens collected by Lady Dalhousie by unknown means. One of his protégés, botanist Philippe Marie Mercier (1781–1831), built a huge herbarium containing Dalhousie's specimens (likely from de Candolle), which were later acquired by Philipp Barker Webb (1793–1854), whose herbarium serves as the foundation of the Florence Herbarium; her Canadian material there is dated 1827–29, but she left Canada in 1828.

58 A selection of specimens collected by Lady Dalhousie at the Herbarium of the *Museum des Sciences naturelles de Paris* reveals that Hooker sent these as duplicates of his collection in the years 1840 and 1850.

59 Archibald, "Dalhousie Castle & Gardens," 255.

60 Hooker, *Exotic Flora*, vol. 2, 145.

61 Hooker, "On the Botany of America," 126.

62 Pringle, "Canadian Botanical Specimens," 16.

63 Whitelaw, ed., *The Dalhousie Journals*, Vol. II.

BIBLIOGRAPHY

Archives

Dalhousie, Countess of. *The Countess of Dalhousie Herbarium, 1823*. Canadian Museum of Nature, Gatineau, Quebec.

– Letters to William Jackson Hooker, Royal Botanic Gardens Kew: Archives: *Director's Correspondence*, DC 44: ff. 56, 59, DC 53, f. 38 [three letters from Château St Louis: 31 October n.d. but likely 1825; Coulstown (East Lothian, Scotland), 27 February 1829; Dalhousie Castle, 4 February 1838].

– *Plants Native of Canada Sorell 1823*. Nova Scotia Museum, Halifax, Mss. 85.119.34.

Perceval, Anne Mary. *Specimens of Canadian Plants, Presented by the Hon. Mrs. A.M. Percival, of Spencer Wood, near Quebec, to Wm. Darlington. 1826*. Darlington Herbarium, West Chester, PA.

Books and Articles

Archibald, Joseph. "Dalhousie Castle & Gardens: the Botany of the Neighbourhood." *Gardener's Magazine* 1 (1826): 251–8.

Argus, George W., and Kathleen M. Pryer. *Les plantes vasculaires rares du Canada; notre patrimoine naturel.* Ottawa: Musée canadien de la nature 1990.

Asselin, Alain, Jacques Cayouette, and Jacques Mathieu. *Curieuses histoires de plantes du Canada*, vol. 3. Quebec: Septentrion 2017.

Bernatchez, Ginette. "La Société Littéraire et Historique de Québec (The Literary and Historical Society of Quebec), 1824–1890." *Revue d'Histoire de l'Amérique Française* 35, no. 2 (1981): 179–91.

Blair, Louisa. "Un Orignal en Flammes; L'Histoire de la Literary and Historical Society of Quebec." In Louisa Blair, Patrick Donovan, and Donald Fyson, *Étagères et Barreaux de Fer; Une Histoire du Morrin Centre*, 171–252. Quebec: Septentrion 2016.

Bouchard, André, Denis Barabé, Madeleine Dumais, and Stuart Hay. "Les plantes vasculaires rares du Québec/The rare vascular plants of Quebec." *Syllogeus* 48 (1983): 1–79; 1–75.

Browne, Janet. "Ramsay, Christian, countess of Dalhousie (1786–1839)." In *Oxford Dictionary of National Biography*. Oxford: Oxford University Press, 2004.

Cayouette, Jacques. "Anne Mary Perceval, l'élégante châtelaine de Spencer Wood." *Bulletin de FloraQuebeca* 9, no. 1 (2004): 1–11.

– "Les découvertes botaniques d'Andrew Fernando Holmes autour de Montréal." *Bulletin de FloraQuebeca* 7, no. 1 (2002): 6–7.

– "Enquête sur la *Flora* perdue." *Quatre-Temps* 32, no. 2 (2008): 14–16; "corrigendum," *Quatre-Temps* 32, no. 3 (2008): 2.

– "Pursh dans l'est du Bas-Canada et à l'île d'Anticosti, I: la saga des spécimens d'herbier." *Bulletin de FloraQuebeca* 10, no. 1 (2005): 13–18.

– "Pursh explore le Haut-Canada en 1816." *Bulletin de FloraQuebeca* 5, no. 1 (2000): 6–7.

CDPNQ – Centre de données sur le patrimoine naturel du Québec. *Les plantes vasculaires menacées ou vulnérables du Québec, Ed. 3*. Quebec: Gouvernement du Québec, ministère du Développement durable, de l'Environnement et des Parcs, Direction du patrimoine écologique et des parcs 2008.

Dalhousie, R.H. "Catalogue of Canadian Plants collected in 1827 and Presented to the Literary and Historical Society." *Transactions of the Literary and Historical Society of Quebec* 1 (1829): 255–61.

Elwood, Marie. "The Discovery and Repatriation of the Lord Dalhousie Collection." *Archivaria* 24 (Summer 1987): 108–16.

Ewan, Joseph. "Darlington, William." In *Complete Dictionary of Scientific Biography*, vol. 3, 562–3. Detroit: Charles Scribner's Sons, 2008.

Gallichan, Gilles. "La bibliothèque personnelle du gouverneur Dalhousie." *Les Cahiers des dix* 65 (2011): 92–3.

Gillett, John M. "A Revision of the North American Species of *Gentianella* Moench." *Annals of the Missouri Botanical Garden* 44 (1957): 243–6.

Goldie, John. "On the culture of North American plants." *The Gardener's Magazine* 2, no. 6 (1827): 129–35.

Hedge I.C., and J.M. Lamond. *Index of Collectors in the Edinburgh Herbarium*. Edinburgh: Her Majesty's Stationery Office, 1970, 73.

Hooker, William Jackson. *Exotic Flora*, vol. 1. Edinburgh: William Blackwood, 1823; vol. 2, 1825; vol. 3, 1827.

– *Flora Boreali-Americana; or, the Botany of the Northern Parts of British North America*. London: Henry G. Bohn, 1829–40. 2 vols.

– "Flora of the British possessions in North America." *Botanical miscellany* … 1 (1830): 92.

– *Icones Plantarum*. London: Longman, Rees, Orme, Brown, Green & Longman, 1, plate 105, 1837.

– "On the Botany of America." *Edinburgh Journal of Science* 2 (1825): 126. Reprinted in *American Journal of Science and Arts* 1, no. 9 (1825): 281.

– "*Species Filicum*." London: William Pamplin, vol. 2: 80, 170, 1852 and 1858, 8.

Labrecque, Jacques, and Gildo Lavoie. *Les plantes vasculaires menacées ou vulnérables du*

Québec. Quebec: Gouvernement du Québec, ministère de l'Environnement, Direction du patrimoine écologique et du développement durable, 2002.

Lamoureux, Gisèle, and Patrick Nantel. *Cultiver des plantes indigènes … sans leur nuire*. Saint-Henri-de-Lévis: Fleurbec Ed., 1999.

Lavoie, Gildo. *Plantes vasculaires susceptibles d'être désignées menacées ou vulnérables au Québec*. Quebec: Direction de la conservation et du patrimoine écologique, ministère de l'Environnement du Québec, 1992.

LeMoine, John MacPherson. *Album Canadien, Histoire, Archéologie-Ornithologie*. Quebec: Des Presses Mécaniques du *Canadien*, 1870: 96.

Marie-Victorin, F., and F. Rolland-Germain. *Flore de l'Anticosti-Minganie*. Montreal: Les Presses de l'Université de Montréal, 1969.

Nelmes, E., and W. Cuthbertson. *Curtis's Botanical Magazine Dedications 1827–1927: Portraits and Biographical Notes*. London: Bernard Quaritch, 1932, 26–8.

Pringle, James S. "Anne Mary Perceval (1790–1876), An Early Botanical Collector in Lower Canada." *Canadian Horticultural History* 1 (1985): 7–13.

– "Canadian Botanical Specimens Collected 1826–1828 by the Countess of Dalhousie, Acquired by the Royal Botanical Gardens." *Canadian Horticultural History* 3 (1995): 1–21.

– "*Gentianella*," *Flora of North America*, vol. 14, forthcoming.

Pursh, Frederick. *Flora Americae Septentrionalis*. London: White Cochrane & Co, 1814, but published in December 1813, vol. I and II; ed. 2, 1816.

Shteir, Ann B. *Cultivating Women, Cultivating Science: Flora's Daughters and Botany in England 1760 to 1860*. Baltimore: The Johns Hopkins University Press, 1996.

Shteir, Ann B., and Jacques Cayouette. "Collecting with 'botanical friends': Four Women in Colonial Quebec and Newfoundland." *Scientia Canadensis* 41, no. 1 (2019): 1–30.

Stafleu, Frans A., and Richard S. Cowen. "William J. Hooker." *Taxonomic Literature*, vol. II: H-Le, ed. 2. Utrecht: Bohn, Scheltema and Holkema 1979: 283–301.

Steinberg, Carlo H. "The collectors and collections in the Herbarium Webb." *Webbia* 32, no. 1 (1977): 1–49.

Tardif, Bernard, Gildo Lavoie, and Yves Lachance. *Atlas de la biodiversité du Québec. Les espèces menacées ou vulnérables*. Quebec: Gouvernement du Québec, ministère du Développement durable, de l'Environnement et des Parcs, Direction du développement durable, du patrimoine écologique et des parcs, 2005.

Tardif, Bernard, Benoît Tremblay, Guy Jolicoeur and Jacques Labrecque. *Les plantes vasculaires en situation précaire au Québec*. Quebec: Centre de données sur le patrimoine naturel du Québec (CDPNQ), Gouvernement du Québec, ministère du Développement durable, de l'Environnement et de la Lutte contre les changements climatiques (MDDELCC), Direction de l'expertise en biodiversité, 2016.

Whitelaw, Marjory, ed. *The Dalhousie Journals*. Canada: Oberon Press, 1978–82.

3

Gender, Botany, and Imperial Networks

Reflections on a Letter

VIRGINIA VANDENBERG

In 1829, Christian Ramsay, Lady Dalhousie, wrote to her friend from Quebec Anne Mary Perceval about the long sea journey from her home in Scotland to her husband's new posting as commander-in-chief of the British Army in India. Her letter describes in detail the expeditions and exploration of the local environment of stops in Madeira, Rio de Janeiro, and Cape of Good Hope before their arrival in now Kolkata (was Calcutta). It was Lady Dalhousie's first experience as a botanical observer in a colonial context vastly different from the one she had become accustomed to, and she confessed to Perceval that, while visiting the tropical locations, she was "really bewildered with the richness of botanical treasures."[1] By the time she travelled to India with her husband, Lady Dalhousie was already an avid and skilled botanical collector, part of a global network of correspondents and collectors that botanist William J. Hooker fostered to conduct botanical research abroad in the British Empire. Her passion for and skill in botanical collecting are made clear through her letters, journals, and herbaria. Dalhousie's scientific activities provide historians of empire and science a useful case study for analyzing global networks around botanical collecting because she conducted botanical work in multiple colonial locations and was at the centre of several social circles connected by interests in botany.

This chapter focuses on Lady Dalhousie's letter to Anne Mary Perceval dating from December 1829 and reflects through it on the entanglement of the early nineteenth-century global knowledge economy and the intimate realm of colonial experience. The full text of this letter can be read in the Appendix to this chapter. The original manuscript copy of the letter is not found among Lady Dalhousie's papers but fortunately a typed transcript made by a family descendant in the 1920s can be found in the John Gray Centre Archives in Scotland.[2] The survival of such a letter between two female botanical collectors only as a typed transcript speaks to the challenges of archival work in the fields of women's imperial science histories. My chapter considers how such a letter, and the life experiences of female botanical collectors that it illustrates, can be used to explicate stories of women in science. It uses the letter, and Lady Dalhousie's movement across the globe, to explore recent historiographical trends from global history that illumine interactions and circulation of knowledge in the construction of empire and imperial science. Applying these new trends to studies of women in science is productive for analyzing how women's scientific activities were part of a constructed imperial science culture, sustained through correspondence networks and incorporating colonial appropriation of local knowledge and expertise for the advancement of British science. Reading Lady Dalhousie's letter through these lenses challenges the assumptions that nineteenth-century women's botanical practices were marginal to the projects of the British Empire and botany.

Botanical collecting and illustrating had been a key part of an Enlightened education for elite women during the eighteenth century, and botanical practices were an established part of many women's activities. As British women began to move about the globe in increasing numbers as part of British imperial expansion, the cultural practices around the accumulation and development of scientific knowledge went with them. Letters were not just locations for information about the experiences of women in colonial locations or their scientific practices but were themselves important parts of constructing imagined imperial space or the formation of the scientific community globally.[3] Thinking critically about correspondence between scientifically interested women travelling about the British Empire becomes a valuable tool for examining the role of scientific activities by individual women such as Lady Dalhousie in the larger scope of empire.

Feminist historians contend that women's everyday lives and practices are worth studying but acknowledge the difficulty of finding archival records that too often were ignored or erased in favour of male accomplishment and achievement.[4] Scholars in postcolonial studies similarly acknowledge the challenges inherent in recovering the voices of the subaltern subject in the archives of Western knowledge; Durba Ghosh has demonstrated this, for example, in the literal erasure of names from British colonial records. Postcolonial scholars also have long been interested in reading the silences in the archive, mirroring the work done by feminist scholars to challenge the assumptions made about the types of stories worth telling and the established methods of telling those stories.[5] Both their methodologies analyze the role that power plays in the construction of the archives and the stories told from the archives. Beyond merely trying to tell the stories hidden in the archives, both approaches also help to examine the complicated ways power is enacted in the lives of ordinary people and help as well to re-evaluate the lives recorded.

This chapter uses approaches from feminist and postcolonial studies to read Dalhousie's 1829 letter and reflect on the complex factors at play in the global knowledge economy. Moving beyond trying to recover the stories of participants traditionally omitted from the story of science, my analysis focuses on a single letter to illustrate the ways in which colonial science was navigated by women within the larger, sprawling studies in global histories. In the case of women conducting scientific activities in colonial locations, integrating fields of critical theory such as feminist biography and postcolonial critiques of Western epistemological violence enables us to complicate the story of scientific women.

Reading British women's personal correspondence through these critical lenses allows us to re-evaluate scientific activities as appropriative imperial practices, keeping in mind the multiple levels of power navigated by elite white British women in colonial locations. Historian Adele Perry argued recently that biography is one of the tools that can be used to facilitate critical discussions around colonial processes and thereby to re-evaluate imperialist and nationalist narratives that have erased colonized peoples.[6] This chapter takes up Perry's call as a starting point for considering the complex ways in which empire, science, race, gender, and class shape women's life experiences narrated in their own letters. In this way, my chapter traces the entangled threads of colonial processes, imperial science, and women's activities that can be discerned in early nineteenth-century relationships/networks and appropriative practices, and thereby incorporates critical feminist and postcolonial theories into reading the lives of women in science in Canada and empire.

ANNE MARY PERCEVAL, LADY DALHOUSIE, AND BOTANICAL COLLECTING IN THE EMPIRE

Lady Dalhousie's letter to her friend, dated 15 December 1829, allows us to see how her experiences and relationships in multiple locations of the British Empire had a significant impact on her life and shaped her botanical knowledge and practices. The chapter by Deborah Reid in this volume, using extensive archival materials, shows that Lady Dalhousie, having collected plants as a young girl and gained experience overseeing the gardens at Dalhousie Castle as a young married woman, carried an interest in plants with her to the North American colonial context. While in British North America during the years 1817–28, that interest developed further through her association with several other British women living in Quebec. Shteir and Cayouette have written about the significance of this circle of "botanical friends," notably including Anne Mary Perceval and Harriet Sheppard.[7] Dalhousie's ties to Perceval and Sheppard were intertwined with shared colonial experience, interest in botanical science, and work collecting and preserving plant specimens. These relationships did not end with Lady Dalhousie's departure from Quebec. Indeed, as the letter to Anne Mary Perceval illustrates, relational connections remained a vital way of linking empire and science together.

By the time Lady Dalhousie was writing to Perceval on her journey to India, a strong relationship had already developed between the two botanizing women. Anne Mary Perceval was a like-minded British woman practising botany in North America at the time that Dalhousie was also botanizing. Perceval was born in England in 1790 and, as the daughter of landed gentry, had been well educated in languages and artistic skills. Perceval married Michael Henry Perceval, collector of customs and superintendent of the Port of Quebec in Lower Canada, in 1810. After the Percevals acquired the estate they named Spencer Wood in 1815, Anne Mary Perceval spent time on plant-collecting excursions in the local woods. Perceval also corresponded with many naturalists, including John Torrey, Frederick Pursh, and William Hooker. Her botanical contributions were acknowledged multiple times in Hooker's *Flora Boreali-Americana* and Torrey and Gray's *Flora of North America.*

Perceval's own letters are particularly descriptive of the interwoven connections that existed between Lady Dalhousie, her social circle in Quebec, and "experts" of botany like Hooker, illustrating the interdependence between both female and male correspondents in North America for sharing and developing botanical knowledge and skills.[8] Social capital demonstrated through "name-dropping" was a way of establishing belonging, credibility and authority, and Perceval was eager to demonstrate her connections to other botanists with whom she was corresponding, including John Torrey, whom she describes as "a young man, but of very promising talents, and very highly thought of amongst his own countrymen."[9] John Field notes that social capital is the "intangible resources of community, shared values and trust upon which we draw in daily life,"[10] and women's participation in scientific communities was often dependant on these intangible resources available to only a privileged few.

Perceval herself travelled to various locations to visit botanists she knew, most likely circulating specimens and texts between practitioners in multiple locations as was common among travellers to colonial destinations.[11] An elite woman but lower in status than Lady Dalhousie, Perceval needed to establish that she had the relevant social capital of connections and knowledge to participate in the scientific community.

The acquisition of plant specimens and botanical texts was central to the relationship between Perceval and Dalhousie, as it was for many corresponding practitioners. Perceval was particularly keen about collecting and reported in a letter to Hooker having "often said" to her collecting companion Dalhousie that "if the love of money increase in proportion to the love of flowers, I can thoroughly understand, how a man becomes a miser."[12] Perceval sought to utilize Lady Dalhousie's connections to further her own botanical knowledge, especially in requesting copies of botanical texts through Dalhousie's bookseller.[13] From Hooker himself she received his publication *Botanical Illustrations*, along with the "most beautiful collection of plants," and she promised to deliver "the parcels … immediately to Lady Dalhousie and Mrs. Sheppard."[14] Writing about practices of Joseph Hooker and others in Victorian botany, Jim Endersby notes that "metropolitan gents needed expert collectors to avoid being bombarded with misidentified plants, worthless duplicates, and poorly preserved rubbish. One way to improve your collectors was to send them gifts, especially botanical books."[15] Such texts "invariably stressed the need for practical collecting before the novice even contemplated the more abstruse branches of the subject."[16] The exchange of botanical texts was an important aspect of connecting those practising botany in the colonies to the larger community by providing a kind of long-distance education in botanical practices.

Maintaining connections like these was significant because botanical collecting played an important role in increasing British knowledge of plants in the empire. Botanical experts relied on collectors to go out, find plants, collect and preserve them, label them, and send them to botanists for classification and cataloguing. These collectors could be male or female and come from a wide variety of social backgrounds. Botanical collecting had long been popular among natural history enthusiasts in Britain but began to play an important role in colonial locations in the late eighteenth and early nineteenth centuries.[17] Suzanne Zeller has detailed how botany was a significant part of inventorying natural resources of the British North American colonies, and Maroske has shown that botanical collecting networks formed by Ferdinand von Mueller were similarly critical to shaping natural knowledge of Australia.[18] In addition, Maroske has analyzed how botanical collecting might function as an "agent of cultural change" and argues that collecting in colonial locations demonstrates important relationships formed between colonial actors. Relationships centred around collecting plants became essential to the practice of imperial botany.

Dalhousie's letter to Perceval is a significant document for research in women's botanical histories. Connecting Dalhousie's North American and Indian botanical work, it points to botanical networks maintained across the globe, not just between male scientific correspondents but between female botanical collectors as well. As the only extant letter identified thus far written by Dalhousie to another practising female botanical collector, it is a window onto scientific interests shared with a like-minded friend. Listing for Perceval the stops she and her family made on their journey to India in 1829, Dalhousie gives details about the local area in each port-of-call that are similar to accounts by other travellers about the journey between Britain and India.[19] She especially focuses, however, on the natural landscape and vegetation. In Madeira, for example, they "employed three days in visiting the beauties of the Island," the view having a "beautiful effect mingled with the Banana and aloe … [and] these are interspersed with the native Chestnut forests." In Rio de Janeiro, their next port-of-call, she highlights the beauty of the greenery: "Mountains rising in endless succession of the very wildest and most fantastic forms, many are cones rising singly from their base or springing from the summits of other mountains all wooded to the top, in the luxuriance of tropical vegetation … covered with mimosa, melastoma, etc. while the graceful palm trees wave above."

While in Rio, she first encountered totally unfamiliar flora on an "excursion … through 'virgin forest,'" where "immense trees totally dissimilar to those I am accustomed to were <u>bound</u> together by innumerable 'Lianes' or climbing plants, and so matted with the profusion of underwood as to be totally impenetrable." Their final stop in Cape Colony also included an "excursion of 3 days into the 'Interior of Africa,'" where she described the "wilderness … covered with the most beautiful Ixias, Gladiolus, and all the tribe of bulbous plants, Geraniums, Heaths, Proteas, in short all the most precious contents of our Greenhouses flourish in a Waste!!!" Lady Dalhousie's interest in observing her environment and recording those observations were shaped by the expectations around scientific practices for both men and women in the colonies, especially the leisure activities of botanical collecting, illustration, and travel that were a feature of an enlightened elite back in Britain. Dalhousie, as an imperial subject moving across the empire, carried British scientific interests, skills, and experiences into new locations, in Dalhousie's case to India from British North America.

The majority of the 1829 letter to her friend illustrates Lady Dalhousie's journey into a new botanical world, but the letter also refers to experiences that she and Perceval shared in the colonial landscape of North America. Lady Dalhousie compares what she is seeing to Quebec rather than to her own home environment of Scotland and notes that the beautiful sunsets they saw when on board their ship were "as lovely as the best we had [from their home in] dear Sorel." Furthermore, upon seeing the "sacred mystical mythological Ganges!" she was reminded of the landscape of Quebec: "to my extreme discomposure, instead of 'dark blue waves' the river is the exact colour of the little brook at Spencer Wood while the snow is melting ... the scenery (setting aside the palm trees) strongly reminded me of the St. Lawrence about Nicolet." It is notable that Dalhousie would use her time in Canada rather than her "home" in Scotland as the benchmark for her experiences in other colonial locations, despite the vastly different climates and landscapes. In a letter written to Hooker several years later, she declares that "there is certainly more resemblance to the Flora of Canada in that of the Himalaya than to that of England!"[20] These comparisons highlight the global scope of Lady Dalhousie's frame of reference for her imperial and scientific knowledge.

TOWARD GLOBAL HISTORIES OF SCIENCE

Dalhousie's report on her scientific activities in the letter to her botanical friend focuses on a specific context of her life experience. However, the letter also reveals how imperial science functioned in a larger global frame. This chapter is drawn from a larger global history study of how elite British women like Dalhousie carried with them to the colonies cultural practices of botany that were also shaped within the colonial context; the result was a uniquely imperial science culture that was part of the implementation of British cultural imperialism.[21] The study focuses on the botanical collecting of several British women across the British Empire to illustrate the relationships that formed between botanical practitioners and to understand how global networks linked figures and sustained scientific research in colonial locations. Anna Maria Walker in Ceylon, Annabella Telfair in Mauritius, and even Lady Frances Cole, whom Dalhousie mentions visiting in Cape Town, are women whose botanical experiences can be read as what James Secord terms "'case studies' in the social character of the making of knowledge."[22] Even though travel to India and other British colonies was an experience that many other British officials and their wives undertook, Dalhousie's letter illustrates factors at play in her practice of science in a particularly vivid and focused way.

Global histories of science move us away from science as formal propositions or discoveries and consider instead, as Kapil Raj writes, the "construction, maintenance, extension, and reconfiguration of knowledge, focusing equally on its material, instrumental, corporeal, practical, social, political, and cognitive aspects."[23] As a material object in the construction of botanical knowledge, Dalhousie's letter from 1829 illustrates how this global system of knowledge functioned. Global history works therefore to expand critical perception of the complex factors behind the practice of imperial botany.

One significant context for reading the letter is the movement of people, goods, and ideas across the global space of the British Empire. While Fa-ti Fan warns that "circulation" has begun to function as a buzzword in global history,[24] it remains a central part of the framework of global history and the history of science more specifically. The movement of the letter highlights the mobile nature of science within the global frame. However, for global histories of science, circulation is more than the dissemination of knowledge but rather the "processes of encounter, power and resistance, negotiation, and reconfiguration that occur in cross-cultural interaction."[25] Global knowledge production is therefore not merely the transfer of knowledge from one context to another. In the letter, we see that Lady Dalhousie did not carry botanical knowledge into a "blank space."[26] Instead, her scientific knowledge was being produced in the encounters she had with both the natural world and local peoples as she moved about the globe.

New imperial historians have examined how correspondence and the networks that correspondence travelled along construct an imagined "space" of empire.[27] Letters like the one between Dalhousie and Perceval fashioned imperial science culture and ultimately the empire itself. Such letters played a key role in communication networks that produced imperial knowledge. Cecily Devereux and Kathleen Venema, in their contribution to Klaus Stierstorfer's multi-volume collection of women's letters across the British Empire, *Women Writing Home*, state that letters "represent the context in which British women perform and, more agentially [*sic*], develop racialized and gendered identities in relation to specifically Canada as a location for British imperial expansion."[28] Lady Dalhousie's identity as a white British woman was in constant development through her botanical practices and her correspondence around those practices.

Letters circulating within and across the British Empire were more than mere familiar links across the distance; the letters themselves became important agents that produce empire. Laura Ishiguro writes that personal letters "facilitated networks of information, ideas and affections that made empire possible and sustainable; in so doing, they did not just reveal a different side of empire, but they worked to constitute it."[29] For Ishiguro, letters *are* the empire, constantly being constructed through the "discursive work of making and defining 'colonisers' and 'colonised.'"[30] In this regard, Dalhousie's 1829 letter to Anne Mary Perceval does more than just illustrate the botanical practices of nineteenth-century British women; the letter itself works to act as women's participation in imperial botany. Such a reading of the letter reaffirms the importance of participation in the community for the developing scientific knowledge and culture in the empire.

For global histories of science, communication is also central to the interactions that occur within the circulatory space of imperial or global networks and, therefore, to the process of knowledge construction. Letters, collected plant specimens, and botanical texts become instruments of communication in a global network around imperial science. British women such as Lady Dalhousie become as much a part of the process of creating knowledge as those considered more formally as scientists. The inclusion of the correspondence and ephemera of women active in science, the material objects collected and circulated in the larger space of the British Empire, and the authorship and publication of popular science texts as locations for analyzing imperial science – all these expand the narrow lines of traditional histories of science. Secord points out that "thinking always about every text, image, action, and object as the trace of an act of communication, with receivers, producers, and modes and conventions of transmission" means that global histories of science eradicate the "distinction between the making and the communicating of knowledge."[31] There is a need, he writes, to analyze audiences and readerships to avoid reproducing the notion that science passes from "highly individualized sites of production to an undifferentiated mass public."[32] In challenging the simple flow of knowledge from expert to non-expert, space is opened up to explore the multiple ways that knowledge can be constructed outside formal channels. In this space, authors such as Fan suggest there is rich material for exploring the "trading zone" of multiple forms of knowledge, including those found in domestic spaces and the interactions between Indigenous and Western knowledges.[33] Fan argues for the investigation of reasons and circumstances that fostered or hindered the movement of knowledge or material objects, "to find out how and why certain mechanisms were introduced to control the coming and going of people and things."[34]

Considering the correspondence of scientific women like Dalhousie and Perceval alongside those of male correspondents provides the opportunity to explore matrixes of power within colonial and imperial networks. Historians of science have acknowledged the importance of correspondence between male scientific figures as being critically important to developing and sustaining scientific work across the globe.[35] By contrast, the relative lack of archival evidence left behind by female practitioners has made it difficult to fully analyze the role that women's knowledge, skills, and activities have played in botanical history and British imperial science. Dalhousie's own contributions to botany are only now being fully understood, in part due to practical aspects of archival loss and disorganization.[36] As historians of science re-examine correspondence between women collectors such as Dalhousie and "experts" such as Hooker and continue to find traces of correspondence between female botanizers, new ways of understanding will emerge to show how women functioned as significant parts of scientific networks and facilitated the spread of scientific connections through introductions, recommendations, specimens, and more.

Dalhousie's 1829 letter reflects a transition and connection between the colonial spaces of a global empire as she moved from British North America to India. Her interest in botanical work began to change with her move to India, and she began to further develop her scientific skills and expand her knowledge. The opening of her letter to Anne Mary Perceval, written while still at sea, notes the exact latitude and longitude of her location.[37] She observed astronomy on the sea journey and named constellations and stars with enough familiarity to "confess [to] dearest Mrs. P that the Great Cross of the South disappointed me." She expressed disappointment that while in Rio, "it was the season when few either shrubs or plants are in flower so <u>that</u> part of the beauty was <u>comparatively</u> lost to me tho' there were still multitudes with which I was unacquainted and had neither time nor opportunity to examine." Upon arrival in Calcutta, she became familiar with the local scientific community, writing that "[o]f Society I must not pretend to report," but "there are many clever men here and a good deal of scientific knowledge." While many British women travelling to colonial locations carefully recorded their journeys for family and friends, Lady Dalhousie's letter makes it clear that she was further developing her scientific interests and intended to continue botanical work.

3.1
Later photograph of the Great Banyan Tree at the AJC Bose Indian Botanic Garden (was Calcutta Botanical Garden, India) that Lady Dalhousie describes in her letter to Anne Mary Perceval.

Colonial botanical gardens were a significant attraction for Dalhousie, and she made it a priority to visit botanical gardens in each location. In Madeira she observed "English conservatories flourishing in all the wild luxuriance of Nature" and in Rio de Janeiro made "several excursions to the Botanic Gardens, where Cinnamon, Tea plant and bread fruit tree flourish in profusion; together with every plant of New Holland and Asia."[38] Her love of gardens took especial precedence when they arrived in Calcutta, taking on an almost religious significance for her: "We went on shore and first trod on Asiatic ground in the Botanic Garden!!! At once I plunged beneath a sacred Banian tree! A forest itself! And I saw the Teak and the Caout chouc and Oh! What did I not see?" The Calcutta Botanical Garden was an important site for her and she declared that her first real steps on "Asiatic ground" were there.[39] Although colonial gardens were often where botanists attempted to transplant "exotic" plants from the Americas and Europe, Dalhousie's primary interest was in the local plants, such as the famed "Banian" (Banyan) tree. For Dalhousie, the botanical garden represented the colonial environment itself and presented a significant opportunity for her to "know" the colonial landscape.

During the time that Lady Dalhousie was corresponding with Perceval, both women were part of William Hooker's network of female collectors of plant specimens in colonial locations. Their letters are filled with self-deprecation and confessions of insufficient botanical knowledge, but both women were ambitious in their goals to contribute to the advancement of British botanical knowledge. Historians of science and empire such as Richard Drayton have demonstrated that there was a growing relationship between botanical science and the "improvement culture" of empire in the early nineteenth century, influenced by Enlightened ideals about scientific knowledge that drove imperial expansion and liberal reform.[40] Sciences such as botany, minerology, and geography had important political and economic roles to play, but the leisure activities of British men and women were also important for maintaining that imaginary space of British Empire built on ideals of scientific knowledge and improvement and also of elite culture that valued polite manners and refined aesthetic sensibilities. Although most studies of imperial science focus on the male experts and explorers constituting the formal projects of empire, elite British women with an interest in natural history would have been aware of the connections made between the sciences and the goals of British imperial exploration and expansion, including the construction of a civilized empire. Dalhousie is an example of how British women could develop their scientific interest and knowledge in colonial locations, serving as part of diverse imperial projects.

Participation in scientific societies and associations was one way for elite women to contribute to imperial science, and Dalhousie's experience as a botanist out in the empire rendered her contributions particularly important to the Botanical Society of Edinburgh. Dalhousie presented her Indian Herbarium to Scottish botanist Robert Graham in 1837, and her gift formed a significant part of the collections that would later become the Royal Botanic Garden Edinburgh. Although there is no official number of specimens attributed to Lady Dalhousie in the Herbarium at the Royal Botanic Garden Edinburgh, some sources put the number at 1,200 specimens.[41] Upon receipt of the specimens, Graham and the Botanical Society of Edinburgh "unanimously and warmly expressed [thanks] to Lady Dalhousie for her splendid donation" and proposed that she be elected an honorary member.[42] While she told him that she does not "know either the names of the plant or the Class," Lady Dalhousie demonstrated her familiarity with botanical work through the initial organization of specimens according to the Linnaean system. She asked Graham to "cause them to be arrayed in any other form or according to such system as you consider most useful & conducive to the advancement

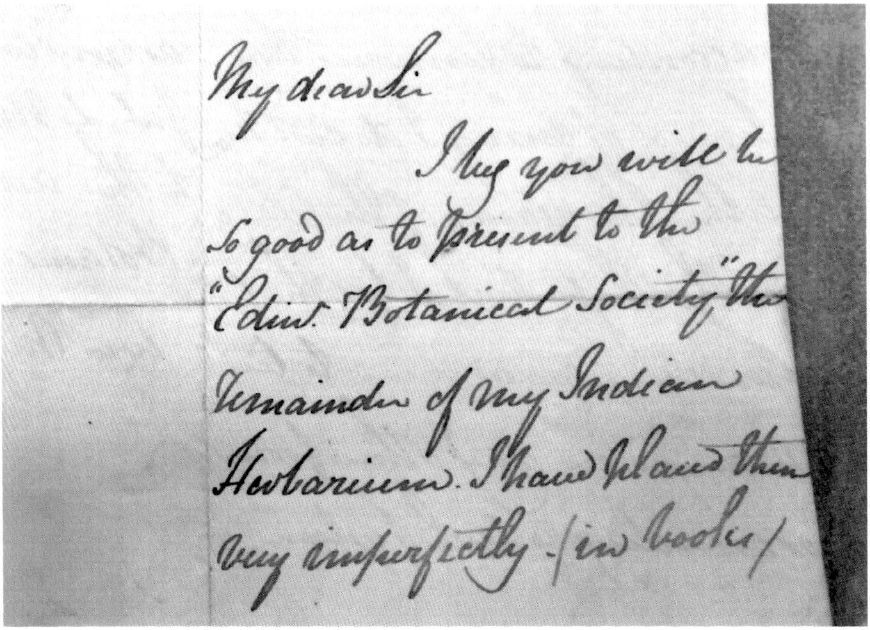

My dear Sir
I beg you will be
so good as to present to the
"Edinr. Botanical Society" the
remainder of my Indian
Herbarium. I have placed them
very imperfectly (in books)

3.2
Section of surviving letter from Dalhousie to Graham, presenting her Indian Herbarium to the Botanical Society of Edinburgh in 1837.

of Botanical Knowledge."[43] Unfortunately, the majority of Graham's correspondence was lost in a fire at the Botanical Society of Edinburgh and this remains the only extant letter from Dalhousie to Graham. A decade earlier, Dalhousie had been named an honorary member of the Literary and Historical Society of Quebec, an associational body founded by her husband in 1824, and had published her botanical findings in the society's transactions.[44] Like colonial collectors of specimens, learned societies such as the Literary and Historical Society of Quebec and the Botanical Society of Edinburgh fostered and developed networks for the circulation of materials and people in service of imperial botany.

Later letters written by Lady Dalhousie, sent to William Hooker and accompanying herbaria specimens, reflect an increased sense of the botanical knowledge and skill she acquired through colonial collecting in India. Although her letters are full of humble statements about her botanical qualifications, such as when she described the "extreme confusion" that "Dabblers in Botany such as I am" experience when encountering the "entirely new & unknown vegetation" of the colonial landscape, she clearly felt confident in her ability to observe and record the environment.[45] Lady Dalhousie's botan-

3.3
A specimen from Dalhousie's donated Indian Herbarium as now found in the Royal Botanic Garden Edinburgh: *Hypericum oblongifolium Choisy,* collected at Simla, 14 August 1831.

ical expertise had increased to the degree that Hooker invited her to contribute to his *Journal of Botany*. She rejected his proposal, stating that it "must require consideration and perhaps more courage than I possess."[46] It was important to Dalhousie to establish scientific authority in her observations and botanical contributions. She offered to show Hooker the Indian Herbarium she had donated to the Botanical Society of Edinburgh, and give him "any explanation in my power, concerning climate [and] situations … of the plants [the herbarium] contains."[47] She later wrote to Hooker in 1838, saying, "It is almost presumption in me to say that if I can answer any question you may wish to ask in regard to elevation, or any thing which has fallen under my own personal observation, it will give me new extreme pleasure to be applied to."[48] These letters demonstrate that Lady Dalhousie's experience in botanical collecting enabled her to consider herself qualified to provide accurate information on the natural history of the colonial landscape.

It is important to appreciate, however, that botanical knowledge about British India was more complex than older narratives of European exploration and discovery have often acknowledged. Lady Dalhousie brought certain practices and skills with her on her travels, but her knowledge was shaped by the experience of botany as practised in the colonial world. In India, botanical practitioners encountered a long tradition of Mughal natural history and botanical art. The expertise of Mughal artists, pre-existing natural history traditions, and Indian interactions with European styles all were part of the intellectual context in which colonial botanical collectors like Dalhousie worked.[49] Like many other botanical collectors, Lady Dalhousie outsourced scientific work and relied on networks to conduct botanical work. Letters to Hooker mention a plant having been "brought to her by a native" and having "drawings by a Native Artist."[50] Outsourcing included the collecting of botanical specimens itself, as well as the tradition of illustrative work being done by local artists, also known as Company artists, many of whom were former Mughal court painters. Dalhousie promised in a 1838 letter that of the 600 specimens she was sending to Kew, "the whole were dried & the greater part gathered by myself."[51] Plants came from a wide range of locations visited by Lady Dalhousie, including in Penang, Cape of Good Hope, Rio de Janeiro, and Madeira, and it is fully possible that some of the specimens she did not "gather" by herself were gifts from local colonial elites or Indigenous "native informants."[52]

Dalhousie's letters, and those of female collectors like her, demonstrate that British women were engaged in constructing and applying scientific knowledge gained through their encounters with the colonial world. Her 1829 letter to Perceval describes one encounter with a local gardener: "None of the European gardeners were to be found, but we [were] attended by a tall handsome Mullah in his white dress and Turban and tho' I could speak no Hindustanee and he no English yet thanks to the language of science we could comprehend each other in regard to names."[53]Although Dalhousie suggests a

3.4
Watercolour entitled *Potentilla nepalensis* "by a Native Artist" for Lady Dalhousie and sent to Hooker in 1833. The Lady Dalhousie (Christian Ramsay née Broun) collection, 1831.

kind of universality to scientific knowledge that enabled her to speak to their attendant in "the language of science," the encounter cannot be read without the context of colonial power dynamics. Although it may not be explicit in Lady Dalhousie's letters, colonial violence, including epistemic violence, played a significant part in the creation of imperial culture through the reconfiguration of local natural knowledge into British botanical science. There is a need to revisit definitions of violence to interrogate the variety of ways in which colonialism enacted its violent processes, including quotidian acts of violence and their role in establishing and maintaining colonial order. Colonial violence in botanical practices can be viewed as one of the mechanisms that worked to prevent or enable different forms of knowledge.

The erasure of the identities of local sources for both knowledge and material specimens in the letters of Dalhousie and other botanical practitioners underscores the nature of colonial epistemological violence in imperial science, even as these women saw themselves as benign figures in their activities.[54] In particular, the racial and class privileges of white collectors were clear in the appropriation of labour and the erasure of names from the records of scientific work that were key components of colonial science. For these British women, undertaking botanical work and participating in networks may not have been an explicit act of imperial aggression, but their activities as practitioners of imperial science and their engagement in and contribution to imperial science culture were no less imperial than those of male science practitioners who rewrote knowledge of the natural world to serve imperial aims of possession and control.[55]

Entangled histories in the global frame provide a way to further trace the complicated ways that power and knowledge worked in British imperial science. Lady Dalhousie's letter illustrates multiple levels of interaction, communication, and control in nineteenth-century imperial science. Anthropologist Nicholas Thomas has employed the concept of "entanglement" to trace the complex economic and political relations around colonialism, debt, and gift exchange in the Pacific. He argues that entanglement is a useful way to both acknowledge the asymmetrical power structures of the colonial project and allow for the "dynamics of local systems, their relative autonomy, and their capacity for resistance."[56] Because asymmetrical power structures are evident in Lady Dalhousie's 1829 letter to Perceval, theories of entanglement help examine the multiple levels of power between male botanists, their female collaborators, and sources of local botanical knowledge in the colonial context.

These power differences reflect the political concerns behind the practice of science in a global imperial world, even for elite British women like Lady

Dalhousie. Kris Manjapra utilized the term "entanglement" recently to explore power differences between European colonizers and Indians colonized in scientific collaborations.[57] Focusing on science and scholarly relationships as the site of entanglement, he examines how intellectual fields and innovations in the nineteenth and early twentieth centuries were reflected in or served the political sphere. While Dalhousie's letter contains no explicit political discussion, the practice of science in the imperial world was by no means free of political meaning. A close reading of her letter reveals how British women like Lady Dalhousie engaged with scientific knowledge within an imperial context, a practice that took place in a space where power dynamics of race and gender were always at play.

A critical examination of the correspondence between scientific women is a productive starting point for in-depth studies of imperial science and its intersections with race, gender, and class. As Adele Perry notes, "the challenges posed to biography by … elite colonial lives … hint at the more radical challenge posed by looking to a colonial archive for evidence that it sought not to record."[58] Thinking critically about the mobility of British women within empire brings a decentred imperial science into view, not merely transplanted onto colonial locations but constructed through encounters and interactions with the colonial world. It is here that we see an entanglement of the intimate and the global; women's scientific contributions and their position within the larger British imperial world become intertwined in less obvious ways than those of male botanists and naturalists. Nevertheless, they are part of imperial science culture on a global scale as they move about the empire and conduct scientific activities.[59] While this chapter has centred around an elite British woman and her connections in British North America and beyond, exciting possibilities remain for considering the many threads tangled in the creation and maintenance of an imperial science culture – not just those of the male experts but those more often obscured in the archives.

APPENDIX: FULL TRANSCRIPT OF THE 1829 LETTER FROM CHRISTIAN RAMSAY, LADY DALHOUSIE TO ANNE MARY PERCEVAL

C. of D. to Mrs Percival Dec. 15.1829. All well.

H.M.S. Pallas. Nov 2nd Lat. 33 13 S Long. 70, 43. E.

Tho' still far distant from Calcutta, I cannot employ more agreeably a forenoon of "quiet" my dearest Mrs Percival than in beginning a letter, to tell you that so far we have proceeded on our voyage with almost unexampled success and good fortune. I wrote last on 2nd July from Portsmouth where we were detained by contrary winds till 15th and again took refige [*sic*] in Plymouth (on 13th) from a heavy S.W. Gale. We finally left England on 20th and on 30th arrived at Madeira. Immediately went on shore to the house of Mr Gordon a wine merchant, and employed three days in visiting the beauties of the Island: the first sketch (as it were) of a southern climate. The mountains rise almost from the beach and the roads are totally inaccessible to any wheel carriage, being narrow, steep and stony in the most extreme degree. I travelled therefore in a sort of Palanquin, carried by two men, who in their wide Moorish trousers, and little blue cloth cap covering merely the top of the head, united with dark complexions and coal black hair, beard and eyes, presented an appearance singularly picturesque. Among the excursions I visited the Covall, 15 miles from Funchal, for this I had six bearers to relieve each other; the gentlemen rode on the little wild looking ponies of the country and each was followed by a portuguese [*sic*] who to assist himself in keeping up with the horses in galloping, seized the animals by the tail. The lower part of the mountains is covered with vineyards, not trained from tree to tree in the classic land of Italy but on low trellises and terraces, which gives no appearance of richness to the general view, but when near, has a beautiful effect mingled with Banana and aloe, and still more when ascending the steep and rocky ridges: these are interspersed with the native Chestnut forests. We passed thro' these, and then the region of heath and grass, until at nearly 4000 feet elevation we came to the edge of a precipice and looked down almost perpendicularly 1760 feet into the narrow valley of the Corrall, but this my dear Mrs Percival you will find excellently described in "Six months in the West Indies." Pray search for it there. Our delightful excursion occupied us 13 hours including resting and refreshing. The Gardens are beautiful in Madeira. English conservatories flourishing in all the wild luxuriance of Nature. We left Madeira 2nd August and on 23rd crossing the Line. In many respects this part of the voyage differed from my preconceptions. The heat was by no means so intense nor had we much of those calms so feelingly described by voyagers – flying fish were abundant, but not so the multitudes of the species I had expected. Two fine Sunsets we had, but in general he [*sic*] disappeared behind thin unmeaning clouds and I have seen as lovely as the best we had here from poor dear Sorell. We had all the usual gambols in crossing the Line, turning the discipline of a well ordered ship into seeming Bedlam for three hours! Proceeding further south the northern constellations gradually disappeared, and those of the other hemisphere replaced them. Shall I honestly confess dearest Mrs. P that the Great Cross of the South disappointed me? The stars are very brilliant which form it, but they are four only, and consequently it does not strike the eye as I had expected. Two very brilliant stars are in Centaur, and Canipus [*sic*] equals Sirius, who is now blazing in the East and with his attendant Orion reminds me of Home! On 4th Sept. we reach Rio Janeiro; this is pronounced by all travellers to be "the most beautiful scene in the world" and truly I believe it: the entrance of the bay is not a mile broad, with high mountains on each side; the Bay extends 30 miles up in the land, and is surrounded by mountains rising in endless succession of the very wildest and most fantastic forms, many are cones rising singly from their base or springing from the summits of other mountains all wooded to the top, in the luxuriance of tropical vegetation; innumerable islands of various sizes stud the bay – these too, covered with mimosa, melastoma etc etc while the graceful palm trees waves [*sic*] above. We were on shore here also at a merchants house and made several excursions to the Botanic Garden, where Cinnamon, Tea plant and bread fruit tree flourish in profusion; together with every plant of New Holland and Asia. The Bay of Bocafuoga is the most fairy spot, enclosed on all sides by mountains, while the tiny waves seem scarcely to break on the silver strand; the falls of Tijuca are situated amid magnificent mountain scenery, but the fall itself is not larger than that of Lorette. We also made a party to an Island about 15 miles above the town, sailing among thousand others, forming beautiful and ever varying groups, and returned by a tropical moon light; but I most admired a ride I took along the Aqueduct which supplies the town, and which ascending a mountain to a height of 2000 feet looks down from a ridge, on one side, on the town beyond which is the bay broken by the Islands into apparent Lakes bounded by the wild range of the organ mountains, and on the other side hangs over the deep narrow valleys of Larengevos and Orange valley. With Boto and its mountains, the mighty ocean stretching beyond all. Part of this excursion lay through the "virgin forest" where immense trees totally dissimilar to those I am accustomed to, were bound together by innumerable "Lianes" or climbing plants, and so matted with the profusion of underwood as to be totally impenetrable. Gigantic ferns 20 feet high were abundant!! Unfortunately it was the season when few either shrubs or plants are in flower so that part of the beauty was comparatively lost to me tho' there were still multitudes with which I was unacquainted and had neither time nor opportunity to examine. The Country is a paradise of beauty, but Man is in his most degraded state; every sort of labour is performed by slaves, even to that of drawing large and laden carts; but even when chained together these light hearted wretches laugh and sing, and the expression of their countenance is far less disagreeable than the dark scowling brows of the portuguese masters.

The Emperor held a Levee at which Ld D. and his Staff and Capt. Fitzclarence and his Officers attended. I went with some others to the Royal Chapel where before the Levee the Emperor with all his Court heard High Mass. The Chapel is handsome and the Choir is said to rival any in Europe, the Emperor being passionately fond of music. In the evening I went to the Opera, the house neat and the preformance [*sic*] respectable. The Imperial box occupies nearly the whole front and extends the height of two tiers. Here again was the Emperor attended by his Courtiers in the most splendid uniforms, really the Coats were masses of gold embroidery, while our uniforms only had the singularity of being scarlet, but from its very singularity attract most attention. Pedro is very dark with black hair and whiskers, not tall but well set up, soldierlike and decided in appearance. He is said to be by no means deficient in abilities but deplorably uneducated and so violent in temper that at times he is little better than a savage. The Levee was held on the anniversary of the declaration of Brazilian Independence and we attended a review of the troops on the Campo St Ana, when a salute of 101 guns was fired! You cannot imagine anything more wretched than the equipages – a strange looking kind of buggy drawn by two miserable mules or horses and for the roads! Suppose that to Lake St Charles. The town with its spires and convents looks well from the ship but the streets are so narrow that two vehicles can scarcely pass and filthy to a degree indescribable! I visited the Museum where are some splendid specimens of native Gold and Silver and a considerable number of beautiful birds, but they are all ill presented and worse arranged! Altogether Brazil is an extraordinary state – everything unsettled! The British merchants are few in number and quarrelling among themselves, nor have they any society whatever with the Portuguese. Society indeed seems nearly confined to the different Corps diplomatique and We have there an Admiral (unluckily a very silly man) and several Men of War. The French have the same and there is an American Commodore Creighton, brother to little Capt. C. of 70th, these are all on very friendly terms. On 16th Sept. we sailed again and arrived at the Cape of Good Hope in 21 days. Here we went to Government House and remained with Lady Frances Cole for 10 days. Unfortunately Sir Lowry was on the frontier but Lady F. is an old acquaintance of mine, a very sensible pleasant person so we passed the time most agreeably. We made an excursion of 3 days into the "Interior of Africa" but first let me say that Capetown is the perfect contrast to Rio, the streets broad and regular with trees planted each side. The houses well built, whitewashed, and everything clean as it is possible to be, lying at the base of the magnificent Table Mountain. We travelled in a waggon [*sic*] drawn by eight oxen "in hand" by a Malay, while another man walked behind with a whip and it is really astonishing to see how they guide a string of horses with the greatest ease. We crossed the "flats", an immense sandy plain, but this wilderness was covered with the most beautiful Ixias, Gladiolus, and all the tribe of bulbous plants, Geraniums, Heaths, Proteas in short all the most precious contents of our Greenhouses flourish in a Waste!!! I was really bewildered with the richness of botanical treasures. We crossed a pass in a very high and rugged range of mountains which Ramsay said was extremely like the Pass of St Gothard, and as fine in stile tho' not rising from so lofty a base. Stellenbosch and the Paare were two beautiful villages, regularly built with fine trees in the streets or surrounding the scattered houses, and one valley is most lovely consisting of farm houses, neat and picturesque in form situated each in its grove of Oak and Orange trees. Some of the latter are 55 feet high, and the branches bending beneath the golden fruit. The Vineyards are not ornamental for kept low they resemble fields of currant bushes; the Oaks are one the American species – introduced – for a singular feature of African scenery is the total want of wood. Bushes "of various flower" there are in profusion but literally not one native tree in the Cape district, tho' a few hundred miles in the interior of the Colony it is said that there are large forests of Mimosa and various other species. We sailed from the Cape of the (??) and now in 14 days we are 2400 miles from it! Without a Gale! tho' the most violent rolling sea! I am so good a sailor that I suffer from nothing. By the bye at the Cape I met Capt. Fitzroy and Lady Mary; he is dark and heavy as formerly; she is grown very fat with as little of "L'air noble" as any one I ever saw. They have 3 boys and 1 girl and seem to live very quietly and prudently. She spoke to me of Canada and enquired with much interest of several persons and among others of you and your family.

Calcutta. Dec. 15th

After the most prosperous voyage that was ever made we arrived [paper torn] Tides up the sacred mystical mythological Ganges! Its banks are low and covered with jungle apparently, to the waters edge, but even there it is teeming with population. To my extreme discomposure, instead of "dark blue waves" the river is exactly the colour of the little brook at Spencer Wood while the snow is melting. We came up the last sixty miles in a steam boat and the scenery (setting aside the palm trees) strongly reminded me of the St Lawrence about Nicolet. We were too early so anchored for two hours to avoid arriving before the troops were assembled, and now my dearest Mrs Percival we went on shore, and first trod on Asiatic ground in the Botanic Garden!!! At once I plunged beneath a sacred Banian tree! A forest of itself! And I saw the Teak and the Chaout chouc and Oh! what did I not see? None of the European gardeners were to be found, but we attended by a tall handsome Mullah in his white dress and Turban and tho' I could speak no Hindustanee and he no English yet thanks to the language of science we could comprehend each other in regard to names! For six miles below Calcutta Garden reach is a succession of beautiful villas, which resemble those on the Thames but surrounded by Palms and everything rich and rare! We landed at this City of Palaces amid a multitude who looked like animated statues, their dark sparkling eyes and rich bronze complexions contrasting with their snowy robes and Turbans. Each Caste has a different dress, the Dandies or boatmen and the Bheesties or Native carriers have no covering to their wild elf locks and no attire beyond a strip of cotton round their waist, yet from their colour, they do not give an idea of Nakedness! We are now at Government House, most kindly received by Lord and Lady William Bentinck: he appears a plain honestlike man. She a nice unaffected person so I hope we shall assimilate well. This house is splendidly magnificent. We dine in an immense marble hall (floor and pillars) about 25 persons with at least 55 attendants, the graceful attitudes and eastern costume, black eyes and beards so very picturesque makes them seem pictures, rather than men. The buildings in the English part of town are all in the Italian style; the native huts are mere mat hovels, but some of the rich Hindoos live in splendid style. I was out riding in a Tonga this morning at six o'clock, it is exactly Ramsay's little Cariole with a top to put up when the sun shines and is carried on the shoulders of four men with an equal number to relieve them; but Lady William and I were carried side by side and conversed easily. We go out before sunrise and may remain till 8 o'clock at this season and again about sunset, and remain in the house all day. At present it is not half so hot as a Canadian summer, but there is something particularly dangerous in the suns rays, as no one thinks of exposing himself to them; this is however the Cold season and I must not judge of Bengal by the thermometer this morning at 55. The Ladies seem to dress very well and are not so fat as I expected; but of the Society I must not pretend to report, there are many clever men here and a good deal of scientific knowledge, but of all these things you shall hear further ere long. I shall send this letter to the care of Sir Charles Flower, very uncertain where it will find you but I trust my dear Mrs Percival you have written and whole history of yourself. And now farewell. God bless you my dear Friend; give my kindest love to all your children and mention each individually. I very truly love you all! All our party are in perfect health and very busy house hunting, horse buying etc etc

Ever your most affectoy [*sic*] attached

C.B.D

Remember me to Mr Percival if with you.

NOTES

1 Lady Dalhousie to Anne Mary Perceval (spelled Percival), 15 December 1829. EL568/3, Colstoun Papers.

2 My thanks to the Broun-Lindsay family for permission to reproduce the 1829 letter from the Colstoun papers for this chapter and for confirming the loss of the original manuscript letter.

3 For scholarship on the role played by correspondence in the British Empire, see Errington, *Emigrant Worlds and Transatlantic Communities*; Gerber, *Authors of Their Lives*; Elliott et al., *Letters across Borders*. For women's imperial correspondence specifically, see Devereux and Venema, *Women Writing Home*, Vol. 3.

4 Booth and Burton, "Critical Feminist Biography;" Lepore, "Historians Who Love Too Much"; Zinsser, "Feminist Biography: A Contradiction in Terms?"

5 Spivak, "Can the Subaltern Speak?"

6 Perry, "Beyond Biography."

7 Shteir and Cayouette, "Collecting with 'botanical friends.'"

8 Like Lady Dalhousie, Anne Mary Perceval had been corresponding with William Hooker for several years about her botanical activities in North America. See Hooker, *Director's Correspondence*, DC/44/116–119.

9 Perceval to William Hooker, 7 October 1826, DC/44/118.

10 For more on social capital, see Field, *Social Capital*.

11 Multiple offers to carry gifts and letters are found in the letters of Dalhousie and other female correspondents in Hooker, *Director's Correspondence*. Further examples of this can be found in the correspondence of other botanical collectors in colonial locations, such as Anna Maria Walker (Ceylon) and Annabella Telfair (Mauritius). DC/53/133 and DC/53/122, respectively.

12 Perceval, 23 October 1825, DC/44/117.

13 Ibid.

14 Perceval, 7 October 1826, DC/44/118.

15 Endersby, *Imperial Nature*, 17.

16 Ibid., 15.

17 Jardine, Secord, and Spary, *Cultures of Natural History*; Shteir, *Cultivating Women, Cultivating Science*.

18 Zeller, *Inventing Canada*; Maroske, "'A Taste for Botanic Science.'"

19 Dalhousie to Perceval, 15 December 1829.

20 Dalhousie to Hooker, 4 February 1833, DC/53/38.

21 Vandenberg, "'Transient Beauties.'"

22 Secord, "Knowledge in Transit," 657.

23 Raj, "Beyond Postcolonialism," 341. See also Roberts, "Situating Science in Global History."

24 Fan, "The Global Turn," 252.

25 Raj, "Beyond Postcolonialism," 343.

26 For more on how older narratives of empire relied on the concept of "blank space" in the production of imperial knowledge, see Kennedy, *The Last Blank Spaces.*

27 For further historiographic discussion of spatial conceptions of networks in empire, see Lester, "Imperial Circuits and Networks," 124–41.

28 Devereux and Venema, *Women Writing Home*, x.

29 Ishiguro, "Relative Distances," 31. See also Ishiguro, *Nothing to Write Home About.*

30 Ishiguro, "Relative Distances," 8.

31 Secord, "Knowledge in Transit," 661.

32 Ibid., 662.

33 Fan, "The Global Turn," 253.

34 Ibid.

35 See for example, Browne, "Asa Gray and Charles Darwin," and Endersby, "'From having no Herbarium.'"

36 See chapter by Jacques Cayouette in this volume.

37 Dalhousie to Perceval, 15 December 1829.

38 The Botanical Garden of Rio de Janeiro was founded in 1808 by King John VI of Portugal and opened to the public in 1822. On the development of the garden, see Marchant, "Dom João's Botanical Garden."

39 For more discussion on the imperial role that the Calcutta Botanical Garden played in the nineteenth century, see Axelby, "Calcutta Botanic Garden"; Baber, "The Plants of Empire."

40 See for example, Drayton, *Nature's Government*; Brockway, *Science and Colonial Expansion*; Bennett and Hodge, *Science and Empire.*

41 Email correspondence with Lesley Scott, Assistant Herbarium Curator, 26 March 2019, Royal Botanic Garden Edinburgh; Horwood, *Women and Their Gardens*, 27.

42 Thursday, 9 March 1837, *Proceedings of the Botanical Society*, 2nd ed. (Edinburgh, 1841): 50.

43 Lady Dalhousie to Robert Graham, 12 April 1837. For help on the Graham correspondence, I thank HJ Noltie, Royal Botanic Garden Edinburgh.

44 Christian Ramsay, "Catalogue of Canadian Plants."

45 Dalhousie to William Hooker, 4 March 1833, DC/58/38.

46 Ibid.

47 Dalhousie to Hooker, 18 July 1833, DC/53/40.

48 Dalhousie to Hooker, 4 February 1838, DC/53/38.

49 On interactions between Mughal court artists and European botanists, see Noltie, *Robert Wight and the Botanical Drawings of Rungiah and Govindoo*; Dalrymple,

Princes & Painters; Archer and Falk, *India Revealed*; and Damodaran, Winterbottom, and Lester, *The East India Company*.

50 Dalhousie to Hooker 4 March 1833, DC/53/38, and 18 July 1833, DC/53/40. Identical copies of a plant illustration attributed to both Lady Dalhousie and Mrs Govan, the wife of colonial naturalist and botanical garden superintendent George Govan, suggest some kind of relationship in India where they had the opportunity to copy a common illustration done originally by a local artist – possibly a friendship like the social circle Dalhousie had been part of in colonial Quebec.

51 Dalhousie to Hooker, 4 February 1833, DC/53/38.

52 On "native informants," see Bayly, *Empire and Information*; Dirks, "Colonial Histories and Native Informants."

53 Dalhousie to Perceval, 15 December 1829.

54 For a discussion of colonial epistemological violence, see Dirks, *Castes of Mind*. For similar discussions of colonial appropriation and erasure in the North American context, see Parrish, *American Curiosity*.

55 Natural history's close association with military expeditions also reinforces the colonial violence at work in botanical collecting. One example was the botanical expedition by Anna Maria Walker into the recently conquered Kingdom of Kandy in Ceylon. (See Vandenberg, "The Walkers and Colonial Communities of Science in Ceylon" in "'Transient Beauties,'" ch. 5, and Schrikker, "The Colonial Project Completed: The Fall of the Kandyan Kingdom," in *Dutch and British Colonial Intervention in Sri Lanka*, ch. 11.) For more on the military connections with British natural history, see Greer, *Red Coats and Wild Birds*.

56 Thomas, *Entangled Objects*, 205–6. For further exploration of the concept of entanglement in the history of science, see discussion of the term in Rieppel, Lean, and Deringer, "Introduction."

57 British imperial scholars such as Cohn have argued that knowledge, especially in the scientific modes of "knowing" the world, has played a notable role in how the British ruled its empire, particularly colonial India. Scholars from the Subaltern School, such as Spivak and Ghosh, have argued that this relationship effectively suppressed other ways of knowing the world and have criticized Western hegemony in the study of historical processes.

58 Perry, "Beyond Biography," 328.

59 For more in-depth exploration of postcolonial critiques of Western science in British India, see Prakash, *Another Reason* and Raj, *Relocating Modern Science*.

BIBLIOGRAPHY

Archives

Dalhousie, Lady Christian Ramsay. 1829 letter to Anne Mary Perceval in Colstoun Papers (1272–1939). John Grey Centre Archives, Haddington, Scotland.

– Lady Christian Ramsay letters to William Jackson Hooker, *Director's Correspondence*. Archives at Royal Botanic Gardens, Kew. DC/44/56-60, 53/38–40.

– Letter to Robert Graham, 12 April 1837, Botanical Society of Edinburgh, W.H. Campbell Correspondence, Aglen Collection. Archives at Royal Botanic Garden Edinburgh.

Hooker, William Jackson, *Director's Correspondence*, Royal Botanic Gardens, Kew.

Perceval, Anne Mary. letters to William Jackson Hooker, *Director's Correspondence*. Archives at Royal Botanic Gardens, Kew. DC/44/116–119.

Books and Articles

Archer, Mildred, and Toby Falk. *India Revealed: The Art and Adventures of James and William Fraser, 1801–35*. London: Cassel, 1989.

Axelby, Richard. "Calcutta Botanic Garden and the Colonial Re-ordering of the Indian Environment." *Archives of Natural History* 35, no. 1 (2008): 150–63.

Baber, Zaheer. "The Plants of Empire: Botanic Gardens, Colonial Power and Botanical Knowledge." *Journal of Contemporary Asia* 46, no. 4 (2016): 659–79.

Bayly, C.A. *Empire and Information: Intelligence Gathering and Social Communication in India, 1780–1870*. Cambridge: Cambridge University Press, 1997.

Bennett, Brett, and Joseph Hodge, eds. *Science and Empire: Knowledge and Networks of Science across the British Empire, 1800–1970*. Basingstoke, UK: Palgrave MacMillan, 2011.

Booth, Marilyn, and Antoinette Burton. "Critical Feminist Biography." Special Issue, *Journal of Women's History* 21, no. 3 (2009).

Brockway, L.H. *Science and Colonial Expansion: The Role of the British Royal Botanic Gardens*. New York: Academic Press, 1979.

Browne, Janet. "Asa Gray and Charles Darwin: Corresponding Naturalists." *Harvard Papers in Botany* 15, no. 2 (2010): 209–20.

Cohn, Bernard S. *Colonialism and Its Forms of Knowledge: The British in India*. Princeton: Princeton University Press, 1996.

Dalrymple, William. *Princes & Painters: In Mughal Delhi 1707–1857*. London: Penguin Books Limited, 2013.

Damodaran, V., A. Winterbottom, and A. Lester, eds. *The East India Company and the Natural World*. Basingstoke, UK: Palgrave Macmillan, 2014.

Devereux, Cecily, and Kathleen Venema. *Women Writing Home, 1700–1920: Female Correspondence Across the British Empire, Volume 3: Canada*. Edited by Klaus Stierstorfer. Cambridge: Pickering & Chatto, 2006.

Dirks, Nicholas B. *Castes of Mind: Colonialism and the Making of Modern India*. Princeton: Princeton University Press, 2011.

– "Colonial Histories and Native Informants: Biography of an Archive." In *Orientalism and the Postcolonial Predicament: Perspectives on South Asia*, edited by Carol A. Breckenridge, Peter van der Veer, 279–313. Philadelphia: University of Pennsylvania Press, 1993.

Drayton, Richard. *Nature's Government: Science, Imperial Britain, and the 'Improvement' of the World*. New Haven: Yale University Press, 2000.

Elliot, Bruce S., David A. Gerber, and Suzanne M. Sinke. *Letters across Borders: The Epistolary Practices of International Migrants*. London: Palgrave MacMillan, 2006.

Endersby, Jim. "'From having no Herbarium.' Local Knowledge versus Metropolitan Expertise: Joseph Hooker's Australasian Correspondence with William Colenso and Ronald Gunn." *Pacific Science* 55, no. 4 (2001): 343–58.

– *Imperial Nature: Joseph Hooker and the Practices of Victorian Science*. Chicago: University of Chicago Press, 2008.

Errington, Elizabeth Jane. *Emigrant Worlds and Transatlantic Communities: Migration to Upper Canada in the First Half of the Nineteenth Century*. Montreal and Kingston: McGill-Queen's University Press, 2007.

Fan, Fa-ti. "The Global Turn in the History of Science." *East Asian Science, Technology, and Society* 6 (2012): 249–58.

Field, John. *Social Capital*. London: Routledge 2004.

Gerber, David A. *Authors of Their Lives: The Personal Correspondence of British Immigrants to North America in the Nineteenth Century*. New York: New York University Press, 2006.

Ghosh, Durba. "Decoding the Nameless: Gender, Subjectivity, and Historical Methodologies in Reading the Archives of Colonial India." In *A New Imperial History: Culture, Identity, Modernity, 1660–1840*, edited by Kathleen Wilson. Cambridge: Cambridge University Press, 2004.

Greer, Kirsten A. *Red Coats and Wild Birds: How Military Ornithologists and Migrant Birds Shaped Empire*. Chapel Hill: University of North Carolina Press, 2019.

Horwood, Catherine. *Women and Their Gardens: A History from the Elizabethan Era to Today*. Chicago: Chicago Review Press, 2012.

Ishiguro, Laura Mitsuyo. *Nothing to Write Home About: British Family Correspondence and the Settler Colonial Everyday in British Columbia*. Vancouver: UBC Press, 2019.

– "Relative Distances: Family and Empire between Britain, British Columbia and India, 1858–1901." PhD diss., University College of London, 2012.

Jardine, N., J.A. Secord, and E.C. Spary, eds. *Cultures of Natural History*. Cambridge: Cambridge University Press, 1996.

Kennedy, Dane. *The Last Blank Spaces: Exploring Africa and Australia*. Cambridge, MA: Harvard University Press, 2013.

Lepore, Jill. "Historians Who Love Too Much: Reflections on Microhistory and Biography." *Journal of American History* 88, no. 1 (2001): 129–44.

Lester, Alan. "Imperial Circuits and Networks: Geographies of the British Empire." *History Compass* 4, no. 1 (2006): 124–41.

Manjapra, Kris. *Age of Entanglement: German and Indian Intellectuals across Empire.* Cambridge, MA: Harvard University Press, 2014.

Marchant, Anyda. "Dom João's Botanical Garden." *The Hispanic American Historical Review* 41, no. 2 (1961): 259–74.

Maroske, Sara. "'A Taste for Botanic Science': Ferdinand Mueller's Female Collectors and the History of Australian Botany." *Muelleria* 32 (2014): 72–91.

Noltie, Henry J. *Robert Wight and the Botanical Drawings of Rungiah and Govindoo*, 3 vols. Edinburgh: Royal Botanic Gardens Edinburgh, 2008.

Parrish, Susan Scott. *American Curiosity: Cultures of Natural History in the Colonial British Atlantic World.* Chapel Hill: University of North Carolina Press, 2019.

Perry, Adele. "Beyond Biography, Beyond Canada." *The Canadian Historical Review* 98, no. 2 (2017): 321–37.

Prakash, Gyan. *Another Reason: Science and the Imagination of Modern India.* Princeton: Princeton University Press, 1999.

Ramsay, Christian. "Catalogue of Canadian Plants, Presented to the Literary and Historical Society by the Right Hon. The Countess of Dalhousie." *Transactions of the Literary and Historical Society of Quebec*, vol. 1 (Quebec, 1829): 255–60.

Raj, Kapil. "Beyond Postcolonialism … and Postpositivism: Circulation and the Global History of Science." *Isis* 104, No. 2 (2013): 337–47.

– *Relocating Modern Science: Circulation and the Construction of Knowledge in South Asia and Europe.* London: Palgrave MacMillan, 2007.

Rieppel, Lukas, Eugenia Lean, and William Deringer. "Introduction: The Entangled Histories of Science and Capitalism." *Osiris* 33 (2018): 1–24.

Roberts, Lissa. "Situating Science in Global History: Local Exchanges and Networks of Circulation." *Itinerario* 33 (2009): 9–30.

Schrikker, Alicia. *Dutch and British Colonial Intervention in Sri Lanka, 1780–1815: Expansion and Reform.* BRILL, 2007.

Secord, James A. "Knowledge in Transit." *Isis* 95, No. 4 (2004): 654–72.

Shteir, Ann B. *Cultivating Women, Cultivating Science: Flora's Daughters and Botany in England, 1760–1860.* Baltimore: The Johns Hopkins University Press, 1996.

Shteir, Ann B., and Jacques Cayouette. "Collecting with 'botanical friends': Four Women in Colonial Quebec and Newfoundland." *Scientia Canadensis* 41, no. 1 (2019): 1–30.

Spivak, G.C. "Can the Subaltern Speak?" In *Marxism and the Interpretation of Culture*, edited by C. Nelson and L. Grossberg. Urbana: University of Illinois Press, 1988.

Thomas, Adrian P. "The Establishment of Calcutta Botanic Garden: Plant Transfer, Science and the East India Company, 1786–1806." *Journal of the Royal Asiatic Society* 16, no. 2 (2006): 165–77.

Thomas, Nicholas. *Entangled Objects: Exchange, Material Culture, and Colonialism in the Pacific.* Cambridge, MA: Harvard University Press, 1991.

Torrey, John, and Asa Gray. *A Flora of North America.* New York: Wiley and Putnam, 1838–[1843].

Vandenberg, Virginia M. "'Transient Beauties': Early Nineteenth Century British Women and the Construction of a Global Imperial Science Culture." PhD diss., Queen's University, 2019.

Zeller, Suzanne. *Inventing Canada: Early Victorian Science and the Idea of a Transcontinental Nation.* Montreal and Kingston: McGill-Queen's Press, 2009.

Zinsser, Judith P. "Feminist Biography: A Contradiction in Terms?" *Eighteenth Century* 50, no. 1 (Spring 2009): 43–50.

PART TWO

Collecting and Its Contexts

4

"I dare not say Botanical … Mine is a real love for flowers"

Mary Brenton in 1830s Newfoundland

ANN SHTEIR

In September 1830, Mary Brenton, living in British colonial Newfoundland, sent William Jackson Hooker, then regius professor of botany at the University of Glasgow, a collection of plants for his *Flora Boreali-Americana* (1829–40), an ambitious compilation of botanical information about some 5,000 plants found across "the Northern Parts of British America." Mary Brenton became Hooker's principal collector in Newfoundland and Labrador, and each year until 1838 shipped him plants, many as dried specimens but others as "living plants … very flourishing and easily cultivated in bog mould."[1] Hooker cites "Miss Brenton" in the *Flora Boreali-Americana* as the source for a wide variety of plants, including ferns, woody shrubs, grasses, asters, cranberries, orchids, and thistles. Three plants among these were new to botanists at that time, and one of them, a gentian currently known as "American spurred gentian," was named *Halenia brentoniana* in her honour. Brenton's actual specimen of this gentian is extant in the Herbarium at the Royal Botanic Gardens, Kew.[2]

Along with Christian Broun Ramsay (Lady Dalhousie), Anne Mary Perceval, and Harriet Sheppard, Mary Brenton is one of only four women named as contributors to Hooker's "Canadian" Flora.[3] Who was Mary Brenton, and how did she come to collect plants in the early nineteenth century from her far eastern part of British North America? What personal, familial, social, and cultural circumstances led to the contributions she made to knowledge of nature? Biographical details are sparse, yet letters, archival records, textual citations, and print resources, as well as plant specimens and material artifacts, make it possible to reconstruct her activities within British cultures of collecting and offer evidence of how she thought about her work with plants. As this chapter will show, Mary Brenton was an elite woman in a transatlantic

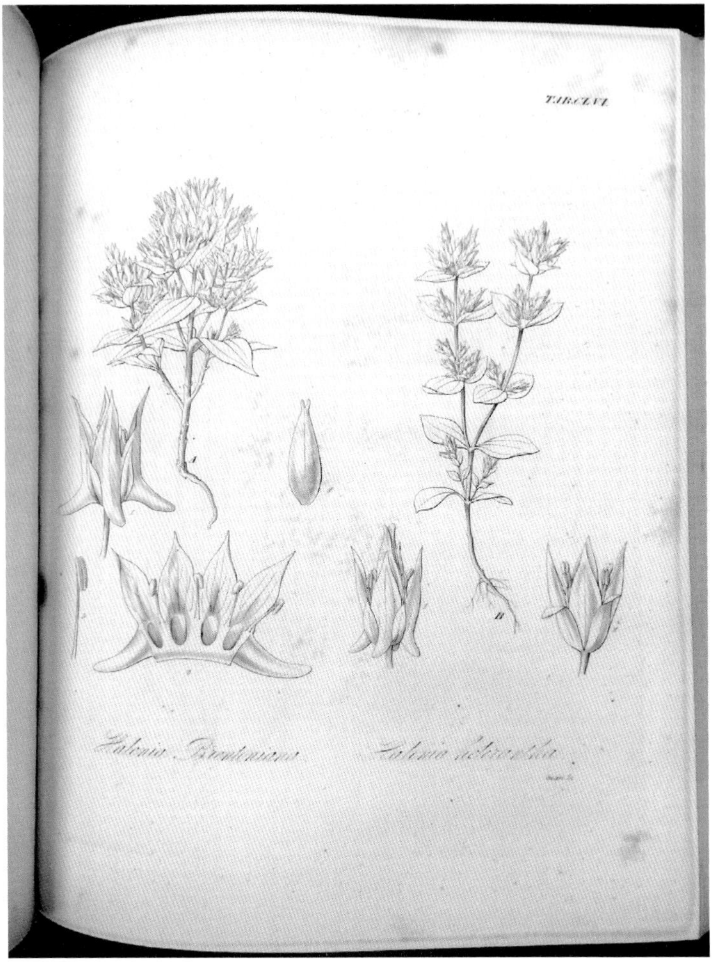

4.1
Halenia brentoniana, Flora Boreali-Americana Tab. CLVI.

world as well as a colonial daughter and imperial botanical recruit. Her experiences as a plant collector in 1830s Newfoundland, while individual to her, were also part of histories of involvement by women in British colonial and imperial science that continue to be mapped.[4]

Even as scholars in many fields and over several decades have challenged the presuppositions that obscure women's activities and achievements, it still is startling to see how much of women's presence in historical records depends on their inclusion in the letters, publications, papers, and profiles of notable men. Mary Brenton has textual visibility because her letters to William Hooker, dating from 1830 into the late 1840s, are archived in his institutional correspondence at the Royal Botanic Gardens, Kew, as a legacy of his appointment

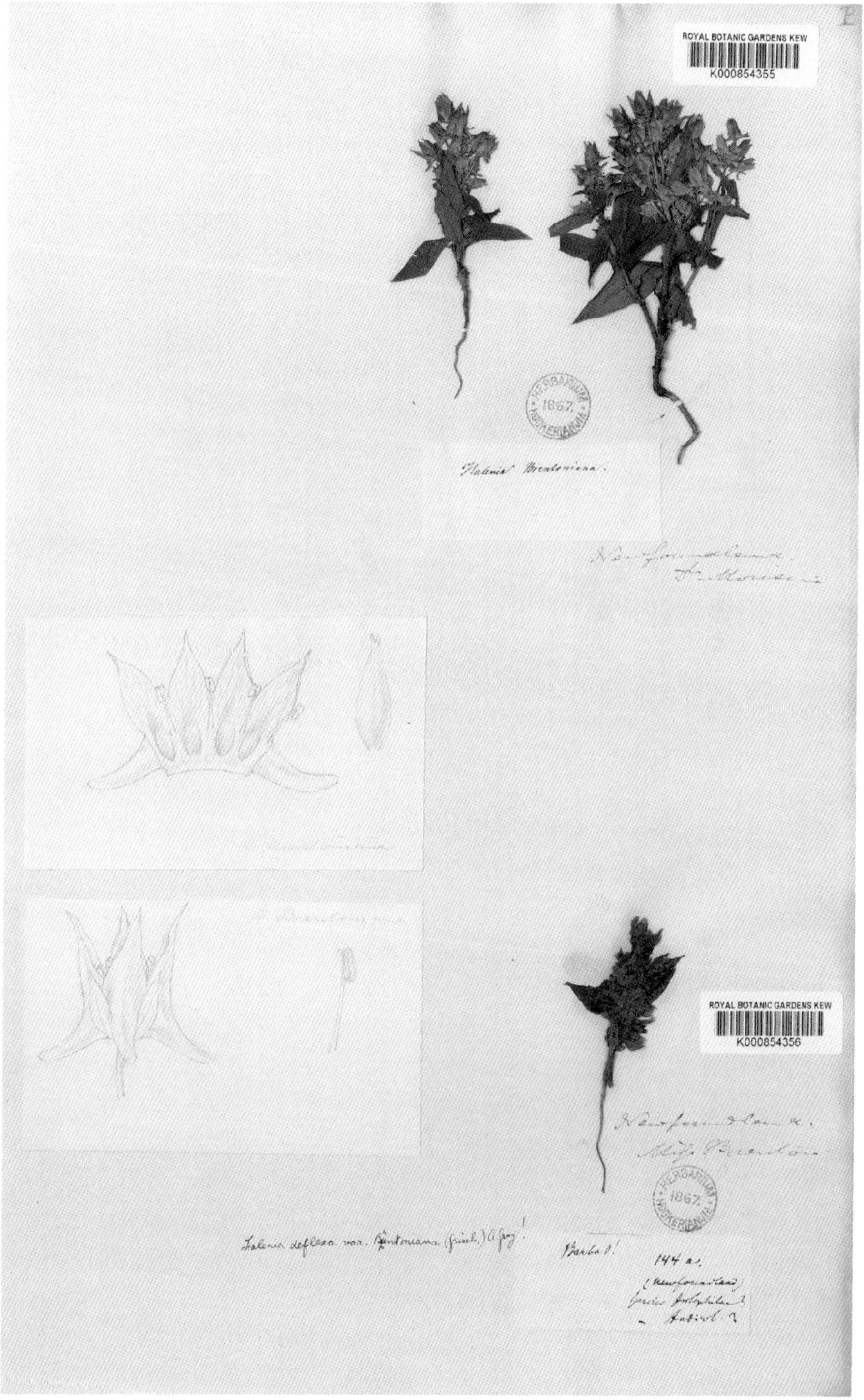

4.2
Mary Brenton's specimen of *Halenia brentoniana*, Herbarium, Royal Botanic Gardens, Kew, K000854356.

to the position of director in 1841.[5] This is a one-sided correspondence because Hooker's replies to Mary Brenton are not extant. Nonetheless, her letters abound in details for recovering her botanical activity and they also are rich in tone about her work in relation to Hooker the "botanist." Mary Brenton also comes into view within the archival *fonds* of Newfoundland's colonial Governor Sir Thomas Cochrane, where references place her within specific social and political worlds. Contemporary documents like these are central to the evidence base for exploring and characterizing Mary Brenton's botanical activities here. They are resources that can continue to bring women's cultural work into the historical foreground.

COLONIAL SEEDBED

Mary Elizabeth Brenton (1791–1884) was in all likelihood enculturated to British practices well before she began collecting plants for Hooker in 1830s Newfoundland. She was born into an old elite family from Newport, Rhode Island, that had settled in Nova Scotia before the American Revolutionary War.[6] Her father, Edward Brabazon Brenton, trained as a lawyer in London and served the Crown in colonial legal, official, and administrative positions during years when Halifax, as Britain's primary naval base in North America,[7] was a garrison town for British military and naval forces. Brenton was a pre-Loyalist allied closely with established power in Nova Scotia and its legal, religious, and social institutions. Mary Brenton would have grown into adulthood in that heartland of British America, with values, tastes, and practices that were geographically colonial but otherwise much like those of her counterparts in Georgian England. As a daughter of privilege, young Mary would have been brought into activities, ceremonial occasions, and public events among her social equals. In May 1809, as reported in *The Nova-Scotia Royal Gazette*, four hundred guests gathered at "an elegant Ball, and supper" given by the "Gentlemen of Halifax" to celebrate Britain's military conquest over the French colony of Martinique during the Napoleonic Wars and the return of Lt General Sir George Prevost from "triumph … over His Majesty's enemies."[8] Events like this with their "elaborate social rituals" of dress, manners, and dancing were one way that elite women and men in early British North America asserted their rank within colonial and imperial hierarchies.[9] Mary Brenton was eighteen years old at that time, and she and her mother, Catherine Taylor, may well have been among the guests, along with Lady Prevost and the "Ladies," who, in the words of that public notice of the event, "honoured us with their Company."

Halifax was a bastion of British values during those years, and a very active colonial print community consolidated this by reporting on current British events and foregrounding social, intellectual, and cultural topics that affirmed Britain's primacy in colonial life.[10] The monthly *Nova-Scotia Magazine and Comprehensive Review of Literature, Politics, and News* carried extracts from British periodicals as well as moral and educational essays that promoted an Enlightenment spirit of improvement.[11] Articles about agriculture and natural history were one regular feature. A long piece from 1792 "On the Study of Botany," for example, signed by "Philo-Botanices," discusses the benefits of botanical knowledge for health and safety and also aims to promote interest in the plants of Nova Scotia, where "many new plants are undoubtedly yet to be discovered."[12] "On the Study of Botany" reflects for the magazine's readers developments in the burgeoning science culture of later eighteenth-century England when audiences for knowledge of nature were expanding. In that spirit, a notice in *The Nova-Scotia Royal Gazette* for 23 August 1808 advertised instruction at a school for "young Ladies and Gentlemen" that included arithmetic, astronomy, geography, and natural philosophy." For the most part, however, as Gwendolyn Davies has shown, female education in colonial Halifax emphasized "the domestic and drawing room values that conservatives demanded as part of woman's sphere."[13] Unlike sons of the late eighteenth-century colonial elite in Nova Scotia who had access to a classical education at King's College preparatory to taking their place in professions, girls from the middle and upper levels of society received a curriculum that was differently gendered and less substantive. Mary Brenton would have been educated at home by her mother, or perhaps she was sent to one of numerous private schools in Halifax offering girls instruction in writing, French, and dancing as well as "a strict observation in their morals and manners."[14] Yet, even though she was raised within a family culture that gave priority to stability and established values, she was born into the 1790s when political turmoil and intellectual ferment challenged old assumptions about women's capacities and life-paths. At the start of that decade, *The Nova-Scotia Magazine* carried a piece by "a Lady" who critiqued "the tyranny of custom" and ideas about the "distinction of the sexes"; the author called instead for the "[e]qual advantage [that] may be hoped from the instructions of our sex in the present century, since we have women who excel in the sciences of commerce, government, poetry, and history; and in the various branches of the polite arts."[15] Might expansive ideas like those have echoed for Mary Brenton during her girlhood?

Early years in colonial Halifax were a plausible seedbed for Mary Brenton's attention to plants. Her parents had a substantial estate outside of town, with

Just Received

From LONDON, and for Sale at the Store of

JOHN WITHAM,

A LARGE Assortment of GARDEN, GRASS & FLOWER SEEDS, warranted fresh—Likewise, Hemp and Canary Seed, Split Peas, and a few bushels of the best Poland Oats for Seed, a few casks of excellent Burton Ale, English Cheese, Fish Sauces and Pickles, and a large assortment of Earthen ware, and many other articles. May 4.

4.3
Newspaper advertisement for seeds, *The Weekly Chronicle*, Halifax, Nova Scotia, 22 June 1810.

meadows and woods for walks and wandering[16] and perhaps a garden for growing flowers. Her interest might have been piqued by a newspaper advertisement in June 1810 for "A LARGE Assortment of GARDEN, GRASS, & FLOWER SEEDS," "Just Received from London," and "warranted fresh."[17] Planting flowers and creating gardens with seeds "Just Received from London" were one way that colonials, recent settlers, and imperial sojourners might connect to Britain, family, and "home." Mary Brenton might also have been aware of botanical activities among some elite women in Halifax at that time. Katharina Pyndar, wife of Nova Scotia's Lieutenant Governor Sir John Sherbrooke, wrote in her diaries from those years about long summertime walks in gardens, fields, and woods, finding "some beautiful moss" and gathering "some sweet Fern," and listed forty-eight "Plants found at Birch Cove growing wild," each identified by botanical name.[18]

Mary Brenton's location within the colonial elite linked her immediate private world to wider social and intellectual opportunities. As discussed in part 1 of this volume, Christian Ramsay, Lady Dalhousie, came to British North America because of her husband's imperial appointments. In like manner, Mary Brenton journeyed outward from Halifax because of her father's official government positions. In 1811 Edward Brenton went to the city of Quebec as civil secretary and aide-de-camp to Sir George Prevost, newly appointed as governor-in-chief of British North America and commander of British forces in North America. Details from this time are elusive, but one can surmise that Mary Brenton and her mother were with Edward Brenton for the next several years.[19] Documents after 1815 place the Brentons in England, with Edward Brenton resident in London at one point and Mary Brenton and her mother living or visiting in southeast Hampshire.[20] Their extended British family included naval officers and other imperial appointees, and through them the Brentons would have come into sociable activities in both city and country. Perhaps Mary Brenton had opportunity to visit landscapes such as a large park in Hampshire whose "shrubberies," "wood-walks," and gardens were described in 1817 as "planned with great judgment, and furnished with a pinery, hot-house, green-houses, and stoves."[21] A woman of her age, class, and station in early nineteenth-century England could hardly have missed exposure to books, magazines, flower shows, and horticultural societies that promoted attention to plants.[22] Interests like these would have come with Mary Brenton to Newfoundland when she and her parents arrived in St John's from London in September 1827.

IMPERIAL RECRUIT

In contrast to the scant direct resources about Mary Brenton's Halifax years, the letters sent to Hooker about plant collecting during the 1830s give plentiful material for placing her and her work geographically and socially. Edward Brenton was appointed a judge on the Supreme Court of Newfoundland in 1827, and for the next ten years Mary Brenton and her parents were part of the circle around Sir Thomas Cochrane, governor of the colony. As before in Halifax, they lived within the comforts of the elite, including an elegantly furnished house in town and a summer residence with "highly cultivated and fruitful Gardens." The privileged lives of Mary Brenton and her mother during the 1830s differed considerably, of course, from those of labouring women in St John's and women in fishing communities in the outports.[23]

The tone for British life in the colony of Newfoundland was set by Governor Cochrane, a colourful figure who assiduously crafted a British presence in St John's. Cochrane expended energy and funds on behalf of activities and causes that interested him and promoted structures for municipal and representative government, liberal agricultural policies, and pauper relief.[24] He instigated and oversaw construction of a new official residence and entertained lavishly in Government House throughout his tenure. A French visitor to

Newfoundland in 1828 took note of the vivid social scene in St John's: "the Governor displayed great pomp and ceremony, like a miniature court of the King of England. The society members, it is said, are hand-picked; the ladies are fresh and pretty and the most pleasant season one can spend there is winter, in spite of its harshness."[25] Thomas Cochrane's journals document public balls and concerts and also private entertainments – playing billiards, reading aloud to small groups, and going sailing with members of his close social circle.[26] Mary Brenton and her parents were among the guests at numerous formal dinner parties as well as multiple evenings-at-home in Government House, and "Judge Brenton and Mrs. Brenton" and "Miss Brenton" are listed often in Cochrane's itemized "Dinner Books."[27] Cochrane was also an avid theatre goer, and the Brentons were invited to join him in the governor's box for performances that would manifest Britain's cultural presence in the colony.

Mary Brenton also was part of excursions to Cochrane's "Cottage," his retreat from administrative duties, where he socialized and enjoyed his garden, often with the "Ladies" of his social circle.[28] Cochrane had a keen eye for plants, flowers, and trees. He notes in his journal in late April 1830 that "the Buds of the Phlox & Willows … have made a considerable push" and comments soon after on how much is growing – "the Lombardy poplar, Willow, Mountain Ash, Birch … Wild Pear, Larch."[29] It hardly seems coincidental that Mary Brenton's correspondence with Hooker dates from this period. She came into botanical collecting through Cochrane and other social and imperial connections who knew that Hooker was seeking contributions of plants from Newfoundland and Labrador. In line with social practices of the time, Cochrane would have vouched for Mary Brenton as a potential contributor, and he probably wrote the letter of introduction that accompanied her first shipment of plants to Hooker.

Thomas Cochrane mediated Mary Brenton's entry into plant collecting, but William Hooker was her botanical instructor. Author, illustrator, teacher, and editor, as well as diligent correspondent and genial enthusiast with social skills and connections,[30] Hooker entered into correspondence with Mary Brenton because he sought more specimens from Newfoundland and Labrador than had come from an earlier collector.[31] Through letters, compliments, and gifts, he encouraged Mary Brenton and guided her work. She wrote this to him in September 1831: "Miss Brenton, in sending her second Collection of dried Plants to Dr. Hooker, has first to thank him for his most flattering letters, both to Sir Thomas Cochrane and to herself and also for his very gratifying present of 'The Floral Illustrations of the Seasons' both of which she values more than she can express."[32] *Floral Illustrations of the Seasons* (1829) is an illustrated publication about gardens and floriculture that seeks to encourage "a taste for botanical pursuits" among women.[33] Written by Margaret Roscoe, daughter-in-law of the founder of Liverpool's Botanic Garden, the book presents horticultural and botanical information about fifty-five plants that could be grown in English gardens. Many of these plants had been introduced into Britain from seed received from North America, and Roscoe provides a coloured engraving of each, based in her own drawings. Hooker believed in the power of visuals in science, drew plants himself, and taught students how to look at plants.[34] By sending *Floral Illustrations of the Season* to Mary Brenton, he was providing text and visuals to cultivate her eye and be a resource for finding plants in British North America that would excite gardeners at "home."

Continuing his mentorship of Mary Brenton, Hooker also sent her his *British Flora* (1830), a book written with several different audiences in mind. He explains in the introduction that he wanted "to provide the young Student with a description of our native plants, arranged according to the simplest method; and … to afford to the more experienced Botanist, a manual, that should be useful in the field as well as in the closet." At the same time that he aimed to "advanc[e] the cause of Botanical Science," he also was writing for "the attention of the many, who are still apt to look upon Botany as a dry and profitless employment, a system of hard words, destitute of any real utility to mankind."[35] To bridge this gap, he presents botanical collecting and description as activities that can offer sociability and congeniality and bring together people of shared interests and enthusiasms. The diagnostic entries in *The British Flora* would have been too technical for Mary Brenton, but other material would have sharpened her attention and helped her look for plants for Hooker.[36]

Mary Brenton was able to seek out and collect native plants of Newfoundland for Hooker in part because she travelled with her father when Edward Brenton's duties as a circuit court judge took him to districts outside St John's. She records collecting on trips two hundred miles northwards, likely up past the Bonavista Peninsula, and the same distance south along parts of the Avalon Peninsula. She travelled to Cape St John on the far side of Notre Dame Bay and up to the Bay of Exploits in the northeast of the colony and also southwest from St John's to locations in St Mary's Bay and Placentia Bay. Official trips like these were probably on board a vessel outfitted for the governor of Newfoundland and his judiciary, but travel in the coastal waters was often inhospitable. Weather, the seas, and the seasons impeded her fieldwork. The summer of 1832 was particularly fierce, she wrote to Hooker, "such as has not

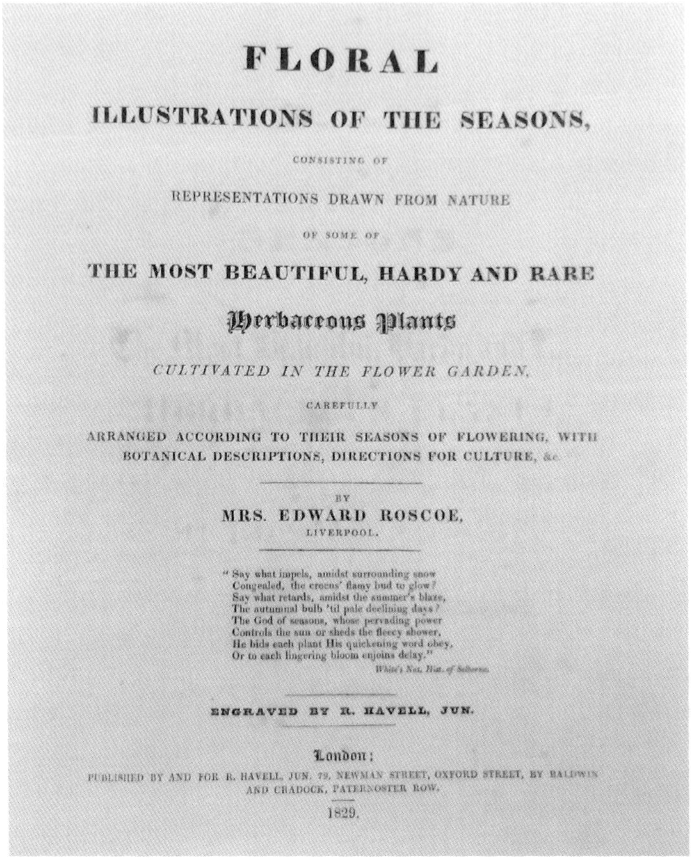

FLORAL
ILLUSTRATIONS OF THE SEASONS,
CONSISTING OF
REPRESENTATIONS DRAWN FROM NATURE
OF SOME OF
THE MOST BEAUTIFUL, HARDY AND RARE
Herbaceous Plants
CULTIVATED IN THE FLOWER GARDEN,
CAREFULLY
ARRANGED ACCORDING TO THEIR SEASONS OF FLOWERING, WITH
BOTANICAL DESCRIPTIONS, DIRECTIONS FOR CULTURE, &c.

BY
MRS. EDWARD ROSCOE,
LIVERPOOL.

"Say what impels, amidst surrounding snow
Congealed, the crocus' flamy bud to glow?
Say what retards, amidst the summer's blaze,
The autumnal bulb 'til pale declining days?
The God of seasons, whose pervading power
Controls the sun or sheds the fleecy shower,
He bids each plant His quickening word obey,
Or to each lingering bloom enjoins delay."
White's Nat. Hist. of Selborne.

ENGRAVED BY R. HAVELL, JUN.

London:
PUBLISHED BY AND FOR R. HAVELL, JUN. 79, NEWMAN STREET, OXFORD STREET, BY BALDWIN AND CRADOCK, PATERNOSTER ROW.
1829.

4.4
Mrs Edward Roscoe, *Floral Illustrations of the Seasons* (1829).

been known in this Island for half a century. Winter and ice lingered till the beginning of June, and since that period cold rains and fogs have hindered any flowers from coming to perfection and made it impossible for me to go any distance in search of them."[37] The next summer was similarly "so extremely cold and wet that few flowers came to perfection and the perpetual rain and fog destroyed the plants as soon as I had dried them. They were sadly knocked about in the small vessel we were cruising in during the autumn gales."[38] Impediments to her collecting work were material in other ways too. Although there is no specific evidence for how Mary Brenton developed or acquired the craft practices that collectors needed in order to preserve specimens for shipment to botanists, she might have had access to Hooker's "Directions for Collecting and Preserving Plants in Foreign Countries," which circulated in a magazine at that time. She understood what was needed to ensure adequate delivery of plants and apologized to Hooker in a letter dated 20 October 1832 for not having "proper paper" to dry the specimens she seeks to send.[39]

Other limitations to Mary Brenton's botanical work were physical and gendered. About 20 per cent of Mary Brenton's contributions to Hooker's *Flora Boreali-Americana* refer to plants collected in bogs or fens, and the challenges of this work were appreciable. She explains to Hooker in October 1832 that "the best flowering plants usually grow in swamps" but that "it is difficult for a lady to reach them." Moreover, "I can find but few persons who have enthusiasm sufficient to induce them to penetrate into the bog up to their knees in water in search of what they may not find after all."[40] The following year she reports further restrictions as well. Her walks were "generally so limited having but a short time to scramble about on shore as my father has leisure from his official duties to accompany me."[41] Mary Brenton was forty-two years old at that time, and the constraints upon her may have had to do with safety or other factors. But it is hard not to contrast the enforced caution and her chaperonage with the opportunities and risk-taking of others who searched for plants across British America on behalf of the *Flora Boreali-Americana*, especially Sir John Franklin and John Richardson who "as Commander and Naturalist of two separate expeditions to the Polar Seas," "collected under circumstances of singular difficulty, hardship, and danger," and to whom Hooker dedicated his book.[42]

Although the letters show that Mary Brenton felt hampered in what she could accomplish, the historical circumstance of being in Newfoundland in the 1830s took her into new terrains. Relative geographical isolation gave her opportunities like those of women in earlier colonial America who, also working within conventional mixed messages about female enquiry into nature, collected plants, shells, birds, and butterflies, and were part of transatlantic natural history correspondence networks.[43] Despite difficulties, Mary Brenton was a receptive pupil with an appetite for knowledge and evident zeal for botanical work. In addition to collecting plants, she became interested in geographical distribution and the effects of climate upon plants, and, noting regional differences in the geographic locales of plants when on a trip north in 1833, she reported to Hooker finding "the same plants … as grow in [St John's], only more luxuriant and the colours brighter [and] the trees much larger."[44] Several years later she wrote about a flower found at a site in St George's Bay on the southwest coast of Newfoundland whose blossoming precedes the arrival of the salmon. Comparing the "Salmon flower" to a plant familiar to her from England, she described it as "in appearance very like the Auriculas – the same stiff leaf, the same farina on the petals of the flower growing on a stalk as high as the English cowslip."[45] She had curiosity and knew how to observe flowers and pursue research about them.

LANGUAGES OF NATURE

Mary Brenton's letters to Hooker put her experiences in cultures of nature back into history. Brenton and Hooker were British subjects of the same generation, based in different locations but living the possibilities of empire. She enjoyed colonial social status, but he had expertise and intellectual standing. She seems to have had no prior technical experience as a collector, but he needed contributions that, with the right cultivation, she might provide. Letters from the early 1830s show Mary Brenton positioning herself in relation to Hooker the "Botanist." She apologized in September 1830 and again in September 1831 that, "from not being a Botanist, [she] has been unable to give all the Botanical names and but few of the Vernacular" and expressed regret "that her want of Botanical Knowledge has rendered her collections so little worthy of Dr. Hooker's attention."[46] Her vocabulary of apology can be read as rhetorically conventional in women's letter writing of her time, but it was not misplaced – at least not at the start of her work for Hooker. Like many other collectors for botanical projects of her time, women and men alike, she was new to this kind of task. But Mary Brenton also seems to have had little interest in formal features of plant description, nomenclature, or classification; this contrasts with Lady Dalhousie, discussed earlier in this volume, whose plant lists show clear knowledge of Latin botanical terminology.

Later, when her work as a plant collector in Newfoundland was winding down, Mary Brenton thanked Hooker in a letter from 1836 for encouraging what she termed her "floral researches." She then continued: "I dare not say Botanical for I have not [the] slightest knowledge of that interesting science. Mine is a real love for flowers."[47] In this statement Mary Brenton may be voicing yet another gendered expression of modesty, or she may be acknowledging her own limited botanical knowledge in comparison to Hooker's expertise. However, her demarcations between "botany" and "flowers" and between "knowledge of that interesting science" and "real love" suggest another, and further, reading in which Mary Brenton gives her work with flowers a value in her own mind that while different from Hooker's is equivalent to his activities with plants. This is a striking and resonant observation about cultures of nature at that time.

Hooker's gift of Margaret Roscoe's *Floral Illustrations of the Seasons* to Mary Brenton was an apt choice for his female botanical recruit in 1830. The book is an amalgam of the horticultural, botanical, literary, religious, and artistic languages of nature that were flourishing in publications for women at that time, and it combines botanical information with poetic epigraphs and pious phrasings in ways that touch many gendered bases. In that same decade, botanists were developing the disciplinary turn toward the "scientific" that would come to have priority as a way to know nature. Mary Brenton's statement in her letter to Hooker in 1836 indicates an awareness of this turn and registers differences in types of plant-related activities and the orientation of those who take part in them. Whereas Hooker's "Botanical" work is a series of intellectual practices, her own interest in plants is about emotive connection – "real love" as the impetus for her "floral researches." Botany may be "that interesting science," but "love for flowers" represents for her a different way of knowing and one that she particularly values.

In a study of women in networks of botanical correspondence in nineteenth-century America, Tina Gianquitto has reflected on motivations that can become apparent in and through letters. In particular, she identifies the importance that an exchange of scientific letters offered as companionship for women living in geographical and intellectual isolation in nineteenth-century America who had few others with whom to share scientific interests and discoveries. "The epistolary pursuit of such interests," she writes, enabled these women to expand their parameters beyond the home, to bond with others over a shared love of the natural world, and to create networks that encouraged and sustained all parties through long years."[48]

Mary Brenton's letters reflect her sense of welcome into a social and intellectual world. Her activities on behalf of the *Flora Boreali-Americana* brought friendship and the acknowledgment of shared interests as well as opportunities for intellectual enquiry and adventure outside her immediate sphere. Hooker's invitation to her, and his attention to her, drew her into activities of standing that in turn led to connection with others across borders. At one point, Hooker conveyed funds to Mary Brenton for distribution, she wrote, "to such persons as have assisted me in my floral and other collection,"[49] but her own involvement had nothing to do with financial considerations.

Botany connected Mary Brenton to Britain and collecting plants and sending them "home" were ways for a colonial-born daughter of empire to express this. Her activities mirrored her father's service to empire as a public servant as well as Thomas Cochrane's work on behalf of British culture and power in a remote Atlantic location. Mary Brenton was smart to welcome involvement in plant-related interests that connected her in these ways. She was a learner with a good eye and her decade of collecting work on behalf of William Hooker's Flora of plants found in "the Northern Parts of British America" also earned her a place in the history of botanical practices by women in nineteenth-century Canada.

Mary Brenton had arrived in St John's, Newfoundland, as a woman in her mid-thirties, living with aging parents. In a society where women conventionally married, was "Miss Brenton" a spinster by choice, a lack of opportunity, or family circumstances? The letters indicate that her health was not always sturdy and in 1836 was, she wrote, "so indifferent … that I have been unable to leave home as usual and to walk any distance."[50] She lived on, however, for another nearly five decades after the Brenton family left Newfoundland for England in 1838. St John's was her father's last imperial posting, and both parents died within a few years after that.[51] Might Mary Brenton have blossomed in the years that followed, like British botanical artist and traveller Marianne North (1830–1890), daughter and devoted companion to a widowed father, who set out after his death to fulfill her dream "of going to some tropical country to paint its peculiar vegetation on the spot in natural abundant luxuriance"?[52]

After Mary Brenton's contributions to the *Flora Boreali-Americana* came to an end in the late 1830s, a social relationship of favours and compliments continued when Hooker became director of the Royal Botanic Gardens, Kew. One letter to Hooker requests permission for entry to Kew at times other than the one day weekly designated for public visits.[53] Another seeks advice about "a pretty little fern" that Brenton understands has been successfully cultivated there, reflecting an interest shared with others in England at that time who were drawn into the "fern mania" of the 1840s.[54] In a letter written near the end of the decade, she reports having been "quite exhausted with my last visit to the Botanical Garden" and continues, "I went through all the Houses and saw the flower you mentioned which is beautiful and curious. The Palm House is worth going miles to see."[55] Clearly, the relationship to Hooker, addressed in that letter as "My Dear Sir William," continued to matter to Mary Brenton, though it likely was more significant from her side than from his.

BIOGRAPHY AND HISTORICAL IMAGINATION

This chapter has taken a research-oriented biographical approach to Mary Brenton and her early nineteenth-century work with plants. "'I dare not say Botanical … Mine is a real love for flowers'" looks beyond "the usual records" of botanical history and seeks to reconstruct her work as a plant collector. It pieces together extant details about family and networks from available documents and traces the kinds of practices that brought her into plant work. In the absence of specific evidence, it extrapolates from contexts at times and offers conjecture.

Mary Brenton collected plants in colonial Newfoundland during the 1830s, but her activities open up histories of women and science beyond her time and place. Her visibility in historical records began with her inclusion in Hooker's imperial Flora and then was augmented by research into letters, archival materials, periodical writings, and records that situate her within colonial and transatlantic social, geographical, and cultural settings. Her work in Newfoundland is part of a history of plant study that dates back to seventeenth- and eighteenth-century floristic explorations,[56] and her contributions join activities that Suzanne Zeller has mapped in *Inventing Canada* as foundational to nineteenth-century nation building and science. Research into Mary Brenton also participates in the project of writing women into Canadian history. Historians Catherine Carstairs and Nancy Janovicek call for extending chronological lenses back farther than has been customary in feminist scholarship on women in Canada over the past generation: "[w]e need to better understand women's lives and gender relations in earlier periods in order to construct more complete portraits of recent times."[57] The intersecting circumstances of an elite colonial woman such as Mary Brenton can add considerable texture to accounts of women's lives and opportunities in British North America. They also can offer material for historians studying region, class, and empire in ways that are specific to Atlantic Canada.[58]

Research on Mary Brenton in 1830s Newfoundland has benefited not only from print, manuscript, and digital sources but also from help by archivists eager to bring forward evidence of early nineteenth-century interest in plants of the colony. To them we owe access to the *Log of the* HMS *Royal George* [1808], a logbook from a late eighteenth-century warship of that name. With nautical entries dating from 16 March to 14 September 1808, the *Log of the* HMS *Royal George* records data about the ship's course and its signals and depth soundings, as well as meteorological notes on winds and weather.[59] At that time HMS *Royal George* was the flagship of Vice-Admiral Sir John Duckworth, whose

long career in the British Navy had taken him into significant naval battles of the American and French Revolutions and the Napoleonic Wars.[60] In 1810 Duckworth was appointed governor of Newfoundland and commander-in-chief of the naval squadron, and the logbook may have come to him from the captain of his ship.

Sometime after, the *Log of the* HMS *Royal George* was repurposed as a plant press and album, in service not of war but of science, and more than sixty plants were placed between its pages. A few plants are identified by name – several mosses, a wild rose, and a lichen, and also a "Golden Rod." Goldenrod is a familiar native species still found widely throughout Newfoundland,[61] and the specimen found in the *Log* provides an evocative link to Mary Brenton and her activities as a botanical recruit and collector in early nineteenth-century British America. In the *Flora Boreali-Americana* Hooker cites "Miss Brenton" for two species of goldenrod sent from Newfoundland. One of them was a bog species then termed *Solidago serotina* that was later described under the name *Solidago terrae-novae* as new in botanical literature.[62] Mary Brenton's own specimen is extant in the Herbarium at Royal Botanic Gardens, Kew.[63] Botanist Jacques Cayouette has reconstructed the possible circumstances be-

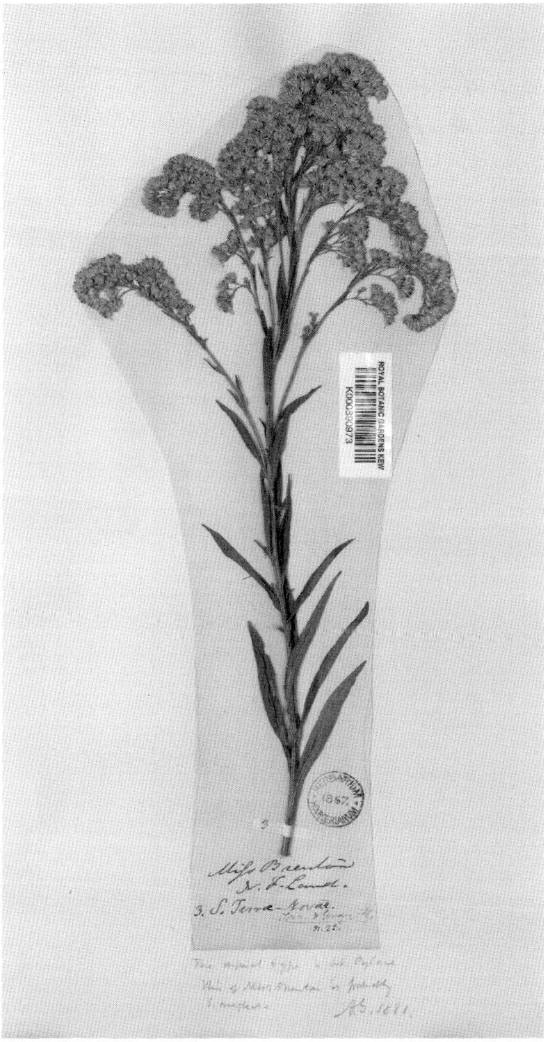

4.5 *Opposite*
"Goldenrod," *Log of the HMS Royal George* [1808]. This goldenrod is probably *Solidago rugosa* var. *rugosa* (Semple).

4.6 *Left*
Specimen of goldenrod collected by "Miss Brenton" in Newfoundland, now known as *Solidago juncea*, Herbarium, Royal Botanic Gardens, Kew, K000890973.

hind Mary Brenton's discovery of this goldenrod, which is now known as *Solidago juncea*. In a letter to Hooker from October 1832, Mary Brenton wrote of her desire to find something new for him despite the physical difficulties of collecting in bogs. She reported having "gathered one or two Plants that I hope you have not seen before," and Cayouette conjectures that the *Solidago serotina* may have been the reward for Mary Brenton's persistence.[64]

Until recently, the *Log of the HMS Royal George* lay deep in the storage vaults of The Rooms, Newfoundland's Provincial Archives. This artifact is believed to have come into their collection, via the Newfoundland Museum, from Government House in St John's, where Governor Thomas Cochrane hosted Mary Brenton and her parents many times at events both public and private during the decade of their imperial sojourn. The Brenton family even lived in Government House for a while after arriving from London in September 1827.[65] Might the *Log* have played a role in the history of Mary Brenton's botanical work for Hooker? Two of the specimens in the *Log of the HMS Royal George* carry dates of collection, "May 1, 1825" for one and "August 1828" for the other. It is possible in an act of historical imagination to picture Mary Brenton coming upon the repurposed *Log* on a shelf or a table in Government House, set out there for browsing or reference. Might the pressed and named specimens of unfamiliar Newfoundland plants in the album have aroused Mary Brenton's interest in the invitation from Hooker to collect plants for his *Flora Boreali-Americana*? It is similarly possible to picture her in conversation at one of Cochrane's evening parties, sharing enthusiasms about Newfoundland plants and flowers, and reporting on her own botanical quests. Might Mary Brenton even have used the *Log of the HMS Royal George* as a plant press herself and placed a specimen of "Golden Rod" between its pages?

ACKNOWLEDGMENTS

Toward locating Mary Brenton in Halifax, I thank Garry Shutlak at the Public Archives of Nova Scotia and Sheila Johnson Kindred, Janet Chute, Michèle Raymond, and G. Brenton Haliburton. In St John's, Melanie Tucker at The Rooms, The Newfoundland Provincial Archives, brought the *Log of the HMS Royal George* and its history to my attention. Jacques Cayouette has been my guide on botanical matters, and historian Bettina Bradbury oriented me at key moments. I thank Sara Maroske for reading drafts and record here my gratitude to Jane Couchman and Joan Gibson for joining with me in learning about Mary Brenton.

NOTES

1 Mary Brenton to William Jackson Hooker, *Director's Correspondence* DC 61: 52, 10 September 1831 with addendum 19 October 1831; DC 63: 47, 15 March [1838]. References to letters from Mary Brenton to William Jackson Hooker in Hooker's *Director's Correspondence* will appear henceforth as "Brenton, DC."

2 *Halenia brentoniana* Grisebach is recognized in recent floras as *Halenia deflexa* var. *brentoniana* or subsp. *brentoniana*. (pers. comm. J. Cayouette) "Spurred gentian" is a small wildflower found on peaty coastal headlands in Newfoundland (Boland, *Wildflowers and Ferns*, 370). The specimen Mary Brenton sent to Hooker is available in digitized format (URL apps.kew.org/herbcat/ ID K000854356). The other

new specimens Mary Brenton found were *Halenia heterantha* and a bog species of *Solidago*.

3 Shteir and Cayouette, "Collecting with 'botanical friends.'"

4 On activities in natural history and science by women in the colonial Atlantic world, see Schiebinger, *Plants and Empire*, ch. 1, and Parrish, *American Curiosity*, ch. 5.

5 Brenton, *Director's Correspondence*, DC 12: 95, DC 21: 120, DC 23: 76–9, DC 61: 51–6, DC 62: 47–49, DC 63: 47.

6 On Nova Scotia history during the years 1749–1820, see Conrad, *At the Ocean's Edge*, esp. ch. 7–9. On Halifax within the global history of British subjects during and after the American War of Independence, see Jasanoff, *Liberty's Exiles*, ch 5, and Fingard et al., *Halifax*, ch.2.

7 Conrad, *At the Ocean's Edge*, 190.

8 *The Nova-Scotia Royal Gazette*, 2 May 1809. In June of the following year, while "All the great People" dined at an exclusive men's club, Mary Brenton's mother was among the "Ladies" with Lady Prevost at a "very pleasant" afternoon party (Kindred, *Jane Austen's Transatlantic Sister*, 55).

9 McKenna, "Class, Race, and Gender Roles," 117–19.

10 See Eamon, *Imprinting Britain*, esp. part 1, "Print as Sociability."

11 Davies, "'Good Taste and Sound Sense.'"

12 *The Nova-Scotia Magazine*, vol. 5, No. 3 (March 1792): 141–2. Accessed 7 March 2018, http://eco.canadiana.ca.ezproxy.library.yorku.ca/view/oocihm.8_0. "Philo-Botanices" may have been William Cochran, editor of *The Nova-Scotia Magazine*. Cochran had a keen interest in the plants of Nova Scotia and over many years compiled a "List of Plants Indigenous to Nova Scotia" with the scientific and vernacular names and time of flowering of more than two hundred plants (Haliburton, "Of the Objects of Natural History").

13 Davies, "Private Education for Women," 10.

14 Advertisement by "Mrs. Goudge," *The Nova-Scotia Royal Gazette*, 15 August 1809.

15 "On the Supposed Superiority of the Masculine Understanding," *The Nova-Scotia Magazine* 3, no. 3 (September 1790), 196. Accessed 15 March 2018, http://eco.canadiana.ca.ezproxy.libraryyorku.ca/view/oocihm8_0.

16 On Brenton family properties, see Chute, "Halifax's New South End."

17 *The Weekly Chronicle*, Halifax, Nova Scotia, 22 June 1810.

18 Haliburton, *A Colonial Portrait*, 55–6, 73, 208–10.

19 Cahill, "BRENTON, EDWARD BRABAZON"; Burroughs, "PREVOST, Sir GEORGE."

20 "E.B. Brenton" appears on roll 308, London, England tax roll, 1820, for house rental at Dean Street, Hanover Square, St George, Westminster. Accessed via Ancestry.com 30 August 2017. I thank Bettina Bradbury for bringing this to my

attention. Letters from Edward Brenton to Sir Thomas Cochrane in early 1827 are dated from Havant, Hamps., where the Brentons may have been staying with family (Cochrane *Fonds*, ms. 2269).

21 Bingley, *Topographical Account of the Hundred of Bosmere*, 41–2.

22 Shteir, *Cultivating Women, Cultivating Science.*

23 Keough, *The Slender Thread.*

24 Thompson, "COCHRANE, Sir THOMAS JOHN"; Cadigan, *Newfoundland & Labrador*, ch. 5.

25 Ney, "The English Shore," 66.

26 For selections from Cochrane's diaries and journals, see Perkins, "Thomas Cochrane and Newfoundland in the 1820s."

27 Cochrane, *Fonds*, ms. 2353 (January 1829–May 1833), ms. 2358 (December 1829–March 1830), ms. 2359 (1831), ms. 2360 (1832-1833), ms. 2361 (1834).

28 Perkins, "'Fair Daughters of Terra Nova.'"

29 Cochrane, *Fonds*, ms. 2593, 28 April 1830 and 1 June 1830.

30 For an overview of Hooker's career from within nineteenth-century perspectives, see J.D. Hooker's "A Sketch of the Life and Labours." See also Drayton, *Nature's Government*, ch. 5.

31 Cayouette, *À la Découverte du Nord*, 39–40.

32 Brenton, DC 61: 52, 10 September 1831.

33 Roscoe, *Floral Illustrations of the Seasons*, preface. On Roscoe and early Victorian illustrated books for women about plants, see Zytaruk, *Nature on the Page*, 102–9.

34 On William Hooker and the role of visuality in nineteenth-century botanical study see Secord, "Botany on a Plate."

35 Hooker, *The British Flora*, vii–viii.

36 To help Mary Brenton collect mosses for him, Hooker sent material in 1833 from his writing about Cryptogams (Shteir and Cayouette, "Collecting with 'botanical friends,'" 21).

37 Brenton, DC 61: 54, 20 October 1832.

38 Brenton, DC 61: 55, 1 November 1833.

39 Brenton, DC 61: 54, 20 October 1832. On material practices of field collecting, see Endersby, *Imperial Nature*, 54–83.

40 Brenton, DC 61: 54, 20 October 1832.

41 Brenton, DC 61: 55, 1 Nov 1833.

42 Hooker, *Flora Boreali-Americana*, vol. 1, ii.

43 Parrish, *American Curiosity*, 189–200.

44 Brenton, DC 61: 55, 1 Nov 1833.

45 Brenton, DC 62: 49, 13 June 1836. On the "Salmon flower" (*Primula laurentiana*) see Shteir and Cayouette, "Collecting with 'botanical friends,'" 23.

46 Brenton, DC 61: 51, September 1830; DC 61: 52, 10 September 1831.

47 Brenton, DC 62: 49, 13 June 1836.

48 Gianquitto, "'My Dear Dr.,'" 438.

49 Brenton, DC 62: 48, 28 November 1834. Mary Brenton refers to funds for an unnamed woman in Labrador, "a Planter's wife," who "sent me the fruits of her summer" and "promises a further supply" in "the next season."

50 Brenton, DC 62: 49, 13 June 1836.

51 Edward Brenton's will, naming his "beloved daughter" Executrix, was proved in London 31 March 1845 "by the oath of Mary Elizabeth Brenton spinster" (PRO, National Archives: Prob 11/2013).

52 North, *Recollections of a Happy Life*, I: 39. Marianne North travelled to Brazil, Borneo, Ceylon, and India and was briefly in Canada during the 1870s.

53 Brenton, DC 23: 76, August 22 [no year]

54 Brenton, DC 23: 78, 18 October [no year]. On the popularity of collecting and cultivating ferns across mid-nineteenth-century Britain, see Whittingham, *Fern Fever*.

55 Brenton, DC 23: 79, 1 November [no year]. The Palm House in the Royal Botanic Gardens, Kew, was completed in 1848.

56 Pringle, "The History of the Floristic Exploration of Canada," 292–5.

57 Carstairs and Janovicek, *Feminist History in Canada*, 13. See also McKenna, "Class, Race, and Gender Roles."

58 See Glassford, "It's (Still) Complicated."

59 The *Log of the* HMS *Royal George* came to the Newfoundland Museum from an unrecorded source in 1971 and was later transferred to The Rooms, the Newfoundland Provincial Archives.

60 HMS *Royal George* was in the class of naval vessels known as "First Rates of 100 guns and above" that took part in battles in the Channel and Mediterranean (Winfield, *British Warships*, 1–5). On Duckworth see Whiteley, "DUCKWORTH, Sir JOHN THOMAS."

61 The goldenrod in the *Log of the* HMS *Royal George* is probably rough-stemmed goldenrod (*Solidago rugosa var. rugosa*) (Semple). I am grateful to John Semple for help with identification.

62 For Hooker's botanical description of the specimen of goldenrod collected by "Miss Brenton," see *Flora Boreali-Americana*, vol. 2, p.2. Now known as *Solidago juncea* (Bog Goldenrod), it "commonly occurs in peatlands, damp barrens, gravels, meadows, and woodlands throughout Newfoundland" (Boland, *Wildflowers and Ferns*, 274).

63 Mary Brenton's specimen in the Herbarium, Royal Botanic Gardens, Kew is ID K000890973. I thank Philip Bartling, Royal Botanic Gardens Victoria, for his assistance.

64 Brenton, DC 61: 54, 20 October 1832; pers. comm., 27 October 2016.

65 The arrival of the Brentons was reported in *The Newfoundlander*, 26 September 1827. Thomas Cochrane recorded their arrival in his personal journal and added that they would stay in Government House until their own house was ready (MS 2591, p.166)

BIBLIOGRAPHY

Archives

Brenton, Edward Brabazon. PRO, National Archives: Prob 11/2013.

Brenton, Mary. Letters to William Jackson Hooker, Royal Botanic Gardens, Kew: Archives: *Director's Correspondence*, DC 12: 95, DC 21: 120, DC 23: 76–7, DC 61: 51–6, DC 62: 47–9, DC 63: 47. Ten letters dating September 1830–20 October 1837 are from St Johns, Newfoundland, and six letters with partial dates but no year(s) are from London and Leamington Spa, England.

Cochrane, Thomas John. *Fonds*, 8 microfilm reels, National Archives of Canada (now Library and Archives Canada), 1956; acquired by Provincial Archives, St John's, Newfoundland 1957, MG 206.

Log of the HMS *Royal George* [1808]. The Rooms, Provincial Archives Division, Newfoundland Museum #975.35.

The Nova-Scotia Magazine and Comprehensive Review of Literature, Politics, and News, 5 vols., 1789-92. Public Archives of Nova Scotia. Nova Scotia Historical Newspapers.

The Nova-Scotia Royal Gazette, 1808–09. Public Archives of Nova Scotia. Nova Scotia Historical Newspapers.

The Public Ledger and Newfoundland General Advertiser, 1838.

Royal Gazette and Newfoundland Advertiser, 1828.

The Weekly Chronicle, Halifax, Nova Scotia. 22 June 1810. Public Archives of Nova Scotia. Nova Scotia Historical Newspapers.

Books and Articles

Bingley, Rev. W. *Topographical Account of the Hundred of Bosmere, in Hampshire*. Havant: Havant Press, 1817.

Boland, Todd. *Wildflowers and Ferns of Newfoundland*. Portugal Cove – St Philip's: Boulder Publications, 2017.

Burroughs, Peter. "Prevost, Sir George." In *Dictionary of Canadian Biography*, vol. 5, University of Toronto/ Université Laval, 2003-, accessed 16 May 2015, http://www.biographi.ca/en/bio/prevost_george_5E.html.

Cadigan, Sean T. *Newfoundland & Labrador: A History*. Toronto: University of Toronto Press, 2009.

Cahill, J.B. "Brenton, Edward Brabazon." In *Dictionary of Canadian Biography*, vol. 7, University of Toronto/Université Laval, 2003-, accessed 12 May 2015, http://www.biographi.ca/en/bio/brenton_edward_brabazon_7E.html.

Carstairs, Catherine, and Nancy Janovicek, eds. *Feminist History in Canada: New Essays on Women, Gender, Work, and Nation.* Vancouver: UBC Press, 2016.

Cayouette, Jacques. *À la Découverte du Nord: Deux siècles et demi d'exploration de la flore nordique du Québec et du Labrador.* Quebec: Editions MultiMondes, 2014.

Chute, Janet. "Halifax's New South End, the North West Arm Land Company and a Parkland Legacy." *Journal of the Royal Nova Scotia Historical Society* 3 (2000): 33–53.

Conrad, Margaret. *At the Ocean's Edge: A History of Nova Scotia to Confederation.* Toronto: University of Toronto Press, 2020.

Davies, Gwendolyn. "'Good Taste and Sound Sense': *The Nova-Scotia Magazine* (1789–92)." *Essays on Canadian Writing* 31 (Summer 1985), 5–22.

– "Private Education for Women in Early Nova Scotia: 1784–1894." *Atlantis* 20, no. 1 (1995): 9–19.

Drayton, Richard. *Nature's Government: Science, Imperial Britain, and the 'Improvement' of the World.* New Haven & London. Yale University Press, 2000.

Dunlop, Allan C. "Brenton, James." In *Dictionary of Canadian Biography*, vol. 5, University of Toronto/Université Laval, 2003–, accessed 12 May 2015, http://www.biographi.ca/en/bio/brenton_james_5E.html.

Eamon, Michael. *Imprinting Britain: Newspapers, Sociability, and the Shaping of British North America.* Montreal and Kingston: McGill-Queen's University Press, 2015.

Endersby, Jim. *Imperial Nature: Joseph Hooker and the Practices of Victorian Science.* Chicago: University of Chicago Press, 2008.

Fingard, Judith, Janet Guildford, and David Sutherland. *Halifax: The First 250 Years.* Halifax: Formac Publishing Co., 1999.

Gianquitto, Tina. "'My Dear Dr.': American Women and Nineteenth-Century Scientific Corresponence." In *The Edinburgh Companion to Nineteenth-Century American Letters and Letter-Writing*, edited by Celeste-Marie Bernier, Judie Newman, and Matthew Pethers, 435–49. Edinburgh: Edinburgh University Press, 2016.

Glassford, Sarah. "It's (Still) Complicated: Region and Gender in Recent Works of Atlantic Canadian Women's and Gender History." *Acadiensis* 44, no. 1 (2015): 145–54.

Haliburton, G. Brenton, ed. *A Colonial Portrait: The Halifax Diaries of Lady Sherbrooke 1811–1816.* Raleigh, NC: Lulu Press, 2011.

Haliburton, Thomas Chandler. "Of the Objects of Natural History in Nova Scotia." In *An Historical and Statistical Account of Nova Scotia*, 405–13. Halifax: John Howe, 1829.

Hooker, Joseph Dalton. "A Sketch of the Life and Labours of Sir William Jackson Hooker." *Annals of Botany* (1902), OS 16 (4), ix–ccxxi.

Hooker, Sir William Jackson. *The British Flora; comprising the Phaenogamous, or Flowering Plants, and the Ferns*. London: Longman, Rees, Orme, Brown, & Green, 1830.

– "Directions for Collecting and Preserving Plants in Foreign Countries." *Supplement to the Nautical Magazine*, 10 (1832): 575–8.

– *Flora Boreali-Americana, or, The Botany of the northern parts of British America.* London: H.G. Bohn, 1829–40. 2 vols.

Jaing Hong. "Angel in the House, Angel in the Scientific Empire: Women and Colonial Botany during the Eighteenth and Nineteenth Centuries." *Notes and Records of the Royal Society*. doi:10.1098/rsnr.2020.0046.

Janovicek, Nancy, and Carmen Nielson, eds. *Reading Canadian Women's and Gender History*. Toronto: University of Toronto Press, 2019.

Jasanoff, Maya. *Liberty's Exiles: American Loyalists in the Revolutionary World*. New York: Vintage Books, 2012.

Keough, Willeen G. *The Slender Thread: Irish Women on the Southern Avalon, 1750–1860*. New York: Columbia University Press, 2008.

Kindred, Sheila Johnson. *Jane Austen's Transatlantic Sister: The Life and Letters of Fanny Palmer Austen*. Montreal and Kingston: McGill-Queen's University Press, 2017.

McKenna, Katherine M.J. "Class, Race, and Gender Roles in Early British North America." In *Reading Canadian Women's and Gender History*, edited by Nancy Janovicek and Carmen Nielson, 110–29. Toronto: University of Toronto Press, 2019.

Ney, Eugène, "The English Shore" (1828). In *French Visitors to Newfoundland: An Anthology of 19th-century Travel Writings*, translated and edited by Scott Jamieson and Anne Thareau, 65–7. St John's, NL: ISER Books, 2013.

North, Marianne. *Recollections of a Happy Life, being the Autobiography of Marianne North*. Edited by Mrs John Addington Symonds. New York: Macmillan, 1894.

Parrish, Susan Scott. *American Curiosity: Cultures of Natural History in the Colonial British Atlantic World*. Chapel Hill: University of North Carolina Press, 2006.

Perkins, Pam. "'The Fair Daughters of Terra Nova': Women in the Settler Cultures of Early Nineteenth-Century Newfoundland." In *Transatlantic Women Travelers, 1688–1843*, edited by Misty Krueger, 81-94. Lewisburg, PA: Bucknell University Press, 2021.

– "Thomas Cochrane and Newfoundland in the 1820s." *Newfoundland and Labrador Studies* 29, no. 1 (Spring 2014): 117–68.

Pringle, James S. "The History of the Floristic Exploration of Canada." *The Canadian Field-Naturalist* 109 (1996) ["1995"]: 291–356.

Roscoe, Mrs Edward. *Floral Illustrations of the Seasons: consisting of representations drawn from nature of some of the most beautiful, hardy and rare herbaceous plants cultivated in the flower garden*. London: R. Havell, 1829–31.

Schiebinger, Londa. *Plants and Empire: Colonial Bioprospecting in the Atlantic World*. Cambridge, MA: Harvard University Press, 2004.

Secord, Anne. "Botany on a Plate: Pleasure and the Power of Pictures in Promoting Early Nineteenth-Century Scientific Knowledge." *Isis* 93 (2002): 28–57.

Shteir, Ann B. *Cultivating Women, Cultivating Science: Flora's Daughters and Botany in England, 1760 to 1860*. Baltimore: Johns Hopkins University Press, 1996.

Shteir, Ann B, and Jacques Cayouette. "Collecting with 'botanical friends': Four Women in Colonial Quebec and Newfoundland." *Scientia Canadensis* 41, no. 1 (2019): 1–30.

Thompson, Frederic F. "Cochrane, Sir Thomas John." In *Dictionary of Canadian Biography*, vol. 10. University of Toronto/Université Laval, 2003–, accessed 20 May 2015, http://wwwbiographi.ca/en/bio/cochrane_thomas_john_10E.html.

Whiteley, William H. "Duckworth, Sir John Thomas." In *Dictionary of Canadian Biography*, vol. 5. University of Toronto/Université Laval, 2003–, accessed 16 November 2018, http://www.biographi.ca/en/bio/duckworth_john_thomas_5E.html.

Whittingham, Sarah. *Fern Fever: The Story of Pteridomania*. London: Frances Lincoln, 2012.

Winfield, Rif. *British Warships in the Age of Sail 1793–1817: Design, Construction, Careers and Fates*. London: Chatham Publishing, 2005.

Zeller, Suzanne. *Inventing Canada: Early Victorian Science and the Idea of a Transcontinental Nation*, ed. 2. Montreal and Kingston: McGill-Queen's University Press, 2009.

Zytaruk, Maria. *Nature on the Page: The Print and Manuscript Culture of Victoria Natural History*. Toronto: The Thomas Fisher Rare Book Library, University of Toronto, 2019.

5

Baron Ferdinand von Mueller's Plant Collectors

At Home with the Australian Flora

SARA MAROSKE

INTRODUCTION

On 5 November 1883, Sarah Brooks (1850–1928), resident of a remote coastal community called Israelite Bay in the colony of Western Australia, sent a batch of dried flowers to German-born Baron Ferdinand von Mueller (1825–1896), government botanist of the colony of Victoria. The specimens were a response to a letter by Mueller published in the *West Australian* newspaper, appealing for settlers to assist him in a grand undertaking to produce a definitive scientific account of Australia's plants. "Perhaps I may not live many years to carry on my investigations," he declared, "and I should like so much to give the finishing stroke for the elaboration of the rich and varied flora of Western Australia before I pass away." The newspaper's editor added that this was a task especially likely to attract ladies in "far distant parts" who found themselves bereft of the intellectual resources to which they had once been accustomed, and that they would derive much interest "from actively aiding our great Australian botanist in his valuable scientific researches."[1] Brooks evidently agreed, and writing to Mueller said of her specimens that "she hoped [they] may prove useful."[2]

The botanical work referred to by Mueller was part of a series initiated in the middle of the century by Sir William Jackson Hooker (1785–1865) of the Royal Botanic Gardens, Kew, with the aim of producing "floras" of all British dominions and possessions.[3] Each flora consisted of descriptions of the plant species from a given region, in a standard format, usually with citations of the particular collections upon which the descriptions were based. The series was conceived of as an outsider-looking-in initiative, intended to facilitate the

5.1
Ferdinand von Mueller, photographed by J.W. Lindt, c. 1880.

exploitation of colonial vegetable resources by British visitors, help build a notion of the British Empire, and raise the status of the commissioning institution. The Australian flora, however, proved to be problematic. All aspects of its production were contested locally, and even before the final volume appeared, the centre of Australian botany had decisively moved from Kew to Australia or more specifically to the Melbourne Phytologic Museum (now National Herbarium of Victoria) – the institution of Ferdinand Mueller.

This story of contested scientific authority has been well covered by historians of Australian science,[4] although its larger cultural ramifications have not. Nevertheless, researchers who have turned their attention to the correspondence and specimens left by the collecting networks of scientists like Hooker and Mueller have begun to reveal the important role that collectors played in the generation of community knowledge.[5] Tina Gianquitto argues that such "transnational networks" created much more than scientific outcomes, including "a sense of civic engagement and shared responsibility for the explication of natural phenomena."[6] This process had implications for questions of national identity and agency, especially in regions such as the Australian colonies where First Nations peoples and indigenous plants and animals were distinctive and alien compared to those of the countries of origin of most of the European settlers.[7]

Records relating to Mueller's collecting network suggest that Sarah Brooks, and hundreds of his other plant collectors, did not see themselves as contributing to William Hooker's series of floras but rather to the fulfillment of a cherished dream of "their" botanist, who was a household name in the Australian colonies. In this sense, the Australian flora project was defined in opposition to imperial knowledge claims and also in denial of those of First Nations peoples whose occupation of the land preceded that of Mueller's collectors by tens of thousands of years.[8] In the context of their relationships to Mueller, these collectors embedded their new knowledge of native plants into the practices of everyday life, learning to identify species growing nearby, displaying, growing, and selling native plants and seeds, and depicting local vegetation in the decorative and literary arts. As Kate Hill observes, these processes were especially facilitated by women, for whom, collecting "often came out of and reinforced family relationships and domestic practices."[9]

The accumulated experiences of the collectors who formed the end points of hundreds of lines of communication radiating out around Mueller challenge the construction of grand narratives of practitioners such as William Hooker and his colleagues at Kew, who saw the floras of Great Britain's colonies as ornaments of empire. In no small degree, the collectors also contributed to Mueller's eventual success in claiming Australian botany as the proper undertaking of "settler" Australia. As Mueller's collectors turned their attention to the Australian flora, they gave more than specimens to their botanist, helping to transform Kew's imperial project into one that was both national and cultural.

THE COLONIAL FLORAS

The first in Kew's series of colonial floras, although only identified as such retrospectively, was William Hooker's partial flora of Canada, *Flora Boreali-Americana*, issued in two volumes in 1829 and 1840. Another sixteen colonial floras followed before the close of the century, with the flora of Australia, *Flora Australiensis*, written by George Bentham (1800–1884), "assisted by" Ferdinand

Mueller, appearing in the middle of the series.[10] The first group of Kew's floras was uncontroversial, with William Hooker, his son Joseph Dalton Hooker (1817–1911), and colleague George Bentham deciding on the format and authorship of the volumes, and with most of the cited specimens lodged at Kew.[11]

Writing a flora of Australia was a large-scale scientific project, covering a natural biogeographical region made up of six independent British colonies (New South Wales, Queensland, South Australia, Tasmania, Victoria, and Western Australia). In the British Empire, only India and southern Africa were as rich and diverse as Australia in plant species, with Australian vegetation types ranging from rain forest, to alpine meadows, grasslands, heathlands, woodlands and desert.[12] The Australian flora was also relatively poorly known, with the likelihood of thousands of new species waiting to be described. Moreover, to early scientific visitors it seemed to comprise a curiously "primitive" type of vegetation that promised to shed light on major philosophical preoccupations in Western botany about the origin and distribution of world plants.[13]

British scientific interest in Australian plants started with Sir Joseph Banks (1743–1820), who sent a series of paid collectors to the continent after his own visit in 1770.[14] William Hooker initially followed Banks's example in paying collectors in New South Wales but quickly moved to a system where he relied on personal contacts to build a modest network of volunteers and used his influence to secure friendly appointees to early Australian botanical positions.[15] In addition, William's son, Joseph, visited and collected in Tasmania and New South Wales, as part of James Clark Ross's *Erebus* and *Terror* expedition, 1839–43, going on to publish a flora of Tasmania.[16]

While aware of British claims on Australian botany, Mueller drew on a separate tradition of German scientific inquiry in his quest to study the vegetation. This tradition was in large part inspired by the great polymath Alexander von Humboldt (1769–1859), whose travels in Europe and the Americas were devoted to finding laws that united the natural sciences. As Rod Home observes, Australia was a significant beneficiary of the Humboldtian tradition,[17] and was visited by eminent German-speaking scientists such as botanist Baron von Hügel (1795–1870), explorer and naturalist Ludwig Leichhardt (1813–1848), and geophysicist Georg von Neumayer (1826–1909). These Continental Europeans did not automatically defer to scientists in Great Britain and felt fully equal to the task of authoring their own contributions to science.

Mueller joined the flow of German migrants to South Australia in 1847, already having familiarized himself with as many books and specimens relating to Australian botany as he could. After the discovery of gold in Victoria in

1851, he moved to its capital city, Melbourne, and contacted the scientifically minded lieutenant-governor of the colony, Charles Joseph La Trobe (1801–1875). Recognizing Mueller's talent, La Trobe appointed him the first government botanist in Victoria in 1853, and in 1857 Mueller was also appointed the first director of the Melbourne Botanic and Zoologic Garden.[18] Thereafter Mueller used these positions to build up a formidable collection of botanical resources in the form of a herbarium, library, living plant collection, and contacts with botanists and botanic gardens around the world.[19]

In his first letter to William Hooker as government botanist, Mueller declared that he wanted to write a flora of Australia. Hooker was cautiously encouraging in reply,[20] but in subsequent correspondence made it clear that Mueller was expected to work with Kew and to visit Kew's herbarium. Used to being independent, Mueller proceeded to publish hundreds of new Australian taxa in his own journal, *Fragmenta Phytographiae Australiae*, which proved fatal for his reputation at Kew. Mueller was shocked and disappointed in 1858, when George Bentham announced that he had agreed to write a flora of Australia, but from the Hookers' point of view it was the only sensible outcome. Desperate not to lose all contact with his cherished project, Mueller accepted the role of Bentham's "assistant" and agreed to ship his own Australian herbarium to Kew.[21]

MUELLER'S NETWORK OF COLLECTORS

By downgrading Mueller's role in the flora of Australia, Kew was ostensibly able to co-opt the specimens accumulated by Mueller and those of his rapidly expanding network of collectors. Such a view, however, cannot be sustained in the face of the surviving voices of individual collectors who developed full and creative lives for themselves beyond the sight or influence of imperial scientists.

For many years after Mueller's death, collectors existed mainly as surnames in *Flora Australiensis* or on sheets of herbarium specimens, with perhaps the addition of a collecting locality and a date. In recent decades, mass digitization projects have enabled many more details to be filled in. The three most important new resources are "The Correspondence of Ferdinand von Mueller Project" that has located, transcribed, and translated about 15,000 letters to or from Mueller,[22] the "Australasian Virtual Herbarium" that provides access to data associated with millions of specimens of plants, algae, and fungi in Australia and New Zealand,[23] and "Trove," a search engine that brings together content from Australian libraries, museums, archives, and other collecting organizations, most notably including a corpus of text searchable newspapers.[24]

It is now possible to establish that Mueller's collectors totalled around 1,800 individuals, residing across the Australian continent and surrounding islands and representing a broad range of occupations, ages, and commitments. Over the half-century in which Mueller established his network, the European and other settler population of Australia rose from about 440,000 to about 3,200,000 (while the First Nations population fell from at least 750,000 to less than 200,000).[25] A significant proportion of Mueller's collectors (about 12 per cent of the total) were women and girls, a much higher figure than for Kew's network of collectors. This reflected his deliberate policy of seeking all comers in constructing a network and his use of a variety of methods to recruit them.[26] Together Mueller's volunteers left few Australian localities completely uncollected, although the inland regions remained the province of a handful of intrepid missionaries and explorers, who often relied on the assistance of First Nations people to find rare plants.[27]

Mueller established a network of collectors to provide him with examples of as many species of plants that grew in Australia as possible, from which information on geographical range and species limits could be worked up for a flora and associated publications. Nevertheless, in the same way as other scientists with data-collecting networks, Mueller always envisaged his relationship with collectors as one of exchange, with his contribution being, at the very least, authoritative information on the scientific names of plants. Beyond that, the collectors were in charge of how they responded to this information.

While each collector enhanced his or her understanding of Australian plants through Mueller, from a cultural point of view this was especially important when the resulting knowledge became embedded in a domestic setting. As parents and children brought new appreciation of Australian plants into the family home, they folded it into the structures of their everyday life. Four examples of the way in which this occurred are considered here: "Naming and Possessing," "Plants as Culture," "Plants as Income," and "A Bush Classroom," although these categories are by no means exhaustive or mutually exclusive. Collecting specimens for Mueller helped individuals to create connections between how they saw their flora and how they saw themselves. In their mediated responses to native plants they began to shape a new notion of identity that was at once distinctive and national.[28]

NAMING AND POSSESSING

> Whoever wishes to become scientifically acquainted with the native plants of his vicinity or of localities otherwise accessible to him, can obtain the species names, if a duplicate set is retained, in which the specimens are numbered correspondingly to those of the transmitted set.
>
> – F. Mueller, circular, 1876[29]

In 1876, Mueller published a circular calling for volunteers across Australia to send him dried plant specimens for "the completion of an universal work on Australian indigenous plants."[30] George Bentham had, in fact, by now published all but one of his seven volumes of *Flora Australiensis*, but Mueller was still busy soliciting material for possible supplementary volumes. He made his pitch to citizens who wished to become "scientifically acquainted" with their local native plants, by which he meant learning their scientific names, or binomials, made up of a genus and species, to which Mueller added an abbreviation of the author's name.

In spite of the quaint wording of his offer, Mueller understood that there was significant interest among settlers to give names to the plants that grew around them, and not just a common or colloquial name but a standardized and authoritative one. Over the course of his career as government botanist, Mueller received thousands of letters from settlers in the various Australian colonies asking him to identify enclosed plant fragments. While only several hundred of these letters survive they readily give the impression of a community who saw Mueller as something akin to an encyclopedia that they could freely consult to fill in knowledge gaps in their local landscapes.

In September 1876, Mueller's circular elicited a response from Alexander Paul (c. 1816–1891), a miner from the Inglewood gold diggings in central Victoria. In a naive hand, Paul asked, if it were "not too late" for the flora, to send a specimen of a creeper that he had first found ten years ago "in the Scrub in this neighbourhood." Delighting in its ornamental properties, Paul transplanted the creeper into his garden, and it now covered "all the front fences." "I do not think it is common," he reflected, "for every body [*sic*] admires it, but no one can tell me its name, perhaps you would favour me with it as no doubt you have come across it in your travells [*sic*]."[31] Although Mueller's reply to Paul does not survive, Paul's specimen is held by the National Herbarium of Victoria. Mueller identified it as *Clematis microphylla* DC.,[32] commonly known as Native Clematis, or Old Man's Beard due to the feathery tails on its

76.04.02

+ New Guinea

For the completion of an universal work on Australian indigenous plants, it is desirable to obtain additional collections of plants in a pressed and dried state, particularly from districts far inland or recently settled. It is an important aim by these means to trace out the exact geographic limits of the many thousand species, which constitute the original Vegetation of Australia, in order also that all observations on their respective utilitarian value, whether for pastoral culture, medical or industrial purposes, may become recognized and applicable to the widest extent. Moreover it is necessary to study still further the degrees of variability, to which all kinds of plants are more or less subject, with a final view of circumscribing the exact characteristics of each species. It is to be impressed on those, who may feel interested in the promotion of such researches, not to exclude from local collections any plants, merely because they appear frequent or insignificant. The process of drying plants for permanent collections is simple and easy in the extreme, it needs hardly any explanation, beyond perhaps the remark, that the parcels of paper, containing any recently gathered plants, after a few hours pressure, should be divided into thin sets, and be spread out on a dry or warm place, to facilitate and to speed the exsiccation, and to lessen also the requirements of shifting plants from paper, which became moist, into dry paper. Small plants should be gathered with their roots, and all not merely in flower but in fruit also, as indeed from the latter generally the main characteristics are derived. Water weeds, rushes, sedges, mosses, lichens, fungi (and on the seacoast also algae), even if ever so small, should not be passed in collecting. Transits are best effected early after the preparation of the specimens, in small parcels closely packed, by successive mails. Whoever wishes to become scientifically acquainted with the native plants of his vicinity or of localities otherwise accessible to him, can obtain the specific names, if a duplicate set is retained, in which the specimens are numbered correspondingly to those of the transmitted set. An intimate knowledge of the indigenous vegetation, while it largely indicates climatic and geologic circumstances, tends also to afford an insight not only into the natural vegetative resources of any tract of a country, but also into much of the pastoral or cultural capabilities of the respective localities. Researches of these kinds become furthermore the sources of educational works, and unfold to well-trained and intelligent minds pure recreative and healthful pleasures, inexpensively everywhere within reach.

FERD. VON MUELLER,
M. & Ph.D., F.R.S.

Melbourne, April 1876.

The names of finders of new or rare plants are always recorded in Baron Von Mueller's volumes.

5.2
F. von Mueller, [Circular], 1876.

5.3
"Australian Flowers" by Samuel Begg, *Illustrated Australian News*, Melbourne, Victoria, Australia, 27 July 1881, 145.

fruits. A search in Trove reveals that this plant slowly gained in popularity over the last decades of the nineteenth century, with more than one hundred references to "Native Clematis" in Australian newspapers before 1900 and another thirty or so to its scientific name, *Clematis microphylla.*

While the "bush," or regions of Australia inhabited by First Nations people, were seen as the natural home of native flora, Paul's letter indicates that individual species, such as *Clematis microphylla*, could cross boundary fences into "settler" gardens. In 1881, the artist Samuel Begg depicted a young Australian woman gathering Native Clematis for an illustrated newspaper, with viewers tacitly invited to compare the natural beauty of each subject in the lithograph.

First Nations names for native plants, too, were sometimes taken up by collectors but in a way that redefined them as words in settler culture, stripping them of their particular tribal meanings. Sarah Brooks, the Mueller collector in Western Australia, called her cottage, "Waratah,"[33] after *Teleopea speciosissima*, a shrub with striking red flowers that was not native to her colony. The word "Waratah" originally came from the Eora of the Sydney area on the east coast of Australia.[34] John Brooks, Sarah's brother, applied the word "Mulga" to stands of *Acacia aneura*,[35] although it was originally a First Nations word from southeastern Australia for a shield made from the wood of this species. As settlers found *A. aneura* throughout arid Australia, they called it all "Mulga."[36]

Some collectors also adopted scientific names into their vernaculars. In 1888, a resident of alpine Victoria, Annie McCann (1838–1924), published a volume of poetry including verses about plants. Living in a remote and recently settled gold-mining town, Annie had "to snatch" time for writing "from the busy hours of life," and it speaks to the importance of her relationship with Mueller that she was artistically inspired by the specimens that she came to know with his guidance. In "To the Australian Fern," she singled out a number of native plants for special praise, including *Clematis* and *Banksia*, the latter being a genus of "curious blooms" with ties to the British discovery of Australia. In a footnote attached to the poem McCann quoted Mueller's account of how this plant was collected in 1770 by Joseph Banks at Botany Bay in New South Wales and named after him by Carl Linnaeus the Younger (1741–1783). In "*Anguillaria australis*," McCann delighted in the scientific name of a plant that was self-consciously linked to her new country through its specific epithet. Commonly known as the "Early Nancy," McCann declared that this small flowering herb was a "Sweet harbinger of Austral spring" and a "Meek snowdrop of these southern skies!"[37]

While Mueller's stated intention was to give collectors the scientific names of their specimens, his most valued outcome was when they found plants that he could not identify. This became a great inducement for collectors to keep botanizing. "I have not had much time lately looking after specimens," Robert Issell Perrott (1822–1895), a farmer from New South Wales told Mueller in 1871, "and be assured I never lose an opportunity of searching about for fresh and new things [–] it affords me a pleasure I cannot describe."[38]

Perrott did not discover any new species, but about 280 of Mueller's other collectors were fortunate enough to do so, and Mueller named a significant proportion of these new species after them.[39] Sarah Brooks was commemorated twice, first in *Scaevola brookeana* F.Muell., a heart-leaved fan-flower, and again in *Hakea brookeana* F.Muell., a representative of a genus with its highest diversity in southwestern Australia where she lived.[40] Botanical eponymy conferred a special kind of possession on Mueller and his collectors. It co-joined their names in formal botanical citations that were intended to be permanent and would attach to the Australian vegetation beyond the lives of the individuals themselves.

PLANTS AS CULTURE

For Mueller, the main purpose of establishing a network of collectors was to obtain specimens, although he also encouraged individuals to submit drawings or sketches of difficult groups to aid in identification. Notably, although not exclusively, it was female collectors who answered this call.[41] As in Great Britain, middle- and upper-class Australian women were expected to be accomplished in the decorative and literary arts (including drawing, music, fancywork, poetry, and prose), but cut off from the cultural institutions and iconic landscapes of "Home," many turned to the Australian bush for inspiration.[42]

The example of Mueller's own nieces reveals that much of this creativity took place around domestic hearths and also how important Mueller's involvement could be to the outcome. Mueller's sister, Clara Wehl (1833–1901), raised twelve children to adulthood at "Ehrenbreitstein," a farm near Millicent in South Australia (her husband, also a Mueller collector, died in 1876). The property abutted uncleared scrub, rich in plant species.[43] Mueller encouraged his nieces to collect and illustrate specimens of this vegetation and Clara supported the project, writing of her daughters in an unpublished poem:

> They must learn to paint something more than cattle,
> Landscapes, flowers, shrubs, even down to nettle.[44]

A number of Mueller's nieces took up painting wildflowers, with at least three writing identifications received from their uncle beneath images in their folios. Marie Magdalene Wehl (1862–1962) practised every day to acquire competency as an illustrator, and with her uncle's encouragement used her paintbrush to depict freshly collected fungi. This was of great importance to scientists in Europe as the appearance of fungi was often completely altered by the process of drying for placement on specimen sheets.

As Mueller did not work actively on fungi himself, he sent Marie's paintings and specimens to British mycologist Mordecai Cooke (1825–1914). In the mid-

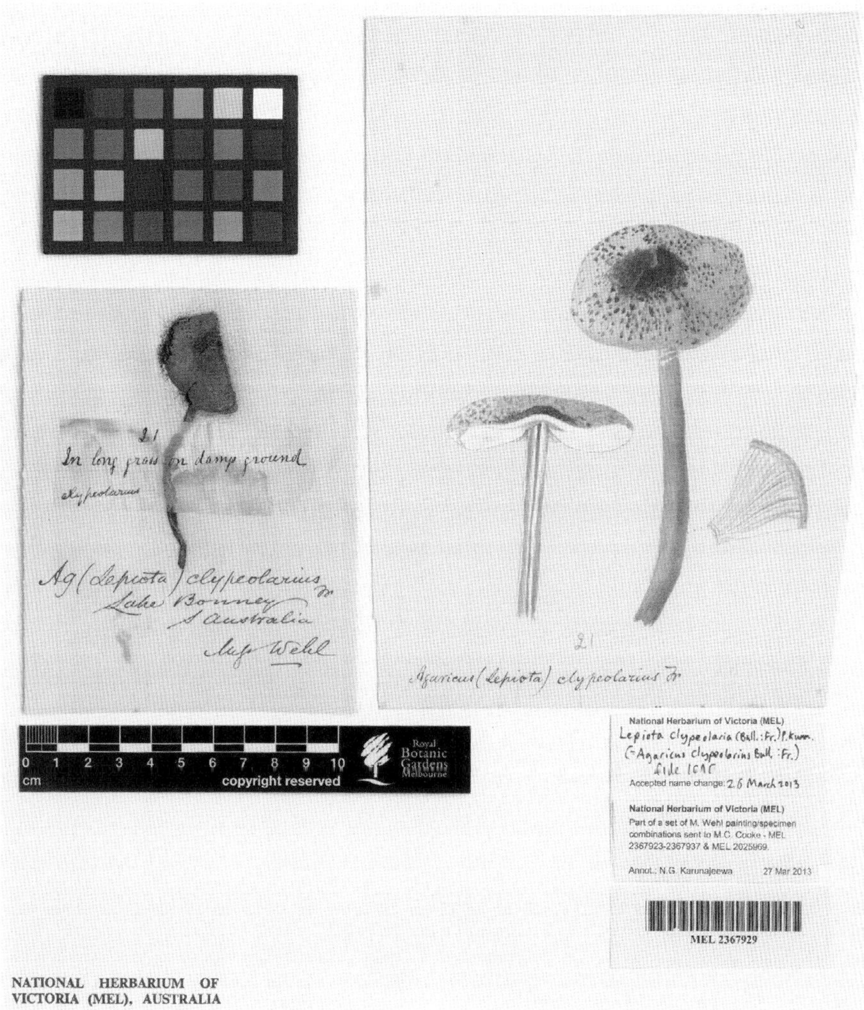

5.4
Specimen and watercolour of *Lepiota clypeolarius* (Bull.) P.Kumm. by Marie Wehl as *Agaricus clypeolarius* Fr., Lake Bonney, South Australia (MEL 2367929).

1880s, Cooke provided identifications and cited Marie's specimens and illustrations in descriptions of new species. Mueller's pride in his niece's efforts led him to step outside his usual field of expertise and name *Agaricus wehlianus* F.Muell. ex Cooke in her honour.[45] Marie's feelings about her labours do not survive, highlighting the frustrations of working with an incomplete documentary record. Later, in what was to be a long life, after marrying and having children, Marie wrote to an aunt, "I get time to do a little painting of which I am very fond."[46]

A handful of Mueller's female collectors became well known for their artworks, which they published and exhibited, helping to develop a broader sense of community pride in the Australian flora as a distinctive subject of creativity. These women included the Scott sisters (Harriet, 1830–1907, and Helena, 1832–1910), who provided illustrations for scientific works on plants, shells, snakes, and mammals;[47] Louisa Atkinson (later Mrs Calvert) (1834–1872);[48] Louisa Anne Meredith née Twamley (1812–1895);[49] and Ellis Rowan née Ryan (1848–1922).[50] Mueller corresponded with all these women, identifying their plant specimens, encouraging them to display their art before the public, and using his authority to give it scientific credibility.[51]

The most famous artist among Mueller's collectors was Ellis Rowan, who became known nationally and internationally as "The Australian flower-painter."[52] She won gold medals at international exhibitions, published several books on her travel adventures, and provided illustrations for two popular field guides of North American plants by Alice Lounsberry.[53] For three decades Mueller communicated with Rowan, identifying her specimens and the species depicted in her artwork. "I used frequently," Rowan recalled later, "in my Australian wanderings, to meet with varieties of shrubs and plants new to science: and these I invariably sent to the late Baron von Mueller to name."[54] So fine was her art that Mueller was reputedly able to recognise new taxa from her drawings alone. The scientist Walter Baldwin Spencer (1860–1929) observed, "a greater tribute to Mrs Rowan's delicate and faithful work could not be paid."[55]

As Mueller and Rowan's fame increased individually they also added lustre to each other and to the flora of Australia. In 1895, Queen Victoria inspected Rowan's portfolios and selected three pictures for the royal private apartments.[56] In conversation with Rowan, Victoria made special mention of Mueller's contributions to Australian botany, an acknowledgment that he regarded as "one of the greatest triumphs of my life!"[57] This was not so much a domestic led transformation of Australian culture as an international one, with Australians learning to appreciate the distinctiveness of their native vegetation because it was valued by overseas observers. It gave a flush of pride to settlers in the Australian colonies (and to their botanist) to know that depictions of endemic Australian taxa were worth hanging on royal walls.

A search in Trove for "Mueller" and "Rowan" produces hundreds of hits before 1900, indicating how far this pairing entered Australian popular culture. While their names mostly occur together in articles about art exhibitions, in 1890 "Baron von Mueller" and "Mrs Rowan" made an unexpected appearance in a Christmas story written for a Western Australian newspaper. Titled "How I Defended the Camp," the tale follows "a 'Jackaroo' of the greenest hue" on an inland journey.[58] While mostly worrying about encounters with deadly

5.5
Aneilema calandrinioides F.Muell. in Queensland, watercolour painting by Ellis Rowan, nla.obj-138801317.

creatures, this would-be bushman is given pause for thought when suddenly coming across the rare sight of the desert in bloom. "It was a revelation of beauty to me," he declared, and trying to convey a full appreciation of what he saw added, "I can fancy Baron Von Mueller's ecstacsies [*sic*] if he were there, while Mrs Rowan would be so spell-bound as to be quite unable to use her clever brush."[59]

PLANTS AS INCOME

Mueller earned his living from botany, as did a handful of other mid to late nineteenth-century scientists in the Australian colonies,[60] but native plants also provided an income, or supplementary income, to a number of Mueller's collectors. The largest group of these were nurserymen and seedsmen, who were motivated to find new showy plants by the prospect of lucrative sales.[61] As revealed by surviving nursery catalogues and advertisements in Australian newspapers, the horticultural trade had an important role to play in moving native flowers from a bush setting into colonial and European gardens and glasshouses.[62] In this way nurserymen attached a monetary value to iconic Australian plants that derived from their beauty and distinctiveness.

In 1870, future Mueller collector Carl Heinrich Hartmann (1833–1887) established the Range Nursery at Toowoomba about 130 km inland from Brisbane in Queensland.[63] The surrounding country was made up of species-rich pockets of wet and dry rainforest, and within a couple of years Hartmann was sending specimens to Mueller for identification. "In two days," Hartmann wrote in a rare surviving letter in 1872, "I have climbed [the mountains] three times, have torn shirt, trousers and shoes, the hands and knees grazed," but added, "imagine the wonderful view."[64] Hartmann's passion was orchids, and Mueller described a new species – *Sarcochilus hartmannii* F.Muell. – after him in 1874. "This very rare plant," Mueller declared in the dedication, was intended as a tribute "to the discoverer to whose friendship I owe many other singular plants from the eastern capricornian regions."[65] Hartmann's *Sarcochilus* proved relatively easy to cultivate and slowly established a reputation as "one of the most beautiful of our native flowers."[66] International gardeners were introduced to the orchid's delicate beauty in 1888 when Walter Hood Fitch (1817–1892) depicted it for *Curtis's Botanical Magazine*, a world-famous journal of botanical illustrations.

Hartmann and other nurserymen usually covered their own travelling expenses, turning to Mueller only for identifications, but Mueller also hired his own collectors and botanical artists. These were almost exclusively men, and Mueller paid them anything from a full wage for a fixed term, to a price per specimen.[67]

The wives and children of male collectors also benefited from Mueller's payments and were especially entitled to do so when collecting was in fact a family business. This was the case for the Webbs in the port town of Albany in Western Australia. William Webb (1834–1897) came to the colony as a convict but after serving his sentence married a Noongar woman (despite apparently leaving a wife behind in England) and had seven children. The Webbs mainly collected natural history specimens for passengers on ships of the Peninsular and Oriental Steam Navigation Company that stopped to provision at Albany on their way to and from Europe and the eastern Australian colonies.[68]

Mueller recruited the Webbs into his collecting network after the publication of *Flora Australiensis* and cited their specimens in his own publications. With money from travellers and from Mueller, the Webbs were able to rent a free-standing cottage in Albany, and William became known locally as "our naturalist" or, more colloquially, "old beetles and bees." These were substantial achievements for an ex-convict with a Noongar wife and children, indicating

5.6
Sarchochilus hartmannii F.Muell., delineated and lithographed by W.H. Fitch, *Curtis's Botanical Magazine* 114, series 3(44) (1888): tab. 7010.

that Australian plants had social as well as material value.[69] In gratitude for Mueller's part in the family's support, William and Lucy named their youngest child May Mueller Webb (b. 1892).[70]

A CLASSROOM IN THE BUSH

> Researches of these kinds become, furthermore, the sources of educational works, and unfold to well-trained and intelligent minds pure recreative and healthful pleasures inexpensively everywhere within reach.
> – F. Mueller, 1876.[71]

Mueller knew from personal experience that immersion in nature from an early age could be transformative. In his circular of 1876, he used the promise of "pure recreative and healthful pleasures" to recruit collectors and reiterated the positive outcomes of studying botany in a textbook for elementary students the following year – *Introduction to Botanic Teachings at the Schools of Victoria; through References to Leading Native Plants*. An early contribution to the nature study movement in Australia,[72] *Botanic Teachings* argued that to ensure botany was both "agreeable and lastingly fruitful" to children it should be learned *in situ*. Mueller used Victoria's native plants as his examples as they were freely and readily available everywhere in the colony, "while even the most ordinary garden-plants are not at all times and at every spot available for illustrative purposes."[73]

Although Mueller intended teachers to guide students through the textbook, he placed an open invitation at the end of the preface for enthusiastic readers to contact him directly by letter.[74] Thus, *Botanic Teachings* became another means of recruiting collectors, especially children, into his network and of shaping their responses to nature. About ten per cent of Mueller's female collectors were seventeen or younger when they began volunteering for the Melbourne Herbarium.[75] *Botanic Teachings* can be directly linked to the recruitment of a handful of these individuals, but its influence on how Australian children viewed nature undoubtedly spread much further.

In 1882, alpine collector Annie McCann used a copy of *Botanic Teachings* to home-school her nine children and also opened the classes to neighbours' children. "Words are but weak vehicles," she wrote to Mueller on receipt of the book, "to convey all the thanks I would fain offer to you for your valuable work which you have so kindly sent *the children* – Believe me it shall long be a prized *souvenir* as well as a warm incentive in promoting that study of which

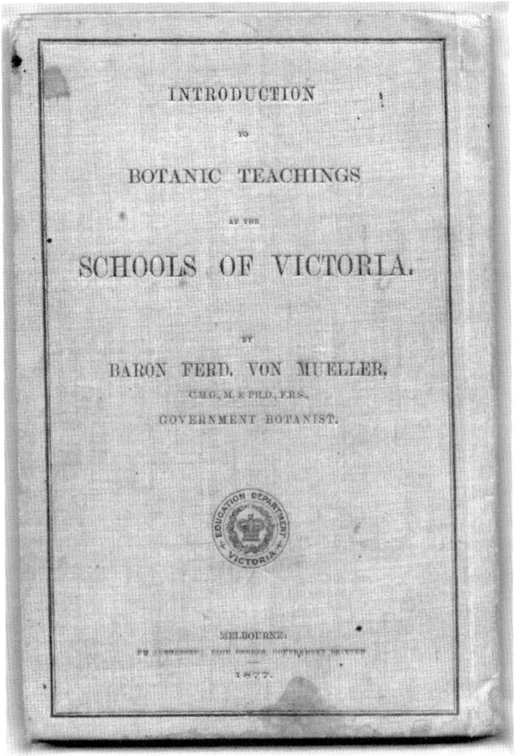

5.7
F. von Mueller, *Introduction to Botanic Teachings at the Schools of Victoria*, 1877.

I am such an ardent worshipper."[76] At least one of Annie's daughters, Mona (c. 1872–1930) collected for Mueller as a child, specializing in ferns, a group that Mueller described in *Botanic Teachings* as forming "delicate specimens when dried for collections."[77]

Few letters from Mueller's child collectors survive, a rare exception being a brief note by thirteen-year-old May Wise (1880–1958) of Sale in Victoria. A student at the local Ladies' High School, May seems to have been introduced to botany by Alexander Purdie (1859–1905), a lecturer at the Sale School of Mines who was himself a Mueller collector. In spring 1895, May and her sister Lilian (1882–1965) used a day off school to collect orchids in nearby scrub. Afterwards May forwarded a specimen of the Greenhood Orchid, *Pterostylis barbata* Lindl., to Mueller, suggesting "it may be a new record for this district."[78] With Mueller's encouragement May, Lilian, and a third school girl, Muriel Bennet, published a "List of Orchids Collected near Sale" in the *Victorian Naturalist*, the journal of the Field Naturalists Club of Victoria.[79] This was a precocious achievement and suggested that Mueller was a powerful mentor.

When *Botanic Teachings* appeared in 1877 the main textbooks in Victoria's elementary schools were imported from Ireland or England and were almost devoid of Australian content.[80] Mueller's textbook was, therefore, a pioneering contribution in the development of an Australian curriculum purposefully intended for settler families and their children.[81] Copies were placed at all Victorian elementary schools, but Mueller's Australia-wide reputation ensured the textbook was noticed in other colonies and action taken to use it there as well. In New South Wales it was published as a serial in the *Sydney Mail*,[82] and in Tasmania the Education Board circulated copies along with a local publication, Spicer's *Handbook of the Plants of Tasmania*.[83] Even in Western Australia, where the vegetation was substantially different to that featured in *Introduction to Botanic Teachings*, requests were made to the government to make the publication available in schools.[84]

CONCLUSION

The production of *Flora Australiensis* has been regarded by historians of Australian science as emblematic of the contest between imperialism and nationalism in colonial science. At the beginning, the Kew botanists were in control of its volumes but by the time the final volume was issued, Baron Ferdinand von Mueller, the government botanist of Victoria, had effectively assumed the mantle of the foremost botanist of Australian plants. His own ambition and drive were important factors in the transition, but I argue here that credit is also due to the efforts of his large and widespread network of collectors. Working in parallel with Mueller, the collectors, in various ways, used their relationships with him to promote Australian botany as an Australian undertaking – defining it in contrast not only to the imperial project of the Kew botanists but also against a backdrop of the dispossession of First Nations peoples.

It has only been possible relatively recently for scholars to appreciate the importance of Mueller's collectors in the story of *Flora Australiensis*. Mass digitization projects such as "The Correspondence of Ferdinand von Mueller," "The Australasian Virtual Herbarium," and "Trove" have made it feasible to trace Mueller's influence on collectors' responses to native vegetation, lifting these individuals out of obscurity and recasting them as agents of cultural change.

Collectors came to embrace the Australian flora by finding out the names of local plants from Mueller and planting them in private gardens. Others turned to the Australian vegetation as an alternative to the high culture of Great Britain and with Mueller's encouragement proclaimed Australian plants as beautiful in words and pictures. Nurserymen and professional collectors benefited from the monetary value placed on rare and showy Australian plants, selling specimens to Mueller or to overseas buyers, and children in several colonies turned the bush into a schoolroom with the assistance of Mueller's *Botanic Teachings*.

Gianquitto points out that by accepting all comers, scientific collecting networks in America promoted the notion that sampling the landscape was a democratic undertaking in which citizens had a shared capacity to participate. Such networks became "the stuff out of which national unity was made."[85] This was undoubtedly also the case in Australia, although First Nations people were excluded from citizenship until 1967. Moreover, in Australia I argue that the domestication of collectors' responses to native plants was an important part of the process of creating a sense of nationhood. In the home, where parents (especially mothers) taught children who they were and what was around them, everyday references to native plants had a powerful role in shaping cultural identity.

Neither Mueller nor his collectors denied being part of the British Empire but sought to initiate and manage their response to Australian botany for themselves – an insider-looking-out approach that also co-opted fragments of First Nations knowledge. While the rationale for the existence of Mueller's network was to produce a scientific account of the Australian flora, the hundreds of individual collectors generated a much more diverse and significant form of botanical knowledge in the context of their relationships with Mueller. They came to regard Australia's vegetation as connected across the various Australian colonies and as distinct in kind from the vegetation of other regions in the world and no less importantly for the development of an "Australian" culture, as beautiful and valuable.

ACKNOWLEDGMENTS

For advice and assistance with this chapter I thank Tom May, Sue Janson, Rusty Shteir, Libby Robin, John Dowe, Sally Stewart, librarian, Royal Botanic Gardens Victoria, Pina Milne, manager collections, Royal Botanic Gardens Victoria, and my Mueller Project colleagues.

NOTES

1 Anon., "Occasional Notes." *West Australian* (Perth, Western Australia), 24 July 1883, 2, http://nla.gov.au/nla.news-title30.

2 Archer and Maroske, "Sarah Theresa Brooks," 190–1.

3 Anon., "Botanical Survey of the Empire," 9–10.

4 Moyal, *'A Bright & Savage Land,'* 149–53. Finney, *Paradise Revealed.* Maroske, "Science by Correspondence."

5 Endersby, *Imperial Nature*, 84–111. Maroske, "'A Taste for Botanic Science.'"

6 Gianquitto, "'My Dear Dr.,'" 435.

7 Robin, *How a Continent Created a Nation*, 213–18, acknowledges that the exceptionalism of Australian nature has played a role in building an Australian cultural identity, although she argues that it has only figured trivially "in the rhetoric of nationhood."

8 Curthoys and Mitchell, *Taking Liberty.* Russell and Olsen, *Australia's First Naturalists.*

9 Hill, *Women and Museums*, 1–2. Opitz et al., *Domesticity in the Making of Modern Science.*

10 Bentham, *Flora Australiensis.*

11 Hooker, *Flora Boreali-Americana.* Hooker, *The Botany of the Antarctic Voyage.* Hooker and Thomson, *Flora Indica.* Harvey and Sonder, *Flora Capensis.* Bentham, *Flora Hongkongensis.*

12 Anon., "Botanical Survey of the Empire," 20.

13 Stafford, "Annexing the Landscapes of the Past," 67–89.

14 Gilbert, "Banks, Sir Joseph (1743–1820)."

15 Gascoigne and Maroske, "Colonial Science and Technology," 446–7. Jackson, "A List of the Collectors," 1–80.

16 Curtis, "Hooker, Sir Joseph Dalton (1817–1911)." Anon., "Botanical Survey of the Empire." *Bulletin of Miscellaneous Information (Royal Botanic Gardens, Kew)* no. 2 (1905): 9–43, esp. 10–11. *Florae Tasmaniae* was part 3 of Hooker, *Flora of the Antarctic Voyage.*

17 Kirchberger, "Naturwissenschaftler im britischen Empire." Home, "Science as a German Export."

18 Home et al., *Regardfully Yours*, 19–38.

19 Maroske, "Science by Correspondence," ch. 2.

20 Home et al., *Regardfully Yours*, 139–42, 144–6.

21 Maroske, "Science by Correspondence," ch. 3.

22 "Correspondence of Ferdinand von Mueller Project," accessed 23 February 2018, https://www.rbg.vic.gov.au/science/herbarium-and-resources/library/mueller-correspondence-project.

23 "The Australasian Virtual Herbarium," accessed 23 February 2018, https://avh.chah.org.au/.

24 "Trove," accessed 23 February 2018, https://trove.nla.gov.au/.

25 Caldwell, "Population," 28, 30.

26 For a detailed discussion of the women and girls in Mueller's collecting network see Maroske, "'A Taste for Botanic Science.'"

27 Maroske and Vaughan, "Ferdinand Mueller's Female Plant Collectors." Clarke, *Aboriginal Plant Collectors*.

28 Robin and Carruthers, "National Identity and International Science," 37–8.

29 Mueller, F., ["Circular"].

30 Mueller distributed the circular to correspondents, but it was also reproduced in newspapers across Australia. See Maroske "'A Taste for Botanic Science.'"

31 Alexander Paul to Ferdinand Mueller. Anon., "Death of an Old Identity," *Bendigo Advertiser* (Victoria, Australia), 5 November 1891, 2, https://trove.nla.gov.au/newspaper/title/346.

32 Alexander Paul, *Clematis microphylla*.

33 Sarah Brooks to Ferdinand Mueller.

34 Moore, *Speaking Our Language*, 7.

35 Brooks, "Mitteilung des Mr. S.P. Brooks."

36 Moore, *Speaking our Language*, 13–14.

37 McCann, *The Poetical Works of Mrs Torrens M'Cann*, 143–4, 193.

38 Robert Issell Perrott to Ferdinand Mueller.

39 Home et al., *Regardfully Yours*, 686–765.

40 *Hakea brookeana* [more correctly spelled *brooksiana*] is now considered a synonym of *H. obliqua* R.Br. by the "Australian Plant Census," accessed 17 February 2021, https://biodiversity.org.au/nsl/services/APC.

41 Maroske, "'A Taste for Botanic Science,'" 85–6. Olsen, *Collecting Ladies*.

42 Jordan, *Picturesque Pursuits*.

43 Maroske and Dowe, "German Farm Gardens," 13–16.

44 Clara Wehl, poem.

45 *Agaricus wehlianus* is now known as *Pluteus wehlianus* (F.Muell. ex Cooke) Sacc., see Dowe et al., "The Wehl family of South Australia," 16.

46 Ibid., 57.

47 Dorey, "The Scott Sisters."

48 Clarke, *Pioneer Writer*.

49 Rae-Ellis, *Louisa Anne Meredith*.

50 Morton-Evans, *The Flower Hunter*.

51 Maroske and Vaughan, "Ferdinand Mueller's Female Plant Collectors."

52 e.g. Anon. "Our Melbourne Lady's Letter." *Daily News* (Perth, Western Australia), 6 June 1896, 2, https://trove.nla.gov.au/newspaper/title/275.

53 Hazzard, "Rowan, Marian Ellis." See also Bazely and McPherson, "Women, Citizen Science, and Botanical Knowledge in Ontario, 1870–1920," in this volume.

54 Anon., "A Flower Hunter." *Argus* (Melbourne, Victoria, Australia), 10 December 1904, 17, https://trove.nla.gov.au/newspaper/title/13.

55 Spencer, "To the Editor of the Argus," 17.

56 Hazzard, *Australia's Brilliant Daughter*, 86–8.

57 Ferdinand Mueller to Ellis Rowan.

58 A "'Jackaroo' of the greenest hue" is an inexperienced labourer on a sheep or cattle ranch.

59 Mann, "How I Defended the Camp."

60 Gascoigne and Maroske, "Colonial Science and Technology."

61 George, "Plant Collectors to 1900," 249–610.

62 Brookes and Barley, *Plants Listed in Nursery Catalogues.*

63 George, "Plant Collectors to 1900," 408–9.

64 Carl Hartmann to Ferdinand Mueller.

65 Mueller, *Fragmenta*, 248.

66 Anon., "Toowoomba Field Naturalist's Club." *Darling Downs Gazette* (Queensland, Australia), 20 September 1910, 5, https://trove.nla.gov.au/newspaper/title/834.

67 Maroske, "'A Taste for Botanic Science,'" 87–8.

68 Little, "Albany History Collection."

69 Anon., "208 Grey St Albany," "Albany Federation Heritage," "Federation House," accessed 3 March 2018, https://federation-house.wikispaces.com/albany+federation+heritage.

70 Anon., "Webb, May Mueller," "Family History," "Births, Deaths and Marriages," last updated 15 August 2017, http://www.bdm.dotag.wa.gov.au/_apps/pioneers index/default.aspx.

71 Mueller, ["Circular"].

72 Kass, *Educational Reform and Environmental Concern*. Kass, "The Nature Study Idea in New South Wales."

73 Mueller, *Botanic Teachings*, 3–4.

74 Ibid., 5–6.

75 Ibid., 73–4.

76 Annie McCann to Ferdinand Mueller.

77 Mueller, *Botanic Teachings*, 132.

78 May Wise to Ferdinand Mueller.

79 Wise and Bennett, "List of Orchids Collected near Sale," 45–7.

80 Gibbs, "Victorian School Books," 57–61, 172–3.

81 Mueller includes some plant names from Indigenous languages but only where they had been adopted as common names by settlers. For example, "Currijong" for *Brachychiton populneum*, and "Waratah" for *Telopea*. Mueller, *Botanic Teachings*, 54, 85.

82 Lucas, Maroske, and Brown-May, "Bringing Science to the Public," 43.

83 Anon., "Education Board." *Mercury* (Hobart, Tasmania, Australia), 11 April 1878, 3, https://trove.nla.gov.au/newspaper/title/10.3. Spicer, *A Handbook of the Plants of Tasmania.*

84 Anon., "Assisted Schools." *Herald* (Fremantle, Western Australia), 20 July 1878, 3, https://trove.nla.gov.au/newspaper/title/386.

85 Gianquitto, "'My Dear Dr.,'" 437, 439.

BIBLIOGRAPHY

Archives

Brooks, Sarah. *Sarah Brooks to Ferdinand Mueller, November 5, 1883.* Letter. RB MSS M1. Library, Royal Botanic Gardens Victoria, Melbourne.

Hartmann, Carl. *Carl Hartmann to Ferdinand Mueller, June 28, 1872.* Letter. RB MSS M1. Library, Royal Botanic Gardens Victoria, Melbourne.

McCann, Annie. *Annie McCann to Ferdinand Mueller, March 1882.* Letter. RB MSS M1. Library, Royal Botanic Gardens Victoria.

Maroske, Sara. "Science by Correspondence: Ferdinand Mueller and Botany in Nineteenth-century Australia", unpublished thesis for the degree of Doctor of Philosophy, Department of History and Philosophy of Science, The University of Melbourne, 2005.

Mueller, Ferdinand. *Ferdinand Mueller to Ellis Rowan, March 3, 1896.* Letter. MS 2206 Ellis Rowan Papers. National Library of Australia, Canberra.

Paul, Alexander. *Alexander Paul to Ferdinand Mueller, September 14, 1876.* Letter. RB MSS M1. Library, Royal Botanic Gardens Victoria, Melbourne.

Paul, Alexander. *Clematis microphylla.* Herbarium specimen, collected 1876. MEL 2208896. National Herbarium of Victoria, Royal Botanic Gardens Victoria, Melbourne.

Perrott, Robert Issell. *Robert Issell Perrott to Ferdinand Mueller, December 5, 1871.* Letter. RB MSS M1. Library, Royal Botanic Gardens Victoria, Melbourne.

Wehl, Clara. Poem, n.d. Millicent Local History Collection, Millicent Public Library, South Australia.

Wehl, Marie. *Lepiota clypeolaria*, c. 1880s. Herbarium specimen. MEL 2367929. National Herbarium of Victoria, Royal Botanic Gardens Victoria, Melbourne, Australia.

William Webb. *William Webb to Ferdinand Mueller, no date, circa 1886.* Letter. RB MSS M1. Library, Royal Botanic Gardens Victoria, Melbourne.

Wise, Mary (May) Isabel. *May Wise to Ferdinand Mueller, October 1, 1895.* Letter. RB MSS M1. Library, Royal Botanic Gardens Victoria.

Books and Articles

Archer, Barbara, and Sara Maroske. "Sarah Theresa Brooks – Plant Collector for Ferdinand Mueller." *Victorian Naturalist* 113, no, 4 (1896): 188–94.

Bentham, George. *Flora Australiensis: A Description of the Plants of the Australian Territory*, 7 vols. London: Lovell Reeve, 1863–1878.

– *Flora Hongkongensis: A Description of the Flowering Plants and Ferns of the Island of Hongkong …* London: Lovell Reeve, 1861.

Brookes, M., and R. Barley. *Plants Listed in Nursery Catalogues in Victoria 1855–1889*. South Yarra: Ornamental Plants Collections Association, 1992.

Brooks, J. "Mitteilung des Mr. S. P. Brooks zu Israelite Bay an Baron Ferd. von Müller, datiert 25. Aug. 1890." *Australische Zeitung* (South Australia, Australia), 19 November 1890, 6.

Caldwell, J.C. "Population." In *Australians: Historical Statistics*, edited by Wray Vamplew, 23–41. Broadway, New South Wales: Fairfax, Syme & Weldon Associates, 1987.

Clark, Dymphna. *Baron Charles von Hügel: New Holland Journal: November 1833–October 1834*. Melbourne: Melbourne University Press, 1990.

Clarke, Patricia. *Pioneer Writer: The Life of Louisa Atkinson: Novelist, Journalist, Naturalist*, Sydney: Allen & Unwin, 1990.

Clarke, Philip. *Aboriginal Plant Collectors: Botanists and Australian Aboriginal People in the Nineteenth Century*. Kenthurst: Rosenberg Publishing Pty Ltd, 2008.

Curthoys, Ann, and Jessie Mitchell. *Taking Liberty: Indigenous Rights and Settler Self-Government in Colonial Australia, 1830–1890*. Cambridge: Cambridge University Press, 2018.

Curtis, Winifred M. "Hooker, Sir Joseph Dalton (1817–1911)." In *Australian Dictionary of Biography*, National Centre of Biography, Australian National University. First published in hard copy 1966. Accessed online 22 February 2018. http://adb.anu.edu.au/biography/hooker-sir-joseph-dalton-3789/text5993.

Dorey, Fran. "The Scott Sisters." Australian Museum, Sydney, New South Wales. Accessed online 22 February 2018. https://australianmuseum.net.au/beauty-from-nature-art-of-the-scott-sisters.

Dowe, John Leslie, Tom W. May, Sara Maroske, and Lucy Therese Smith. "The Wehl Family of South Australia and Their Botanical Connections with 'Dear Uncle' Baron Ferdinand von Mueller." *Swainsona* 34 (2020): 1–79.

Endersby, Jim. *Imperial Nature: Joseph Hooker and the Practices of Victorian Science*. London: The University of Chicago Press, 2008.

Erdos, Renee. "Leichhardt, Friedrich Wilhelm Ludwig (1813–1848)." In *Australian Dictionary of Biography*, National Centre of Biography, Australian National University. Published first in hard copy 1966. Accessed 22 February 2018. http://adb.anu.edu.au/biography/leichhardt-friedrich-wilhelm-ludwig-2347/text3063.

Finney, Colin. *Paradise Revealed: Natural History in Nineteenth-Century Australia.* Melbourne: Museum of Victoria, 1993.

Gascoigne, John, and Sara Maroske. "Colonial Science and Technology." In *The Cambridge History of Australia*, edited by Alison Bashford and Stuart Macintyre, 438–61. Port Melbourne: Cambridge University Press, 2013.

George, Alex S. "Plant Collectors to 1900." In *Australian Botanists's Companion.* Kardinya, WA: Four Gables Press, 2009, 249–610.

Gianquitto, Tina. "'My Dear Dr.': American Women and Nineteenth-Century Scientific Correspondence." In *The Edinburgh Companion to Nineteenth-Century American Letters and Letter-writing*, edited by Celeste-Marie Bernier, Judie Newman, and Matthew Pethers, 435–49. Edinburgh: Edinburgh University Press, 2016.

Gibbs, Desmond Robert. "Victorian School Books; a Study of the Changing Social Content and use of School Books in Victoria, 1848–1948, with Particular Reference to School Readers." Unpublished PhD thesis, University of Melbourne 1987.

Gilbert, L.A. "Banks, Sir Joseph (1743–1820)." In *Australian Dictionary of Biography*, National Centre of Biography, Australian National University. First published in hard copy 1966. Accessed 22 February 2018. http://adb.anu.edu.au/biography/banks-sir-joseph-1737/text1917.

Harvey, W.H., and Otto Wilhelm Sonder. *Flora Capensis; Being a Systematic Description of the Plants of the Cape Colony, Caffraria and Port Natal* …, 3 vols. Dublin: Hodges, Smith, and Co., 1860–65.

Hazzard, Margaret. *Australia's Brilliant Daughter, Ellis Rowan: Artist, Naturalist, Explorer, 1848–1922.* Melbourne: Royal Victorian Institute for the Blind, 1987.

– "Rowan, Marian Ellis (1848–1922)." In *Australian Dictionary of Biography*, National Centre of Biography, Australian National University. Published first in hard copy 1988. Accessed online 1 March 2018. http://adb.anu.edu.au/biography/rowan-marian-ellis-8282/text145133.

Hill, Kate. *Women and Museums 1850–1914: Modernity and the Gendering of Knowledge.* Manchester: Manchester University Press, 2016.

Home, R.W. "Science as a German Export to Nineteenth Century Australia." *Working Papers in Australian Studies* 104 (1995): 1–17.

Home, R.W., A.M. Lucas, Sara Maroske, D.M. Sinkora, and J.H. Voigt (eds). *Regardfully Yours: Selected Correspondence of Ferdinand von Mueller*, vol. 1: 1840–1859. Bern: Peter Lang, 1998.

Hooker, Joseph Dalton. *The Botany of the Antarctic Voyage* …, 3 pts. London: Reeve Brothers, 1844–1859 [i.e. 1860].

– "Sarcochilus Hartmanni: Native of Queensland." *Curtis's Botanical Magazine* 114, series 3 (44) (1888): tab. 7010.

Hooker, Joseph Dalton, and Thomas Thomson. *Flora indica: Being a Systematic Account of the Plants of British India …* London: W. Pamplin, 1855.

Hooker, William Jackson. *Flora Boreali-Americana; or, the Botany of the Northern Parts of British America …* London: Treuttel & Würz, 1833 [i.e. 1829]–1840.

Jackson, B.D. "A List of the Collectors Whose Plants are in the Herbarium of the Royal Botanic Gardens, Kew, to 31st December, 1899." *Bulletin of Miscellaneous Information*, 169/171: 1–80.

Jordan, C. *Picturesque Pursuits: Colonial Women Artists & the Amateur Tradition.* Carlton: Melbourne University Press, 2005.

Kass, Dorothy. *Educational Reform and Environmental Concern: A History of School Nature Study in Australia.* Abingdon, Oxon: Routledge, 2018.

– "The Nature Study Idea in New South Wales: The Role of Charles Tucker Masson (1856–1928)." *Historical Records of Australian Science* 24, no. 2 (2013): 225–41.

Kirchberger, Ulrike. "Naturwissenschaftler im britischen Empire: Die Erforschung der außereuropäischen Welt im Spannungsfeld zwischen deutschem und britischem Imperialisimus." *Historische Zeitschrift* Bd 271, H. 3 (2000): 621–60.

Little, Roy and Beatrice. "Albany History Collection: Persons – Vertical Files Collection: Biographical Summaries." Viewed 3 March 2018. http://history.albany.wa.gov.au/files/persons_vertical_files_biographical_summaries_31.03.2009.pdf.

Lucas, A.M., Sara Maroske, and Andrew Brown-May. "Bringing Science to the Public: Ferdinand von Mueller and Botanical Education in Victorian Victoria." *Annals of Science* 63, no. 1 (2006): 25–57.

McCann [as M'Cann], Mrs [Annie] Torrens. *The Poetical Works of Mrs Torrens M'Cann.* Melbourne: George Robertson, 1888.

Mann, A.C.D. "How I Defended the Camp." *Victorian Express* (Geraldton, Western Australia), 20 December 1890, 2. https://trove.nla.gov.au/newspaper/title/1128.

Maroske, Sara. "Science by Correspondence: Ferdinand Mueller and Botany in Nineteenth-century Australia." Unpublished PhD thesis, Department of History and Philosophy of Science, University of Melbourne 2005.

– "'A Taste for Botanic Science': Ferdinand Mueller's Female Collectors and the History of Australian Botany." *Muelleria* 32 (2014): 72–91.

Maroske, Sara, and John Dowe. "German Farm Gardens in Colonial South Australia." *Australian Garden History* 26, no. 4 (2015): 13–16.

Maroske, Sara, and Alison Vaughan. "Ferdinand Mueller's Female Plant Collectors: A Biographical Register." *Muelleria* 32 (2014): 92–172.

May, T., Sara Maroske, and D. Sinkora. "The Mycologist, the Baron, His Fungi Hunters and the Mystery Artist." *Botanic Magazine* (Melbourne) 6 (1995): 36–9.

Moore, Bruce. *Speaking Our Language: The Story of Australian English.* Oxford: Oxford University Press, 2008.

Morton-Evans, Christine, and Michael Morton-Evans. *The Flower Hunter: The Remarkable Life of Ellis Rowan*. Australia: Simon & Schuster, 2009.

Moyal, Ann. *'A Bright & Savage Land': Scientists in Colonial Australia*. Sydney: Collins, 1986.

Mueller, F. ["Circular"]. Melbourne: Government Printer, 1876.

– *Fragmenta Phytographiae Australiae* 8 (70): 248.

– *Introduction to botanic teachings at the schools of Victoria*. Melbourne: Government Printer, 1877.

Olsen, Penny. *Collecting Ladies: Ferdinand von Mueller and Women Botanical Artists*. Canberra: National Library of Australia, 2013.

Opitz, Donald L., Staffan Bergwik, and Brigitte Van Tiggelen, eds. *Domesticity in the Making of Modern Science*. New York: Palgrave Macmillan, 2016.

Rae-Ellis, Vivienne. *Louisa Anne Meredith, a Tigress in Exile*. Sandy Bay, Tasmania: Blubber Head Press, 1979.

Robin, Libby. *How a Continent Created a Nation*. Sydney: UNSW Press, 2007.

Robin, Libby, and Jane Carruthers. "National Identity and International Science: The Case of *Acacia*." *Historical Records of Australian Science* 23, no. 1 (2012): 34–54.

Russell, Lynette, and Penny Olsen. *Australia's First Naturalists: Indigenous Peoples' Contribution to Early Zoology*. Canberra: National Library of Australia, 2019.

Spencer, W.B. "To the Editor of the Argus." *Argus* (Melbourne, Victoria, Australia), 10 December 1904, 17. https://trove.nla.gov.au/newspaper/title/13.

Spicer, W.W. *A Handbook of the Plants of Tasmania*. Hobart Town: J. Walch and Sons, 1978.

Stafford, R.A. "Annexing the Landscapes of the Past: British Imperial Geology in the Nineteenth Century." In *Imperialism and the Natural World*, edited by J.M. MacKenzie, 67–89. Manchester: Manchester University Press, 1990.

Swan, R.A. "Neumayer, Georg Balthasar von (1826–1909)." In *Australian Dictionary of Biography*, National Centre of Biography, Australian National University. Published first in hard copy 1966. Accessed online 22 February 2018. http://adb.anu.edu.au/biography/neumayer-georg-balthasar-von-4290/text6943.

Vellacott, Helen, ed. *Some Recollections of a Happy Life: Marianne North in Australia & New Zealand*. Caulfield East: Edward Arnold Australia, 1986.

Wise, May and Lilian, and Muriel Bennett. "List of Orchids Collected Near Sale." *Victorian Naturalist* 12, no. 3 (1895): 45–7.

6

Alice Hollingworth, Early Botanical Explorer in Muskoka District, Ontario

JAMES S. PRINGLE

By the standards of the nineteenth century, the District of Muskoka was remote from Ontario's larger cities and from the universities then in existence. Consequently, floristic exploration[1] was late in coming to the district.[2] No concentrated efforts to investigate and catalogue the flora of Muskoka were made until the late R. Emerson Whiting initiated such studies in the 1950s and Dr James P. Goltz joined in these efforts beginning in the 1980s. Early in his study, Whiting found specimens in what is now a component of the herbarium of the Royal Ontario Museum in Toronto that had been collected in the District of Muskoka in 1890 and 1891 by Alice Hollingworth.[3] Only nineteen of these specimens were found at that time, but they were of considerable interest to botanical history.

Before 1890, few botanical specimens had been collected in the District of Muskoka, and those that did exist were from sites in the southern and western part of the district, where summer cottages had been built and resorts had been developed. From 1887 through 1895, Andrew Alexander, of Hamilton, had collected specimens from a few of the Georgian Bay islands and from mainland sites along the Georgian Bay shore near Go Home Bay. Members of the Botanical Club of the American Association for the Advancement of Science, on a three-day excursion in 1889 to Lakes Muskoka and Rosseau, had collected specimens in the vicinities of Gravenhurst and Port Sandfield, including several species of aquatic plants that were new to the known flora of Canada. Blackstone Lake, near Georgian Bay, where Thomas J.W. Burgess collected specimens cited in John Macoun's *Catalogue of Canadian Plants*,[4] is outside the present-day District of Muskoka, and Macoun's three-man "expedition" on the Trent River mentioned in his *Autobiography*[5] may not have

6.1
Alice Hollingworth, ca. 1900.

reached the present boundary of the district. The interior of the district, between the Muskoka Lakes and Bracebridge, a region notable for its mosaic of calcareous and non-calcareous edaphic habitats, with several forest types, rock outcrops, and bogs, fens, and other wetlands, had, until the 1890s, remained unexplored floristically. Until well into the twentieth century, nearly all of the botanical specimens from that part of the District of Muskoka were those collected by Alice Hollingworth.

Moreover, here was a name unfamiliar to historians of botany. When Whiting began his studies, Alice Hollingworth had not been included in any list of persons known to have collected botanical specimens, and nothing on her life, or of her having lived and botanized in Ontario, had been published in the literature of botanical history.

Of the collectors of botanical specimens in what is now Canada during the first half of the nineteenth century, most of those who have become prominent in the literature of botanical history were persons who sent specimens to Sir William Jackson Hooker, who, while regius professor of botany at the University of Glasgow in Scotland, wrote the *Flora Boreali-Americana*,[6] which was an important reference on the Canadian flora for decades. Hooker did not visit North America himself but recruited plant collectors in what was then British North America. Those in the east were mostly high-ranking colonial administrators and British military officers, the wives of some of these appointees, and some of their friends whom they in turn had recruited. They were generally of the upper socioeconomic classes.

Alice Hollingworth was active in botany later than most of the other women mentioned in this book, and times had changed. Chapters in part 1 discuss earlier botanical collectors in British North America, including the Countess of Dalhousie, Anne Mary Perceval, and Harriet Campbell Sheppard, who were active in the 1820s and 1830s, corresponded with botanists in England and Scotland, and contributed specimens for study by British botanists, notably Hooker. By the 1890s Canada had scholarly societies of its own, both national and regional, and several universities had been established in Ontario and elsewhere in Canada.[7] Alice Hollingworth, those who encouraged and guided her, and the societies of which she was a member were all Canadian. Unlike most of the women noted here for their contributions to Canadian floristics, Alice Hollingworth was from a lower socioeconomic class and lived in a remote, newly and sparsely settled farming and logging area rather than in a major population centre.

By 1890, significant botanical explorations had occurred in what is now Canada. The publication of Hooker's *Flora Boreali-Americana* had been completed in 1840. Publication of Macoun's *Catalogue of Canadian Plants* had begun in 1883, and by 1890 only the parts dealing with bryophytes and lichens remained to be published. Nevertheless, much of Canada, even in the eastern provinces, was still poorly known floristically. The professionalization of the sciences was well under way, but amateurs were still prominent among the members of local and regional scholarly societies, which by the 1890s existed in most of eastern Canada's major cities and some smaller population centres, and amateur botanists were making significant contributions to floristic knowledge.

In Canada as elsewhere, collecting objects of many kinds has long been popular, and in the Victorian era, the hobbies of collecting and exchanging natural history specimens were at their zenith.[8] Many persons, mostly among the better educated and more well-to-do members of society, including both women and men, created private herbaria during the latter part of the nineteenth century. Members of botanical exchange clubs sent specimens to fellow participants throughout much of North America and Europe.[9] Collecting plant specimens was a common requirement in North American secondary schools and academies during this time, especially for girls, in whom one objective of this exercise was to promote "gentility." In colleges and universities, systematic botany courses in which plant collections were required were common electives in curricula for both sexes. Relatively little of this avocational or assigned collecting contributed significantly to floristic knowledge, but some amateurs during this time participated in serious botanical explorations, communicated with career botanists, published papers on their discoveries, and contributed specimens to institutional herbaria. Some of their specimens, when exchanged with researchers or when private herbaria were donated or bequeathed to scholarly institutions, provided valuable resources for taxonomic and floristic research.

Despite the popularity of collecting plant specimens in the latter half of the nineteenth century, only a few women contributed significantly to the knowledge of the Ontario flora during this period. Mrs William Roy (as her name appears on herbarium labels), née Jessie Dalrymple Gregg, brought Macoun's attention to the occurrence of rare fern species in the vicinity of Owen Sound and, from the 1860s through the 1880s, collected and exchanged plant specimens from that area that are now in several institutional herbaria.[10] Catharine McGill ("Kate") Crooks (married surname Smart) collected plant specimens in the vicinity of Hamilton and Galt (now part of Cambridge) ca. 1860.[11] Annie Alicia Boyd, while she was a student at Queen's University in 1897 and 1898, collected botanical specimens, mostly from the vicinity of Kingston, some from Ompah, that are now in the herbarium at the Queen's University Biology Station.[12] Among these botanical collectors Alice Hollingworth is notable for having botanized during the 1890s in an area that would not receive much floristic attention from others for many years.

The specimens seen by Whiting in the 1950s had been contributed by Alice Hollingworth to the herbarium of the Canadian Institute (discussed below), which was transferred to the Royal Ontario Museum in 1924 but housed with the herbarium of the University of Toronto until the combined herbaria were moved to the museum in 1980. She had contributed many more than nineteen specimens to the institute's herbarium, but many of them were destroyed by insect infestations that devastated much of the institute's holdings of plant and animal specimens in the mid-1890s.[13]

Fortunately, a much greater representation of Alice Hollingworth's botanizing efforts in Muskoka is extant at the University of Guelph. In 2016, her own herbarium, which had until then remained with the family, was donated to the university by her grandson Donald Webster. This herbarium consisted of nearly 1,200 specimens, including angiosperms, conifers, ferns, bryophytes, lichens, and one charophyte, most which she had collected herself in the 1890s, when she was in her twenties and living in Beatrice, in Muskoka District. The relatively few specimens that she added in the 1920s and 1930s include some from localities near the south end of Georgian Bay, a few from Manitoulin Island, and some from the Niagara Peninsula. There are also specimens from a trip to Saskatchewan, British Columbia, the Yukon Territory, Washington, and Alaska in the summer of 1934.[14]

Unlike those of many amateur naturalists, Alice Hollingworth's herbarium included many grasses, sedges, rushes, asters, and goldenrods. Her identification of plants in these groups indicates not only a keen interest in the Muskoka flora but also considerable skill in plant identification, even with plant groups generally considered "difficult," and an awareness of what species would be considered rare or otherwise noteworthy. What books she had for identifying plants other than *Gray's Manual* (noted below) is not known, but consideration of that topic makes her accomplishments appear all the more remarkable. The early floras of northeastern North America did not include identification keys such as those used by botanists today. Field guides, the earliest of which were published in the 1890s, were limited both in the number of species included and in their illustrations of the species that were included. They would not have sufficed for the identification of rare or inconspicuous plants or grasses and sedges, or for distinguishing among many of the species in large genera.

Alice Hollingworth's discovery of several species that are rare and/or inconspicuous represents enthusiastic and extensive investigation of the Muskoka flora. One especially significant discovery was that of the Shining Ladies'-tresses, *Spiranthes lucida* (H.H. Eaton) Ames, a native orchid rare throughout its range and an obligate calciphile, found at Beatrice in 1890 in an enclave of clayey, calcareous soil in a region where non-calcareous soils predominate.[15] Other notable species that she discovered in Muskoka and reported to the Biological Society of Ontario[16] include Northern Adder's-tongue

Fern, *Ophioglossum pusillum* Raf. [at that time included in *O. vulgatum* L.], Leathery Grape-fern, *Sceptridium multifidum* (S.G. Gmel.) M. Nishida [as *Botrychium ternatum* (Thunb.) Sw.], and Club-spur Orchid, *Platanthera clavellata* (Michx.) Luer [as *Habenaria clavellata* Michx.], all of which are not only uncommon but also small and inconspicuous, and Lesser Round-leaved Orchid, *Platanthera orbiculata* (Pursh) Lindl. [as *Habenaria orbiculata* (Pursh) Torr.], which, although more conspicuous, is not common. Other uncommon and inconspicuous species that she discovered in Muskoka include Water Awl-wort, *Subularia aquatica* L., a minute aquatic; Green Adder's-mouth, *Malaxis unifolia* Michx. [as *Microstylis ophioglossoides* Nutt.], a small, green-flowered orchid; *Calopogon tuberosus* (L.) Britton, Steans & Poggenb. [as *C. pulchellum* (Salisb.) R. Br.] and *Platanthera blephariglottis* (Willd.) Lindl. [as *Habenaria blephariglottis* (Willd.) Hook.], two orchids that are more showy but rare; and Golden Saxifrage, *Chrysosplenium americanum* Schwein. ex Hook., a species with inconspicuous flowers. Her collections from the Niagara Peninsula include such rarities for Ontario as Long-branched Frostweed, *Crocanthemum canadense* (L.) Britton [as *Helianthemum canadense* (L.) Michx.], and American Water-willow, *Justicia americana* (L.) Vahl [as *Dianthera americana* L.]. All of the species mentioned here are represented by specimens from Alice Hollingworth's own herbarium, now at Guelph, except for *Spiranthes lucida,* which is at the Royal Ontario Museum.

Plant morphology, classification, and nomenclature were very much of interest to Alice Hollingworth, and she derived great satisfaction from her study of these subjects, much of it "by candlelight … in the dark hours of night while others slept."[17] She used scientific names on the labels in her herbarium, in her notebooks, and in her 1894 publication cited below. In an address in 1899 she referred to plants by their vernacular names, but she noted that the Pale Corydalis [*Capnoides sempervirens* (L.) Borck.] was "half-brother" to the cultivated Bleeding-heart [*Lamprocapnos spectabilis* (L.) Fukuhara; both species are in the family Papaveraceae, subfamily Fumarioideae], and that both the ground pine and prince's pine were species of *Lycopodium*, as that genus was circumscribed at the time.[18]

In preparing specimens for her own herbarium, Alice Hollingworth followed the usual procedures then and now used by professional as well as amateur botanists. The plants were neatly pressed and dried, showing the diagnostic floral and vegetative characters, then mounted on paper, with the labels in the lower right. The herbarium sheets on which she mounted the specimens are in the size range that had by 1890 become standard in North American and European institutional herbaria, or more precisely in this case,

16¼ × 11 in (91.0 × 28.3 cm). These large sheets of good quality, relatively heavy paper would not have been conveniently – or cheaply – available even in Toronto. This paper, and the printed "Rose Hill Herbarium" labels, must have required considerable effort and a significant expense to acquire and provide further evidence of her enthusiasm and dedication to botanical studies. The data, however, are less detailed than would now be desired, especially with the specimens from her own herbarium, in which the provenance is usually stated no more precisely than "Muskoka." As some of the specimens do bear the names of the Muskoka localities Bracebridge, Mary Lake, and Windemere, it is likely that those with the provenance given only as "Muskoka" were collected in the vicinity of Beatrice. The data on the specimens collected for the Canadian Institute and now in the Royal Ontario Museum herbarium are somewhat better, giving the localities as "Beatrice" or in a few cases "Falkenburg" or "Muskoka Township."

Alice Hollingworth (married surname Webster) was born 2 October 1870 in a log cabin in Muskoka District, Ontario, where the community called Beatrice would develop. She was the ninth of the eleven children of John and Betty Clare (née Bury) Hollingworth, of whom nine survived beyond early childhood.[19] In the words of her daughters Alice Webster Emmett and Beatrice Webster James, "She knew the rigors of pioneer life while her parents, brothers and sisters carved a farm out of woods and rock. She developed a great love for the growth and beauty about her."[20]

The senior Hollingworths had come to Canada from Ashton-under-Lyne, England, where John Hollingworth had been a coal miner. They arrived in Ontario, with their older children, in 1861. After seven years in Toronto, where John Hollingworth was a photographer and dealer in photographic supplies, they went to Muskoka in 1868, when the "colonization roads" were being constructed and settlers were being sought for the lands newly made available for farming.[21] In that year Watt Township and adjacent Stephenson Township were opened for settlement. The Hollingworth farm, acquired under the newly passed Free Grants and Homestead Act, consisted of Lot 4, Concession 1, Watt Township.[22] This homestead, which they called Rose Hill Farm, was located in the present-day Township of Muskoka Lakes, on the north side of Beatrice Town Line Road, about one kilometre west of Manitoba Street/Raymond Road (the name changes at the intersection with Beatrice Town Line Road) or Muskoka Road 4, which approximately follows the route of the southern part of the Parry Sound Colonization Road. In 1876 and 1877, the crops and livestock raised on this farm included white beans, buckwheat, flax, oats, peas, potatoes, rye, swede turnips [rutabagas], wheat, maple sugar, cattle,

6.2
A specimen of Eastern Wild Columbine (*Aquilegia canadensis* L.) collected by Alice Hollingworth in Muskoka District.

and sheep.[23] The present house and other structures have replaced the pioneer cabin, but the property is still in the Hollingworth family at the time of this writing, although it is now a "getaway" residence rather than a working farm. Daffodils planted by Alice Hollingworth in the 1890s still flower each spring along the path from the house to the mailbox.

The natural vegetation in the vicinity of Rose Hill Farm consists largely of northern hardwoods mixed with spruce and jack pine, with tamarack dominant in some of the wetlands. Much more of the land was used for agricultural purposes in the late nineteenth century than at present, but wetlands, rock outcrops, and steep hills prevented some of the land from ever having thus been used.

Beatrice, where Alice Hollingworth grew up, is about thirteen kilometres north-northwest of Bracebridge, near the present-day intersections of Raymond Road with Beatrice Town Line and Camel Lake roads. An 1879 map shows a school, two churches, and a post office in Beatrice. In 1910, Beatrice had an Anglican church, a post office, and thrice-weekly stage service, and soon thereafter had its own telephone company. Otherwise there were only scattered farms, each of which, in most cases, occupied one of the lots of ca. 100 acres (65 ha) granted under the act mentioned above. The first school at Beatrice, like those in most of the rural communities and townships in Muskoka, was built in the 1870s.[24] Alice Hollingworth, according to family tradition, first attended school in 1880, the year she turned ten, and completed the eight grades in four years. In 1884 she was awarded a book, which is still in the family, for her "diligence."[25]

Despite their remoteness from major cities and the limited educational opportunities, the Hollingworths were remarkably active in scientific pursuits. John Hollingworth, Alice's father, was described as "being naturally of a scientific turn," a man who "used his mentality in guiding his efforts and was ever an experimental farmer."[26] He was an associate member of the Astronomical and Physical Society of Toronto, and in 1907 contributed several years' data on precipitation in the Districts of Muskoka, Parry Sound, Nipissing, and Algoma to that society, relating those records to the rapid deforestation that was occurring in those districts.[27] That society, and its successor, the Royal Astronomical Society of Canada, recognized John Hollingworth as a "veteran aurora observer" and noted his correspondence on auroral displays and the sounds associated with them.[28] He served as president of the Farmers' Institute in Muskoka and reported periodically on central Muskoka farm conditions and productions to the Ontario Department of Agriculture, which published this correspondence in the department's *Annual Report*. Three generations of the Hollingworth family contributed climatological data from Beatrice to the Canadian Meteorological Office beginning in 1876 and continuing for more than a century.[29] A meteorological data station, now automated, still exists at Beatrice, on the next property west of Rose Hill Farm.

Samuel Hollingworth, Alice's eldest sibling, had moved to Toronto upon reaching adulthood where he was at first a clerk, then an assistant, and ultimately the proprietor of a homeopathic pharmacy. (The company is still in business at the time of this writing, as Thompson's Homeopathic Supplies.) Alice, while in Beatrice, collected medicinal herbs for the pharmacy, presumably thereby gaining some financial support for her intellectual pursuits. Samuel was a member of the Canadian Institute and its Biological Section and had been the corresponding secretary of the Natural History Society of Toronto prior to that society becoming the Biological Section of the Canadian Institute in 1885. He spearheaded the movement to establish the institute's botanical sub-section in 1890 and served as its chair. He was a botanical collector himself, who contributed specimens to the herbarium of the Canadian Institute's Biological Section. Samuel Hollingworth encouraged Alice's interest in natural history, sent her books, and introduced her to the Canadian Institute and its Biological Section, into which she was accepted as an associate member in 1892, when she was twenty-two years old.[30]

The Canadian Institute (now the Royal Canadian Institute for Science), founded in 1849, was based in Toronto and drew most of its regular membership from that city. For a time, it included several sections, including the Biological Section mentioned above.[31] The Canadian Institute breached Alice Hollingworth's isolation in Beatrice and provided her with an avenue for communication with the amateur and professional scientific community in Ontario's larger cities. Although she was already keenly interested in natural history, especially botany, through the Canadian Institute she met and communicated with several individuals who shared this interest and would provide significant encouragement and guidance. Among these correspondents, the best known to history was Dr William Brodie, of Toronto, who had been president of the Natural History Society of Toronto and chaired the Biological Section of the Canadian Institute. Brodie, a dentist, had wide-ranging interests in biology, geology, and philosophy and amassed large collections of plant and animal specimens. He was called "an inspiration and a wise counselor" to naturalists in the Toronto area.[32] He visited the Hollingworths in Beatrice, and Alice Hollingworth visited the Brodies in Toronto. Brodie's letters to her, written while she was still in Beatrice, included, among other topics, philosophical discussions attempting to reconcile evolutionary concepts with traditional

Christian beliefs about the origin of life and biodiversity. Another noted correspondent in the 1890s was Dr Albert Durrant Watson, a Toronto physician, astronomer, poet, and philosopher. A letter to Alice Hollingworth from Watson[33] suggests that she herself contemplated writing a book on philosophical and/or religious topics in relation to natural history, but no manuscript or other evidence of such an aspiration was seen in this study. Botanists with whom she corresponded included James Fletcher, of Ottawa, the Dominion entomologist and botanist. Another was John Macoun, to whom she sent mosses while he was working on the bryological content for his *Catalogue of Canadian Plants*.[34]

The Canadian Institute member who had the most significant role in Alice Hollingworth's life, in the present context and otherwise, was Charles William Armstrong, secretary of the Biological Section. He was a second-generation member of the institute, the son of Charles Heron Armstrong, who would later be the curator of the Biological Section's collections. Charles W. Armstrong was only 189 days older than Alice Hollingworth, and a close friendship developed. To her, he was "Charley," and he addressed his letters to "Alice," or occasionally "Al" or his private nickname for her, "Cassandra." He encouraged her interest in the native flora and botanical collecting, offering her "all the assistance possible," and gave her detailed instructions for the preparation of quality herbarium specimens. (One of these instructions was that the specimens be accompanied by detailed data on localities and habitats, but, as noted above, she did not follow this advice as fully as one might wish, in that locality data are generally imprecise and information on habitats is lacking.) Armstrong visited the Hollingworths in Beatrice and joined Alice in collecting plant specimens there. He also requested that she collect certain species for the Biological Section's herbarium.[35]

According to Hollingworth family tradition, Alice and Charles may have wanted to marry, but, according to Alice's granddaughter Helen Blackburn, a "class difference" thwarted any such aspirations.[36] Charles W. Armstrong moved to Vancouver ca. 1905, where, after several years as a travelling salesman in pharmaceuticals, he operated Van Alpine Gardens, a nursery specializing in succulents and alpine plants, and acquired a reputation in horticultural circles for his knowledge of cacti.[37] Charles and Alice continued to correspond after his move to Vancouver, and although his letters may have become less frequent, he compensated with their content, each of which ran to about twenty-four pages of news items and chatty comments on diverse subjects, accompanied by photos of cacti, which Alice proudly displayed.[38]

The specimens collected by Alice Hollingworth in Muskoka District that are now at the University of Guelph date from 1890 to 1896. Her collections during the next two years were, with one possible exception, from other Ontario localities. The earliest of her collections now at the Royal Ontario Museum likewise date from 1890, but it seems likely that she had begun contributing specimens to the Canadian Institute at least a year or two earlier, while in her late teens. By 1890, the members of the Canadian Institute had already become so impressed by her enthusiasm and skill in discovering and identifying plants of interest that the institute,[39] at Brodie's recommendation, presented her with a copy of the newly published sixth edition of *Gray's Manual of Botany.*[40] *Gray's Manual*, through its successive editions, was, and would long remain, the primary standard reference for plant taxonomy and identification in northeastern North America. It would have been of immense value to Alice Hollingworth for identifying plants and would have provided her with their current classification and up-to-date nomenclature.

In the report of the Biological Section for 1890–91, Charles W. Armstrong,[41] who was at that time the secretary/curator, noted that most of the section's botanical specimens from outside the vicinity of Toronto had been collected in northern Ontario and contributed by Alice Hollingworth. In the following year's report[42] she was further recognized for "great assistance … in collecting specimens and information on the 'flora' of [Muskoka] District." In 1892 Alice Hollingworth presented an address at a meeting of the Biological Section, entitled "Scientific researches in rural districts."[43] This was the only presentation by a woman at any meeting of the institute or its sections that year. Regrettably, it was not published, nor does a manuscript version exist. In 1896, she contributed ninety-seven specimens from "the Northwest" to the Biological Section's herbarium.[44]

When the Canadian Institute discontinued its division into sections, Brodie, Charles W. Armstrong, and others attempted to perpetuate its Biological Section by founding the Biological Society of Ontario.[45] When this short-lived society launched its *Biological Review of Ontario* in 1894 Alice Hollingworth was among the authors in the first (and only) volume, in which, as noted above, she reported four uncommon plants that she had found in Muskoka District.[46] *Gray's Manual* was being put to good use.

From 1896 through 1902 Alice Hollingworth contributed data on plant phenology (flowering dates and other seasonal phenomena) from the Beatrice area to the Botanical Club of Canada, of which she was a member. This club had been organized in 1891 by George Lawson through the Royal Society of

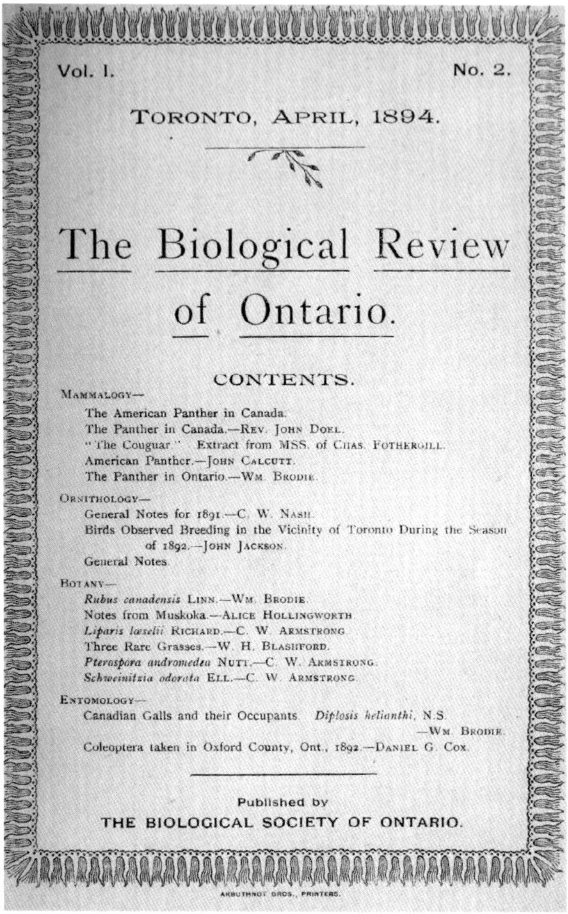

Vol. I. No. 2.

TORONTO, APRIL, 1894.

The Biological Review of Ontario.

CONTENTS.

MAMMALOGY—
The American Panther in Canada.
The Panther in Canada.—REV. JOHN DOEL.
"The Cougar." Extract from MSS. of CHAS. FOTHERGILL.
American Panther.—JOHN CALCUTT.
The Panther in Ontario.—WM. BRODIE.

ORNITHOLOGY—
General Notes for 1891.—C. W. NASH.
Birds Observed Breeding in the Vicinity of Toronto During the Season of 1892.—JOHN JACKSON.
General Notes.

BOTANY—
Rubus canadensis LINN.—WM. BRODIE.
Notes from Muskoka.—ALICE HOLLINGWORTH.
Liparis læselii RICHARD.—C. W. ARMSTRONG.
Three Rare Grasses.—W. H. BLASHFORD.
Pterospora andromedea NUTT.—C. W. ARMSTRONG.
Schweinitzia odorata ELL.—C. W. ARMSTRONG.

ENTOMOLOGY—
Canadian Galls and their Occupants. *Diplosis helianthi*, N.S. —WM. BRODIE.
Coleoptera taken in Oxford County, Ont., 1892.—DANIEL G. COX.

Published by
THE BIOLOGICAL SOCIETY OF ONTARIO.

ARBUTHNOT BROS., PRINTERS.

6.3 *Right*
Front cover of *The Biological Review of Ontario* 1, no. 2 (1894), listing a paper by Alice Hollingworth.

6.4 *Opposite*
Excerpt from *The Naturalists' Directory*, 1898 edition, showing Alice Hollingworth's entry. C indicates that she was a collector; Ex., that she was interested in exchanges; and *, that the compilers had heard from her since the preceding edition was published.

Canada, originally with the primary objective of promoting floristic studies throughout Canada, although the compilation of phenological records soon came to dominate its programs. At the instigation of the club's general secretary, Alexander Howard MacKay, superintendent of education for Nova Scotia, the members were annually supplied with standard lists of events, mostly the first flowering of plant species but also some zoological and climatological phenomena, for which they supplied the dates on which these events occurred in their respective areas.[47] After Alice Hollingworth moved to the vicinity of Creemore, Ontario, data from Beatrice were contributed by her father until the Botanical Club's dissolution in 1910.[48] In recent years, widespread concern about the effects of climate change has led to a resurgence of interest in phenology, and new networks of "citizen scientists" are being recruited to record such data. Like similar data recorded by Henry David Thoreau in Massachusetts in the 1850s, these historic records from across Canada, compiled by MacKay

NATURALISTS' DIRECTORY. 93

2313 Holder, Chas. F., LL.D. Pasadena, Calif. *Zool., Gen. Biol., Arch.*
2314 Holder, Mary H., Berlin, Mass. *Bot.* C.
2315 Holderman, D., Girard, Kansas. *Ethnol., Anth.?*
2316 Holland, Frank, Manchester, N. H. *Mic., Veg. Histol., Protozoa.* Polycistina raw material from all parts of the world desired. C.
2317 Holland, Chancellor W. J., D.D., LL.D., Ph.D., 5th and Bellefield Ave., Pittsburg, Pa. *Biol., Bot., Lepid. of world, Orn.* Large collection, including entire coll. of Mead, W. H. Edwards, Knyvett and Pryer (Japanese Pyrals, etc.) Desires to buy insects from all parts of the world, esp. Africa. Also rare books on Lepid.*
2318 Hollick, Arthur, Ph.B., Columbia, Coll., New York, N. Y. *Bot., Palæo.* C.*
2319 Hollingworth, Alice, Beatrice P. O., Muskoka, Ont., Canada. *Bot., Geol.* C. Ex.*
2320 Hollingworth, Sam'l, 394 Yonge St., Toronto, Ont., Canada. *Bot., Mic.*
2321 Hollis, Frederick S., Ph.D., Newton Highlands, Mass. *Min.*
2322 Holm, Theo., Ass't Pathologist, U. S. Dept. of Agric., Washington, D. C. *Bot: Monocotyledones.* C. Ex. Cyperaceæ.
2323 Holman, David S., Acad. Nat. Sciences, Philadelphia, Pa. *Mic.*
2324 Holman, Silas W., S.B., Prof. of Physics, Mass. Inst. Tech., Boston, Mass.*

and now at the Nova Scotia Museum of Natural History, constitute valuable baseline data, permitting comparison of the dates at which such events occur now with the dates at which the same events occurred in the past.[49]

Alice Hollingworth participated in exchanges of botanical specimens with other naturalists during the 1890s. She was listed in several editions of *The Naturalists' Directory*, a more or less annual publication, of which a major function was to facilitate exchanges of specimens among collectors in many fields of natural history in many countries. She listed her interests as botany and geology, and indicated that she was a collector and was interested in exchanging specimens.[50] Among those in Canada with whom she exchanged specimens, those best known to Canadian botanical history included William Copeland McCalla, then living in St Catharines, Ontario, who contributed specimens from the Niagara Peninsula; William Spreadborough, who contributed specimens from Saskatchewan and Alberta, where he had explored with John Macoun; and the Rev. Arthur Charles Waghorne, whose specimens from Newfoundland included mosses and a few lichens as well as vascular plants.[51] For several years she corresponded and exchanged specimens with Helen Una Grieve in New South Wales, Australia. Helen Grieve was "delighted with [the] beautiful Canadian wildflowers," and, in response, Alice Hollingworth sent her seeds of sixty species.[52] The provenance of some of the specimens from her herbarium indicates that she also exchanged specimens with collectors in the United States, England, Switzerland, and New Zealand.

Material preserved by Hollingworth/Webster family members include manuscript essays and similar writings by Alice Hollingworth, mostly on botanical and agricultural topics. Those in notebooks were probably written for her own purposes, to record and organize both her own observations and what she had learned from reading. Other essays may have been written for her column in *The Farmers' Sun*, discussed below, and still others may have been prepared for talks that she gave to meetings of the Women's and Farmers' Institutes. All or most of these writings appear to date from after 1900, and none of them include accounts of her botanical explorations in Muskoka. The correspondence includes letters that Alice Hollingworth received, notably those from Armstrong, Brodie, and Watson, but none of her own letters or copies of her letters have been located in this study, despite searches of indexes to the archived papers of some of her correspondents.[53]

Correspondents praised Alice Hollingworth for her skill as a botanical artist, who worked with pen and ink. Most of her drawings were sent with letters and specimens, but some remain in the family archives, illustrating a section on weeds within a notebook on various botanical topics.

As the nineteenth century approached its close, a series of events began that would alter many aspects of Alice Hollingworth's life. In 1899 she was twenty-nine years old. If she had indeed dreamt of life as Mrs Charles Armstrong, those dreams had come to naught. Three of her siblings had gone to Toronto, where her brother Samuel was advancing to increasingly responsible positions in the pharmacy and two of her sisters were employed as nurses. Her extant writings contain nothing about her emotions and private thoughts, but one may well imagine that she felt increasing concern about her future. Visits with relatives and others who shared her interest in botany, what she heard from her older siblings, and her attendance at meetings of the Canadian Institute had provided glimpses of an attractive world, with expanded opportunities, beyond the Muskoka interior.

In January through March 1900 Alice Hollingworth took the winter short course in dairying at the Ontario Agricultural College in Guelph (one of the precursors of the University of Guelph). Dairy processing at that time was generally conducted in farm homes and was considered a part of the usual responsibilities of Ontario farm women.[54] This home dairy course was the only course at the OAC open to women at the time. The principal instructor was Laura Rose, supervisor of the Home Dairy Department, who, after graduating from a similar course at the college in 1893, had been hired as the college's first female instructor. Participation in and graduation from the dairy course at the Ontario Agricultural College made Alice Hollingworth aware of

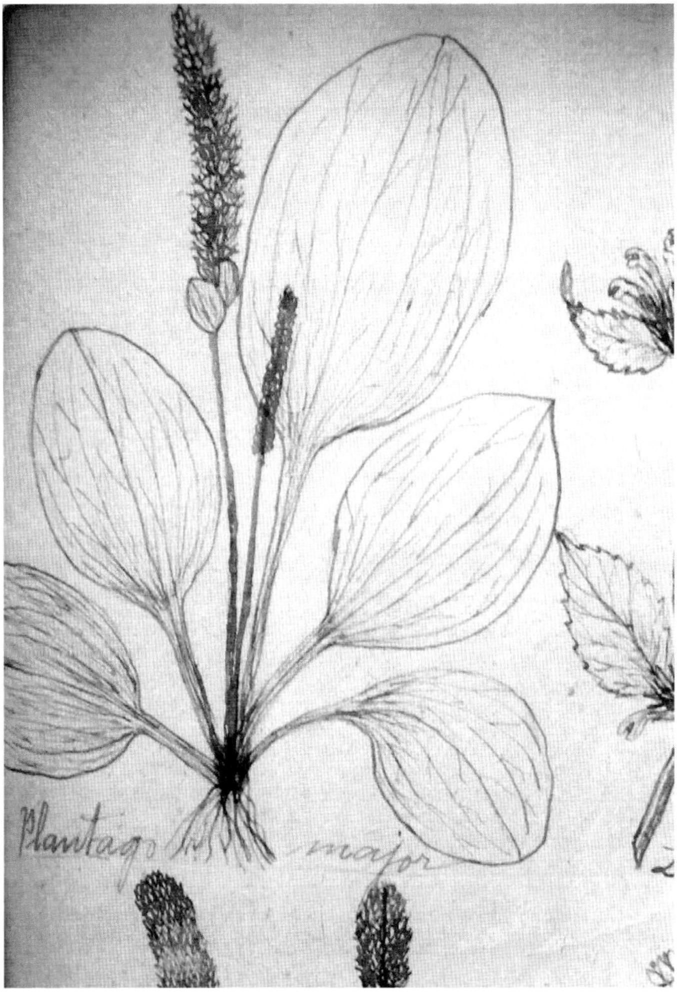

6.5
Drawing of Common Plantain, *Plantago major* L., by Alice Hollingworth. Original in the Hollingworth/Webster family papers.

employment opportunities and provided her with enhanced credentials that would be relevant to such opportunities. Following her graduation from this course, she was the first of five women hired by the Ontario Department of Agriculture as organizers and speakers for the Women's Institutes of Ontario.

Although her participation in the dairy course brought her abilities to the attention of Laura Rose, who was a leader in the formation of the Women's Institutes, it is unlikely that Alice Hollingworth had enrolled in that course with the specific objective of employment on behalf of the Women's Institutes. The concept of the Women's Institutes was so new in 1900 that the position for which she was hired following her graduation may not have been fully envisioned, or not yet announced publicly, when she applied for admission to the course. The Women's Institutes were founded in 1899 with the primary objectives of providing education, both in the school curricula and continuing, in the domestic sciences and arts that the institutes' founder, Adelaide Hunter Hoodless, considered appropriate for women at the time, especially for those in rural communities. Initially, the Women's Institutes were affiliated with the Farmers' Institutes, which had been established in 1892, and were within the purview of the Ontario Department of Agriculture, from which they received substantial support. Later they became independent of the Department of Agriculture and greatly diversified their programs.[55] Alice Hollingworth is credited with having founded, on her own, eighteen of the first forty-four branches.[56] Her efforts, it may be noted, were not met without some resistance from men, one of whom told her "Don't bring the Women's Institute here, or we won't be able to get a woman to milk a cow."[57]

Even before she was employed as an organizer and speaker, Alice Hollingworth had addressed provincial meetings of the Farmers' Institutes, at which both men and women were generally among the speakers. Her first address to a provincial convention of delegates from the Farmers' Institutes and the newly formed Women's Institutes, presented in 1899 and published the following year,[58] dealt with a subject that has become very much of interest to horticulturists and ecologists in the twenty-first century. Drawing from her own experiences at Rose Hill Farm, she promoted the use of native plants in making the environs of farm homes more attractive. "Poverty cannot be pleaded" as an excuse for the dreary-looking environs of some farm homes, she said, "because in the country Nature takes the place of money and provides all that is necessary to make beautiful surroundings." Many species of native trees, shrubs, vines, and herbaceous plants received her praise, with the observation that "Nature has planted with a lavish hand, and it is a shame for any home to be destitute of these attractions." She also recommended the use of native plants in indoor fresh and dried arrangements and discussed arbours, shelters, benches, planters, and other garden structures that could be made from native plant materials. The following year she spoke on butter making.

The use of native species in home landscaping was probably more of a novel idea for those in the audience than might immediately occur to us in the twenty-first century. English and Scottish horticulturists had for some time been interested in the introduction of ornamental species from North America – such plant explorers as David Douglas, Thomas Drummond, John Fraser, and James McNab come to mind – but in rural areas, for immigrants to Canada from the British Isles or the European continent, plants from the "old country" would have cushioned the feeling of separation from what had been familiar and would have represented a component of "civilization" where its amenities must have seemed sparse. Even among those born in Canada there was a tendency to think of native plant species, free for the taking, as "weeds," deeming only plants of exotic origin or purchased from commercial sources in the cities to be worthy of cultivation as ornamentals. And few rural Canadians could afford such luxuries. Alice Hollingworth realized that native plant species could contribute to the beautification of the grounds of the farm home without the expense required for obtaining exotic species and with no question as to their hardiness. Moreover, she appreciated the benefits that attractive landscaping could bring to farm families, by giving visitors and travellers better first impressions of the inhabitants of the farm homes.

Invasive species – exotic species, in many cases introduced as ornamentals, that may spread into natural areas and become abundant, competing with native species, altering ecosystems and reducing biodiversity – had not yet become a widespread concern in 1899, but some of the plants she recommended, including native species of ferns, roses, hawthorns, plums, maples, viburnums, and the wild clematis or virgin's-bower shown here, are now being promoted as alternatives to invasive exotic ornamentals.[59] A manuscript in the Hollingworth/Webster family archives, entitled "An all-Canadian garden," suggests that there may have been other occasions when Alice Hollingworth spoke or wrote articles emphasizing the attractiveness and landscaping value of native species.

On Christmas Day, 1902, on the occasion of her parents' golden wedding anniversary and a family reunion at Rose Hill Farm, Alice Hollingworth married Francis Ernest "Frank" Webster, of Creemore, Simcoe County, Ontario. Frank was president of the Farmers' Institute in Creemore, and Alice had met him while at an organizational meeting of the Women's Institute in Creemore in February of that year. Following their marriage, Frank Webster built a home for Alice, himself, and the children they anticipated in Websterville, a rural community a short distance west of Creemore, which had been founded by Frank's forebears. Alice and Frank Webster had four children, Alice Esther (called Allie), Francis Gordon (called Gordon), Herbert Goldwyn, and Beatrice Mary.[60]

Alice Hollingworth Webster's plant collecting was only one, although probably the most significant, manifestation of her interests in natural history and her penchant for collecting. The Webster home contained many cabinets and drawers housing collections that, her daughters recalled, also included "sea

6.6
Wild Clematis or Virgin's-bower (*Clematis virginiana* L.), an attractive native vine recommended by Alice Hollingworth for use in home landscaping.

shells, minerals, curios and pioneer articles." "Adults and children ... spent many a happy afternoon in her home looking and listening to the history of each article."[61]

Because of her concerns for social reforms, briefly mentioned below, and her advocacy for improvement in living and working conditions for Ontario farm families, the story of Alice Hollingworth Webster's life gains in interest in the broader context of Canadian and Ontario history after 1900, but there is less that is relevant to the botanical theme of this volume. Even before her marriage her involvement with the Canadian Institute had declined, as the

Botanical Sub-section no longer existed. There were fewer opportunities to report – and be recognized for – botanical discoveries, and, after a decade or more of botanizing in Muskoka, she was probably finding fewer rare species that she had not already encountered. Meanwhile, her employment as an organizer and speaker for the Women's Institutes demanded her time and attention.

Nevertheless, Alice Webster's scholarly interests did not so much wane after her marriage and move to the Creemore area as shift in emphasis toward archaeology. The vicinity of Creemore and Collingwood had been the homeland of the Petun or Tionontati Nation during the sixteenth and seventeenth centuries and had already attracted the attention of archaeologists. Alice Webster had become interested in the relics of earlier, Indigenous habitation while she was in Beatrice. She had met and corresponded with David Boyle, an archaeologist and curator of the Ontario Provincial Museum (one of the precursors of the Royal Ontario Museum) and had contributed Indigenous artifacts to the Canadian Institute's museum. Prior to the Websters' marriage, Frank Webster and others in the Creemore area had already become greatly interested in the work of visiting archaeologists. Following their marriage, he, Alice, and later their children and grandchildren contributed local knowledge of archaeological sites and otherwise assisted the archaeologists.[62]

Alice Webster retired from her roles as an organizer and speaker for the Women's Institutes following her marriage, but as one history of the institutes expressed it, "she would continue throughout her life to be an extremely active and beloved member of the organization."[63] Participation in the Ontario Federation of Agriculture and the United Farmers of Ontario enabled both Alice and Frank Webster to take more active roles in promoting improved conditions for farmers and other social reforms.[64] Alice Webster was a founding member and provincial officer of the United Farm Women of Ontario, an auxiliary group of the United Farmers, founded in 1918. After the organization of the Federated Women's Institutes of Canada in 1919, she was also active in that group, which devoted much of its activity to lobbying.[65] As she had been when she spoke on using native plants in landscaping the rural home, she continued to be concerned about the image of farmers and their families among people in other walks of life. Other causes for which she worked included women's suffrage, rural electrification, sex education, and more attention to Canada in the teaching of history in the public schools. She was interested in the history of Muskoka and the Creemore–Collingwood area, on which she wrote articles for the local newspapers.[66] A lasting interest in the native flora was represented in a series of articles collectively entitled

"Wayside Plants," which she wrote for *The Farmers' Sun*, a newspaper published by the United Farmers of Ontario.[67]

Alice Webster served a term as president of the Creemore Horticultural Society, which was founded in 1921. She participated in flower shows and judged both floral and vegetable entries. She also served as a judge at rural school fairs sponsored by the Ontario Department of Agriculture. Her abiding interest in plants found further expression in her flower garden, with a rose bed, a peony border, a rock garden, and many herbaceous perennials, which became a showplace in which she conducted tours.[68]

Alice Hollingworth Webster died 28 January 1954. The minister at her memorial service noted "her interest and activity in local organizations, particularly in the Women's Institute and the Horticultural Society."

Alice Hollingworth was in several respects a pioneer in her botanical pursuits. In the interior of Muskoka, she explored the flora in a part of Ontario that was remote from the major population centres of the time and had not previously been explored floristically. She gave attention to grasses, sedges, and species with inconspicuous flowers, plant groups that had largely been neglected as being too "difficult" by earlier amateur plant collectors in Canada. In addition to documenting the occurrence of species, she was interested in other aspects of plant study. And she contributed directly to organized botanical knowledge in Canada through communicating with Canadian botanists, reporting notable discoveries, contributing specimens to institutional herbaria, and participating in Canadian scholarly societies, which in the 1890s were still largely dominated by men.

Alice Hollingworth's application of her love for and knowledge of the local flora to making rural homes more attractive was one of many aspects of her work toward better living and working conditions on Ontario farms and greater respect for farmers. Her employment on behalf of the Women's Institutes, her marriage, her move from Muskoka to the vicinity of Creemore, the birth of her children, and other changes affected which specific pursuits were favoured by the circumstances of the times, but her diverse accomplishments were manifestations of an integrated whole. Her contributions to archaeology in the Creemore area were essentially a continuation of her earlier pursuits in botany in Muskoka, both representing an intellectual interest in her environment. Likewise, her participation in the scientific and agricultural societies and her advocacy of innovations in education and suffrage for women were among the manifestations of her desire to give women more opportunities to contribute to social progress.

ACKNOWLEDGMENTS

I am very grateful to Helen Emmett Blackburn, of Creemore, Ontario, granddaughter of Alice Hollingworth Webster, for the opportunity to study material in the Hollingworth/Webster family archives, compiled by herself and other descendants of John and Betty Hollingworth, for providing the portrait of her grandmother, and for information that she provided through emails and conversation. I am also grateful to Lois James-Chételat for excerpts from her manuscript *Alice's Daughters*. I am very grateful to Carole Ann Lacroix, of the University of Guelph, for data on the specimens from Alice Hollingworth's herbarium now at Guelph and for scans of specimens from this collection. I greatly appreciate the efforts of Dr James P. Goltz, of Fredericton, New Brunswick, in compiling and providing data from the records compiled by R. Emerson Whiting. I am pleased to have had access to archival resources at the Bracebridge (Ontario) Public Library. The library of the Royal Botanical Gardens (Canada) and the online Biodiversity Heritage Library were also important sources of information in this study.

NOTES

1 In botanical history the terms "botanical exploration" and "floristic exploration" are used routinely in reference to field work where the flora was not well known or well documented, with the objective of determining, recording, and documenting the plant species occurring naturally there. They carry no implication of physical perils and privations or remoteness from modern amenities.

2 On the early history of botanical collecting and floristic studies in Ontario see Pringle, "Botanical Exploration" and "History of the Floristic Exploration of Canada."

3 James P. Goltz, pers. comm. 26 April 2017.

4 Macoun, *Catalogue of Canadian Plants*.

5 Macoun *Autobiography*, 42–3.

6 Hooker, *Flora Boreali-Americana*.

7 Penhallow, "Review of Canadian botany"; Pringle, "History of the Floristic Exploration of Canada."

8 Zeller, *Inventing Canada*; Pringle, "Botanical Exploration" and "History of the Floristic Exploration of Canada"; Keeney, *The Botanizers*.

9 On the botanical exchange networks of this period, with emphasis, respectively, on their operation in the United States and in the British Isles, and in both papers noting the prominent role of women, see Rudolph, "Women in Nineteenth Century American Botany"; Groom et al., "Herbarium Specimens."

10 Pringle, "Botanical Exploration"; Crowder and Taylor, "Mrs. Traill, Mrs. Roy, & Miss Boyd."

11 Logie, "List of Plants Found Growing in the Neighborhood of Hamilton"; David A. Galbraith, pers. comm. 27 September 2017 (in Royal Botanical Gardens archives). On Kate Crooks, see the chapter by David Galbraith in this book.

12 Crowder and Taylor, "Mrs. Traill, Mrs. Roy, & Miss Boyd."

13 C.H. Armstrong, "Report of Curator of the Biological Section"; Hollingworrth/Webster family papers. The latter citation, here and elsewhere in this chapter, refers to assorted correspondence, manuscripts, notebooks, typescripts, photographs, and other unpublished material privately preserved by members of the family, primarily by Helen Blackburn, at whose home in Creemore, Ontario, these items were seen in this study.

14 Later additions to Alice Hollingworth's herbarium included specimens collected by others, which she had acquired through exchanges, noted elsewhere in this chapter. Her son Herbert G. Webster contributed specimens collected in Nottawasaga Township (now Clearview), Ontario, in 1938 and near Woodstock, Ontario, in 1944, these being the most recent additions to the herbarium.

15 James P. Goltz, pers. comm. 11 September 2017.

16 A. Hollingworth, "Notes from Muskoka."

17 James-Chételat, "Alice's Daughters," unpublished manuscript in Ms James-Chételat's possession, excerpts from which were made available for this study.

18 A. Hollingworth, "The Use of Native Trees and Plants."

19 Hollingworth/Webster family papers.

20 Emmett and James, "Mrs. Alice Webster."

21 Spragge, "Colonization Roads in Canada West"; Stewart, "Land Registration in Muskoka."

22 Canada census records, Watt Township (microprint, Archives of Ontario).

23 From the journal of John Hollingworth, in Kirkwood and Murphy, *Undeveloped Lands in Northern & Western Ontario*.

24 Denniss, *A Brief History of Schools in Muskoka*.

25 Helen Blackburn, pers. comm., 11 July 2018.

26 Anonymous, "Death of an aged pioneer. Mr. John Hollingworth, Sr., passed away, aged 94 years, 6 months." Photocopy of a newspaper clipping in the Muskoka Collection at the Bracebridge (ON) Public Library, dated 30 October 1924, probably from *The Bracebridge* [ON] *Gazette*.

27 J. Hollingworth, "Precipitation."

28 Keay, "C.A. Chant and the Mystery of Auroral Sounds."

29 Anonymous, "A century of Weather Observations at Beatrice, Ontario 1876–1975," Ottawa: Environment Canada, Atmospheric Environment Service 1975. (Pamphlet;

includes material from *The Herald-Gazette* [Bracebridge, ON] and *News on the DOT* [Canada Department of Transport]); Britton, "Hollingworth Family Completes a Century of Service."

30 Toronto city directories; Pearce and Armstrong, "Report of the Biological Section of the Canadian Institute"; Boivin, "Survey of Canadian Herbaria"; Helen Blackburn, in conversation 30 August 2017.

31 Wallace, *The Royal Canadian Institute.*

32 Anonymous, "Dr. William Brodie," *The Globe* [Toronto], 21 August 1909, A6; Zeller, "William Brodie"; Littlefield, *Tom Thomson's Toronto Neighbourhoods.*

33 Hollingworth/Webster family papers.

34 James-Chételat, "Alice's Daughters"; Hollingworth/Webster family papers.

35 Hollingworth/Webster family papers.

36 Helen Blackburn, in conversation 30 August 2017. Toronto city directories for 1892 et seq. listed Charles W. Armstrong as a clerk and later a foreman at the Canadian division of Evans & Sons, wholesale druggists and manufacturing chemists, and Charles H. Armstrong as a carpenter.

37 Anonymous, "C.W. Armstrong, Canadian collector," *Cactus and Succulent Journal* (Los Angeles) 23 (1951): 12; Pringle, "Botanical Exploration."

38 Hollingworth/Webster family papers; Helen Blackburn, in conversation 30 August 2017.

39 James-Chételat, "Alice's Daughters."

40 Gray, *Manual of Botany of the Northern United States.*

41 C.W. Armstrong, "Report of the Biological Section."

42 Pearce and Armstrong, "Report of the Biological Section … 1891–1892."

43 Ibid.

44 C.H. Armstrong, "Report of Curator of the Biological Section."

45 Pringle, "Botanical Exploration."

46 A. Hollingworth, "Notes from Muskoka."

47 Penhallow, "A Review of Canadian Botany"; MacKay, "Phenological Observations"; Boivin, "Survey of Canadian Herbaria" and "Botanical Societies in Canada"; Pringle, "Botanical Exploration" and "The History of Floristic Exploration."

48 Pringle, "Botanical Exploration."

49 Beaubien and Hamann, "Plant Phenology Networks"; Primack and Miller-Rushing, "Uncovering, Collecting, and Analyzing Records."

50 Cassino, *The Naturalists' Directory*, 1898, and other editions published during the 1890s.

51 On these Canadian botanists, see Pringle, "Botanical Exploration" and "The History of Floristic Exploration" and references cited therein.

52 A. Hollingworth, "The Use of Native Trees and Plants"; Hollingworth/Webster family papers.

53 I am grateful to Charlotte Chaffey, of the Royal Ontario Museum, for her search of the papers of William Brodie at that institution. Other searches were made using online resources.

54 Cohen, "The Decline of Women in Canadian Dairying"; Steele, *My Heart, My Hands.*

55 Creelman, "Report of the Superintendent," 1900, 1901; Campbell, "Women's Institute and OMAF link"; Steele, *My Heart, My Hands*; Kechnie, *Organizing Rural Women.*

56 Campbell, "Women's Institute and OMAF link"; Steele, *My Heart, My Hands*; Kechnie, *Organizing Rural Women.*

57 Weber, "Women's Institutes."

58 A. Hollingworth, "Notes from Muskoka."

59 See, for example, Cullina, "Alternatives to invasive or potentially invasive exotic species."

60 Anonymous, "Hollingworth family reunion held on old homestead," *The Bracebridge* [ON] *Gazette*, 30 June 1955, 1; Garrad, "Researching the Petun"; Canada Census records; Hollingworth/Webster family papers.

61 Emmett and James, "Mrs. Alice Webster."

62 Garrad, "Researching the Petun."

63 Steele, *My Heart, My Hands.*

64 On these organizations see Zwerver, "Farmers Working for Farmers" and Badgley, *Ringing in the Common Love of Good.*

65 Steele, *My Heart, My Hands.*

66 Badgley, *Ringing in the Common Love of Good*; Blackburn, Letter to the editor; Hollingworth/Webster family papers.

67 This newspaper, which was published weekly from 1892 to 1934, has not been indexed, and a search of issues from the most likely time period did not locate any of these articles.

68 Hollingworth/Webster family papers; Helen Blackburn, in conversation 6 September 2017.

BIBLIOGRAPHY

Archives

Alice Hollingworth Herbarium, University of Guelph Herbarium.

List of specimens from Muskoka, Ontario, collected by Alice Hollingworth, now in the Royal Ontario Museum Herbarium. Compiled by James P. Goltz; copy deposited in the archives of the Royal Botanical Gardens, Burlington, Ontario.

Hollingworth/Webster Family Papers. In possession of Helen Blackburn, Creemore, Ontario.

Books and Articles

Armstrong, Charles H. "Report of Curator of the Biological Section." In "Report of the Natural History Society (Biological Section of the Canadian Institute) for session 1895–96." *Proceedings and Transactions of the Royal Society of Canada*, ser. 2, 2(Proc.) (1896): L (Roman numeral).

Armstrong, Charles W. "Report of the Biological Section." *Transactions of the Canadian Institute* 2 (1892): 72–3.

Badgley, Kerry A. *Ringing in the Common Love of Good: The United Farmers of Ontario, 1914–1926*. Montreal and Kingston: McGill-Queen's University Press, 2000.

Beaubien, Elisabeth G., and Andreas Hamann. "Plant Phenology Networks of Citizen Scientists: Recommendations from Two Decades of Experience in Canada." *International Journal of Biometeorology* 55 (2011): 833–41.

Blackburn, Helen. Letter to the editor, published with the heading "Helen Blackburn has many ties to this newspaper." *The Creemore* [ON] *Echo*, 22 May 2015, 4; pdf online version accessed 23 August 2016.

Boivin, Bernard. "Botanical Societies in Canada." *The Plant Press* (Mississauga) 2 (1984): 103–6.

– "Survey of Canadian herbaria." *Provancheria* 10 (1980): 1–187.

Britton, Ted. "Hollingworth family completes a century of service." *The Herald-Gazette* (Bracebridge, ON), 4 March 1976, 13.

Campbell, S. Lynn. "Women's Institute and OMAF Link Began 91 Years Ago." *Home & Country* (Federated Women's Institutes of Ontario) 54 (1988): 12–13.

Cassino, Samuel E., compiler. *The Naturalists' Directory: Containing the Names, Addresses, Special Departments of Study, etc., of Professional and Amateur Naturalists, Chemists, Physicists, Astronomers, etc., etc., of the United States and Canada*, ed. 14. Boston: S.E. Cassino, 1898.

Cohen, Marjorie Griffin. "The Decline of Women in Canadian Dairying." *Histoire Sociale/Social History* (Toronto) 17 (1984): 307–34.

Creelman, George C. "The Formation of Women's Institutes." In "Report of the Superintendent of Farmers' Institutes of the Province of Ontario, 1899–1900." In *Annual Report of the Department of Agriculture of the Province of Ontario*, 1899. Vol. II (1900): 134–5.

– "Report of the Superintendent." In "Report of the Superintendent of Farmers' Institutes of the Province of Ontario, 1900." In *Annual Report of the Department of Agriculture of the Province of Ontario*, 1900. Vol. II (1901): 78.

Crowder, Adele, and Vivian Taylor. "Mrs. Traill, Mrs. Roy, & Miss Boyd: Plant Collectors in 19th Century Upper Canada." *Occasional Papers, W.D. Jordan Special Collections and Music Library* 001. Kingston: Queen's University Libraries, 2003.

Cullina, William. "Alternatives to invasive or potentially invasive exotic species." Framingham, MA: New England Wild Flower Society 2003. www.newfs.org/docs/invalt2.pdf (accessed 22 February 2018).

Denniss, Gary. *A Brief History of the Schools in Muskoka*. Bracebridge, ON: Herald-Gazette Press, 1972.

Emmett, Alice W., and Beatrice W. James. "Mrs. Alice Webster Was Active in W.I. Activities." *The Creemore* [ON] *Star*, February 1954.

Garrad, Charles. "Researching the Petun." *Ontario Archaeologist* 89 (2010): 3–57.

Gray, Asa, with revisions and additions by Sereno Watson and John Merle Coulter. *Manual of the Botany of the Northern United States, Including the District East of the Mississippi and North of North Carolina and Tennessee, ed. 6, Revised and Extended Westward to the 100th Meridian*. New York and Chicago: Ivison, Blakeman, and Company, 1890.

Groom, Q.J., C. O'Reilly, and T. Humphrey. "Herbarium Specimens Reveal the Exchange Network of British and Irish Botanists, 1856–1932." *New Journal of Botany* 4 (2014): 95–103.

Hollingworth, Alice. "Notes from Muskoka." *The Biological Review of Ontario* 1 (1894): 40.

– "The Use of Native Trees and Plants for Beautifying the Farm Home." In "Report of the Superintendent of Farmers' Institutes of the Province of Ontario, 1899–1900." In *Annual Report of the Department of Agriculture of the Province of Ontario*, 1899. Vol. II (1900): 155–6.

Hollingworth, John. "Precipitation." *Transactions of the Astronomical & Physical Society of Toronto* 1906 (1907): 4–6.

Hooker, William Jackson. *Flora Boreali-Americana; or, the Flora of the Northern Parts of British America*. London: Henry G. Bohn, 1829–1840.

Keay, Colin S.L. "C.A. Chant and the Mystery of Auroral Sounds." *Journal of the Royal Astronomical Society of Canada* 84 (1990): 373–82.

Kechnie, Margaret C. *Organizing Rural Women: The Federated Women's Institutes of Ontario 1897–1919*. Montreal and Kingston: McGill-Queen's University Press, 2003.

Keeney, Elizabeth B. *The Botanizers: Amateur Scientists in Nineteenth-Century America*. Chapel Hill: The University of North Carolina Press, 1992.

Kirkwood, A., and J.J. Murphy. *The Undeveloped Lands in Northern & Western Ontario*. Toronto: Publisher not specified; printed by Hunter, Rose & Co., 1878.

Littlefield, Angie. *Tom Thomson's Toronto Neighbourhoods*. Toronto: Published by the author, 2016.

Logie, Alexander. "List of Plants Found Growing in the Neighborhood of Hamilton, during the Years 1859 and 1860." *Annals of the Botanical Society of Canada* 1 (1862): 90–108.

MacKay, A.H. "Phenological Observations of the Botanical Club of Canada, 1900." *Proceedings and Transactions of the Nova Scotian Institute of Science* 10 (1902): 379–84.

Macoun, John. *Autobiography of John Macoun, M.A., Canadian Explorer and Naturalist.* Ottawa: The Ottawa Field-Naturalists' Club, (1922), 1979.

– *Catalogue of Canadian Plants.* Montreal (parts I–VI), and Ottawa (part VII): Geological and Natural History Survey of Canada 1883–1902.

Pearce, James H., and Chas. W. Armstrong. "Report of the Biological Section of the Canadian Institute, 1891–92." *Transactions of the Canadian Institute* 3 (1893): 56–8.

Penhallow, D.P. "A Review of Canadian Botany from 1800 to 1895." *Proceedings and Transactions of the Royal Society of Canada*, ser. II, 3(IV) (1897): 3–56.

Primack, Richard B., and Abraham J. Miller-Rushing. "Uncovering, Collecting, and Analyzing Records to Investigate the Ecological Impacts of Climate Change: A Template from Thoreau's Concord." *BioScience* 62 (2012): 170–81.

Pringle, James S. "Botanical Exploration of the Canadian Watershed of Lake Huron during the Nineteenth Century." *Canadian Horticultural History* 2 (1989): 4–88.

– "The History of the Floristic Exploration of Canada." *The Canadian Field-Naturalist* 109 (1996 ["1995"]): 291–356.

Rudolph, Emanuel D. "Women in Nineteenth Century American Botany: A Generally Unrecognized Constituency." *American Journal of Botany* 69 (1982): 1346–55.

Spragge, George W. "Colonization Roads in Canada West, 1850–1867." *Ontario History* 59 (1957): 1–17.

Steele, Betty. *My Heart, My Hands: A Celebration of Volunteerism in Canada.* Toronto: Stoddart Publishing, 2000.

Stewart, R. Craig. "Land registration in Muskoka – 127 years of service." *The Ontario Land Surveyor Quarterly*, Winter 1966: 12–13.

Wallace, W. Stewart, ed. *The Royal Canadian Institute: Centennial Volume 1849–1949.* Toronto: The Royal Canadian Institute, 1949.

Weber, Ken. "Women's Institutes: for Home and Country." *In the Hills*: original vol. and pp. not seen; seen online at http://www.inthehills.ca/2009/11/back/womens-institute/ (accessed 2 May 2017), 2009.

Zeller, Suzanne Elizabeth. *Inventing Canada: Early Victorian Science and the Idea of a Transcontinental Nation.* Toronto: University of Toronto Press, 1987. Republished Montreal and Kingston: McGill-Queen's University Press, 2009.

– "William Brodie." In *Dictionary of Canadian Biography*, vol. 13: 112–14. Toronto: University of Toronto Press, 1994.

Zwerver, Harry. "Farmers working for farmers: a brief history of the Ontario Federation of Agriculture." https://ofa.on.ca/uploads/userfiles/files/A%20Brief%20History.pdf (accessed 5 May 2017) 1986.

PART THREE

Natural History "Old" and "New"

7

Catharine Parr Traill

A Natural Historian in Changing Times

MICHAEL PETERMAN

My interest in Catharine Parr Traill's writing career in Canada and her extraordinary passion for natural history culminated in my essay "Splendid Anachronism" (1990). Drawing on personal information from her books and letters, I attempted to document her resolute and steady commitment to the study of Canadian flora. My aim was to situate her sustained botanical commitment within the larger context of the professionalization of botany in Canada during her later decades. To do so, I drew upon contemporary cultural historians Carl Berger and Suzanne Zeller, along with Ottawa botanists John Macoun and James Fletcher who served in the 1880s as Traill's supporters in that city's burgeoning scientific community. Before 1990 little of an academic nature had been written about Traill's botanical writings. While I noted several essays that called attention to her deep interest in plants and ferns,[1] my aim was to investigate her pursuit of scientific recognition in the late nineteenth century. At the same time I was struck by the lack of attention paid by Canadian literary critics to her botanical work.

Shortly after my essay appeared, I became aware that the late Marianne Gosztonyi Ainley had begun to examine Traill's botanical work from a scientific perspective. In her essay "Last in the Field? Canadian Women Natural Scientists, 1815–1965," Ainley noted Traill's prominence among women in the "pre-professional" period of Canadian science and praised both her "true scientific curiosity" and her interest "in the *process* of science."[2] In a subsequent essay, "Science in Canada's Backwoods," Ainley argued that the terms I had applied to Traill – "splendid anachronism" and "struggling amateur" – were inappropriate descriptors of her work "as a pioneering naturalist and popularizer of science."[3] While I had drawn the phrase "splendid anachronism" from Carl Berger, I had to agree that describing Traill as a "struggling amateur" tended to diminish the seriousness of her botanical work.[4]

Having subsequently co-edited Traill's selected letters and edited *The Backwoods of Canada*, I continued to be fascinated by her personalized documentations of her botanical inventory and her unflagging efforts to find a publisher for her floral work. Recent essays by Elizabeth Thompson, Carole Gerson, Rebecca Raglon, Molly Blyth, Angela Byrne, Corinne Bigot, and Jane Stafford have thrown fresh critical light on various aspects of Traill's scientific writing.[5] In response, then, I return to Traill's letters and books in pursuit of a fuller understanding of her struggles as a botanical writer in nineteenth-century Canada.

As Ainley herself noted, Traill (1802–1899) "remained ambivalent … about being called a botanist."[6] Despite "her love of science, her growing knowledge of plant taxonomy and ecology, her eagerness to learn from others, and her increased concern about habitat destruction," she knew that she lacked formal scientific training. Moreover, she had neither the opportunity nor the desire to commit herself to the specialized language and methodology of contemporary botany as she understood it.[7] While she did her fieldwork with passion and precision, she did not aspire to be seen as a professional botanist.

How best, then, to see her today? Initially, I saw her as an amateur committed to her personalized study of Canadian flora. Despite her isolation in the Upper Canadian backwoods and the many challenges she faced in gathering floral information, accessing available scientific studies, and finding publishing opportunities, she carried on resolutely. In using the word "amateur" I did not mean to suggest that she was unprofessional. Nevertheless, the word "amateur" as applied to scientific undertakings by women in the nineteenth century does cast a long shadow. A better choice lies in the vocabulary of Traill's Suffolk youth in early nineteenth-century England. She saw herself as adapting her interests within a well-established way of thinking – that of "the natural historian." She shared this outlook with several of her sisters, particularly Elizabeth (Eliza), the eldest.[8] From her early days as a published writer Catharine had a clear sense of who she was as a writer and the kind of audience she wished to address.

7.1 *Opposite top*
Colour portrait of Catharine Parr Strickland as a young woman in Suffolk (likely painted by Thomas Cheesman).

7.2 *Opposite bottom*
Photo of the elderly Catharine Parr Traill at table.

Despite the lack of literary opportunities in Canada during the pre-Confederation period, she quickly turned her attention to wildflowers and flora, at least partly through Elizabeth's direction. Optimistically she sought to make connections with prominent scientific figures; however, she was well aware that these individuals were geographically and experientially remote from her. Over time she came to see herself, ruefully but accurately, as "a poor country mouse."[9] While this can be viewed as her cagey recourse to the "modesty topos" often evoked by women engaged in scientific fieldwork,[10] the fact was that she lived far from the centres of scientific endeavour in Canada West. She knew no scientists personally and for decades she worked quietly on her own. Moreover, she had no access to the new scientific methods and language being developed and used in nineteenth-century Canada and elsewhere. Instead, Traill drew upon information that was immediately useful to her, and did so on her own terms. She was willing to adapt her natural-history approach in certain ways, but she remained committed to the tradition she learned in rural Suffolk. The audience she sought was not scientific.

THE OLD NATURAL-HISTORY OUTLOOK AND THE NEW WAY OF SCIENCE

Bred to appreciate writers like Gilbert White from the late eighteenth century and his seventeenth-century precursor Izaak Walton, author of *The Compleat Angler*, she was schooled in a Christian view of nature that she learned in the first decade of the nineteenth century from her father, Thomas Strickland. Like Gilbert White (1720–1793), she sought "to pay a more ready attention to the wonders of the Creation, too frequently overlooked as common occurrences."[11] In fact, White, whom she never met, served as her natural-history mentor. Largely overlooked in recent years, his book, *The Natural History of Selborne*, was highly regarded for nearly two centuries. In his 1977 edition, Richard Mabey reported that it had appeared in more than 300 editions or reprintings and had been continuously in print since 1789. He deemed it "the fourth most published book in the English language."[12] White's attention to aspects of natural history, his devotion to his home village, and his engaging literary style impressed generations of English readers.[13] Using Selborne as his laboratory, he assembled a personalized natural history organized around letters he wrote or received.[14]

As a young woman Traill was inspired by White's call to observe the natural world with care and to report her findings to others. She responded keenly to his habit of close observation and his recognition of the healthiness that followed from such efforts. "Under Providence," White had written, natural history "contributed to much health and cheerfulness of spirits, even to old age."[15] White held the view that "natural history" was a distinct pursuit, far from "the dusty cabinets of the taxonomists" on the one hand and different from developing sciences like biology and geology on the other. He relished documenting the natural phenomena around him, especially birds, trees, and local topography. In Suffolk, Catharine adapted his approach to her age, gender, locale, and education; in her later Canadian years she would continue to call attention to his classic text by way of justifying her approach to flowers.[16]

Natural-history study served as an important component of the education of the Strickland girls. Born only twelve years after White's book was published, Catharine came to see her personal efforts as a contribution to an important literary-scientific tradition – whether in Suffolk or abroad. Once settled in Upper Canada, she chose to focus her natural-history lens on native plants and ferns, though her interest in birds and small animals remained keen.[17] She had a few botanical books on hand in her log cabin, but she studied plants from her own feminine and maternal perspective and through the herbal and medicinal knowledge she gained from First Nations friends and immigrant neighbours of her own class. Like White, she adopted a personalized focus. She looked upon the various species she was studying as both the natural phenomena of her new environment and her own offspring. In *The Backwoods of Canada*, her first Canadian book, she offered this playful aside: "I suppose our scientific botanists in Britain would consider me very impertinent in bestowing names on the flowers and plants I meet with in these wild woods: I can only say, I am glad to discover the Canadian or even the Indian names if I can, and where they fail I consider myself free to become their floral godmother, and give them names of my own choosing."[18] Her natural-history nomenclature suited her "new world" environment, even as it reflected her English and eighteenth-century roots.

Like Gilbert White, she was a self-directed generalist who wed the study of natural history to her home place. Growing up with her five sisters and two brothers at Reydon Hall in Suffolk, Traill was arguably her father's keenest student.[19] His curriculum included focusing their attention on the natural world around them. With the exception of her oldest sister Elizabeth, none of her siblings responded to the world of flora quite as purely or intensely as she did. Thomas Strickland's educational plan was unusually prescient for his time – he sought to educate Elizabeth (1794), Agnes (1796), Sarah (1798), Jane Margaret (1800), Catharine Parr (1802), and Susanna (1803) at home, teaching them to think for themselves, to write effectively, and to be well-informed citizens. He urged them as young children to create their own gardens and to take responsibility for their pets. It was their duty to think through the challenges they faced and adopt practical solutions. Only after they had struggled with their difficulties would he offer his advice.[20]

NATURAL HISTORY AND BOOK MAKING

In a late-life sketch entitled "The Five Little Gardeners," Traill remembered the individual gardens created by four of her sisters and herself. Under the eye of the Reydon Hall gardener each was tasked with designing her own garden. Four-year-old "Kate" made a puddle in her small plot and gathered "a lap-full of daisies and buttercups from the meadow which she planted in rows in the soft ground." Only when her flowers withered and died did she realize her mistake. Though embarrassed by her *faux-pas*, she reported in her "Editor's Note" that one might view her naive efforts as a "foreshadow[ing] of her love of the wild-flowers and ferns." That love would lead to what she called "the valuable work she [would do] in bringing our Canadian flora to the knowledge of the world."[21] In her nineties she continued to be proud of the value – nationally and internationally – of her path-breaking botanical work.

During those Reydon Hall years Catharine experienced several losses – the death of her father, the prolonged struggle of her mother and sisters to keep up their manor home, and a failed engagement.[22] Many of her fondest Reydon memories concerned the natural world. Two of her early books make clear her commitment to natural history. *Sketches from Nature; or Hints to Juvenile Naturalists* and *Sketch Book of a Young Naturalist; or, Hints to the Students of Nature* contain stories about her sisters and their experiences with birds and animals. A review of *Sketch Book of a Young Naturalist* in *The Lady's Magazine*, likely written by her sister Elizabeth, called attention to its skilful combination of "information, good sense, and piety, with interesting anecdotes and amusing incidents."[23] In those years, however, Catharine paid less attention to floral matters than to other natural-history subjects.

From the early 1820s four of the Strickland sisters published books for juvenile readers.[24] Indeed, there was a shared Strickland "brand" at play in that sector of London's book market. Recently Jane Stafford has analyzed the shared features of their juvenile books. Arguing that their early writings have been overlooked by critics, she notes that such work was an acceptable "means of obtaining a little ready money" for young females. Writing in the spirit of their father's teachings, the Strickland daughters eschewed Romanticism for the most part. Their work was "rational, scientific and information rich," at-

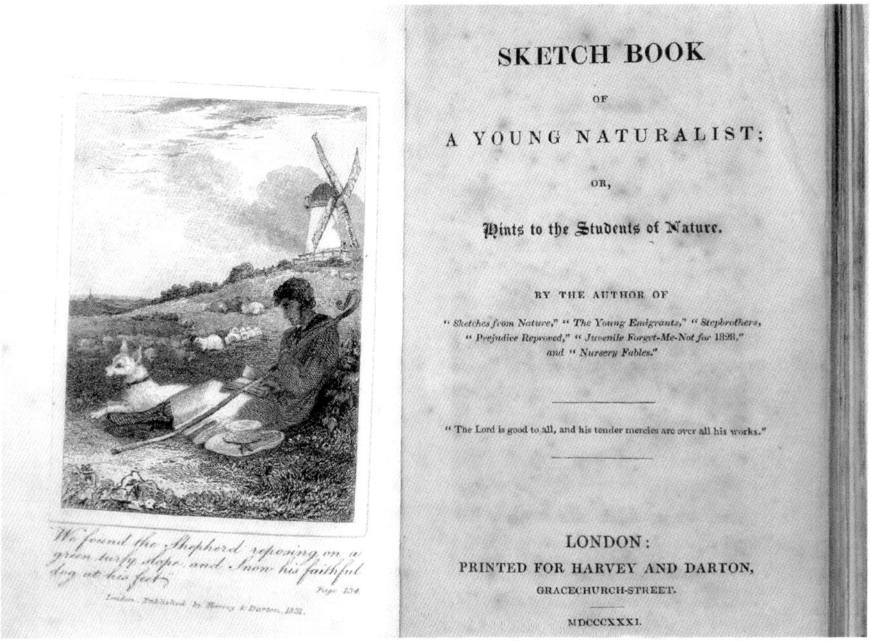

7.3
Title page, *Sketch Book of a Young Naturalist* (London: Harvey and Darton, 1831).

tentive to facts, matters of "civility," and the importance of outward behaviour rather than "interiority." In Stafford's view Catharine was the most productive and scientific of "the resourceful Strickland sisters."[25] When she immigrated to Canada, she brought that rational pedagogy with her. Her later works are rooted in those views.

Stafford's assessment of Catharine's juvenile books counters Northrop Frye's assertion in *The Bush Garden* that her view of nature was "reminiscent of Miss Muffett."[26] He pegged her as a naive practitioner of the sentimental pastoral myth – a Wordsworthian view offering a selective, overly pleasant view of nature that ignored its darker and more threatening sides. In Canada, she presented herself to readers as a more scientifically inclined, non-romantic writer than Frye recognized.

Once she settled in Upper Canada, she began to collect and identify the wildflowers in her wilderness environment. Plants, ferns, shrubs, grasses, and trees all interested her, but wildflowers were her special focus. Rooted in her rational and Christian view of the natural world, she approached her new environment as her personal laboratory.[27] Her position was straightforward and un-ironical. Neither did she falter in her characteristically optimistic outlook, despite experiencing many losses, setbacks, and discouragements over the years. Close observation of the natural world and her "love of natural history" became her mantra in Canada.

BRINGING NATURAL HISTORY TO UPPER CANADA

A newly married emigrant, Traill embraced the dramatic changes she faced on arrival as an Upper Canadian pioneer in 1832. Unlike her younger sister Susanna who immigrated in the same year, she was undeterred by leaving her homeland or by the hardships that lay ahead.[28] Six years earlier in *The Young Emigrants*, she had imaginatively anticipated just such an opportunity of discovering a new country.[29] Within weeks of her own arrival she began to gather floral samples even as she lamented not having taken up Elizabeth's "frequent offers of prosecuting a study which I once thought dry."[30] Now with refreshed eyes she saw botanical study as "highly interesting, and a fertile source of mental enjoyment." For the next seven decades she carefully collected wildflowers and ferns in nearby fields and unbroken lands, making diary entries about her discoveries and accessing information from botanical texts in her possession. Floral study became her disciplinary ballast through years of change and challenge; in effect, Catharine wed its study with her new life in Upper Canada.

Over the years she worried increasingly about the changes she observed as her adopted country emerged from its thinly settled, colonial youth into a confederation of provinces, a vast continental land mass, and an industrializing nation. The consequences of environmental change became an ongoing concern for her. She lamented that the flowers she gathered were "destined sooner or later to be swept away, as the onward march of civilization clears away the primeval forest – reclaims the swamp and bogs, and turns the waste places into a fruitful field."[31] Nevertheless, as a pioneer homemaker and a keen natural historian she made herself more at ease by focusing intently on the study of flora, even as she gathered information about other areas of domestic life then underrepresented in the print outlets of colonial Canada.[32] Often struggling to cope with financial constraints and family problems – in the 1840s, for instance, she suffered her husband's near-bankruptcy, the deaths of two newborn babies, and diseases like the ague, scarlet fever, and typhus affecting the family – she persisted in what spare time and energy she could muster to collect plant specimens and gather information about them.

Though most of her journals were destroyed in a devastating house fire in August 1857,[33] family members managed to save one early diary along with her plant and fern manuscripts. The surviving volume – from 1837 – contains entries made while her family was living on their first bush farm near present-day Lakefield, Ontario. Her entry for 6 August illustrates both her observational skills and her reliance on available scientific literature. On this occasion she was investigating the properties of the herb Impatiens, recommended to her as a remedy for poison ivy:

> Yesterday I was recommended to [wash] the [affected body] part with the juicy stalks of the herb Impatiens or Canadian balsam which is regarded by many as a remedy [but] the application had no good result in the present instance. This flower seems to be a native of Britain as I find an exact description of it in Hull's "British Flora."[34] – It grows in vast quantities along the banks of our crick – The seed pods of this plant seem to be of a sensitive nature [;] on bending down the nearly matured pod it starts from beneath the touch and each membrane curls itself into a ring discharging the seed – I suppose it is from this circumstance it derives its name 'Impatient.' The blossom of this plant is very singularly formed [;] it is deep orange speckled with purple. The nectary forms a sort of head or high pointed cap terminated by a horn at the summit. I find this to be the same plant that is thus described in the book above alluded to –Impatiens Balsamina.[35]

Besides Hull's book, she also mentions Thomas Nuttal's *The Genera of North American Plants*. Through her close friend and backwoods-mentor Frances Stewart,[36] who saw in her a talented student of flowers, she had access to Frederick Pursh's *Flora Americae Septentrionalis; or A Systematic Arrangement and Description of the Plants of North America*. Though it was written in Latin, her husband Thomas had the linguistic training to help her access this crucial resource.[37]

In her early years in Upper Canada, Catharine transformed herself from children's writer to dedicated natural historian. Her first decade in the backwoods was largely a positive experience. The environs of Lake Katchewanook proved a treasure trove of botanical discovery. As the weather and season allowed, she extended her collection of wildflowers and ferns, while closely observing aspects of the local terrain. She hoped that the data she was gathering would someday be of interest to others and useful in her future writing.

The publication of her first Canadian book, *The Backwoods of Canada*, owed much to her sister Agnes's help in England. Agnes took the manuscript to Charles Knight in London who chose to include it in his "Library of Entertaining Knowledge" for the Society for the Diffusion of Useful Knowledge.

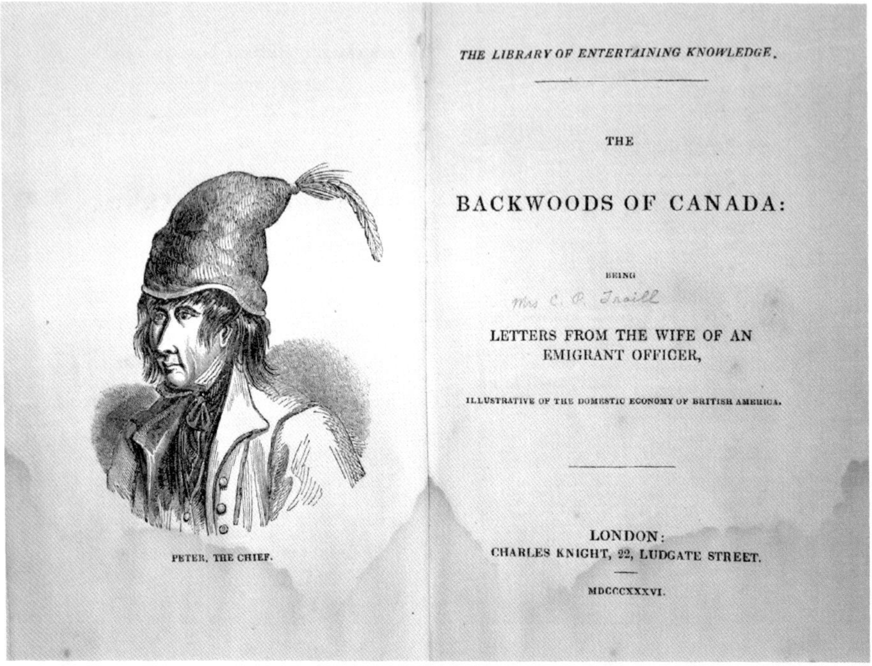

PETER, THE CHIEF.

THE LIBRARY OF ENTERTAINING KNOWLEDGE.

THE

BACKWOODS OF CANADA:

BEING

LETTERS FROM THE WIFE OF AN
EMIGRANT OFFICER,

ILLUSTRATIVE OF THE DOMESTIC ECONOMY OF BRITISH AMERICA.

LONDON:
CHARLES KNIGHT, 22, LUDGATE STREET.

MDCCCXXXVI.

7.4
Title page, *The Backwoods of Canada* (London: Knight, 1836).

That volume, in which Traill identified herself only as "the wife of an emigrant officer," sold steadily and went through numerous editions. From her isolated log cabin, she also cast about for Canadian and American publishing opportunities, mindful of the paucity of outlets and the uncertainties in mail service.

After five years in "the backwoods," Thomas Traill sold their bush farm and moved the family to nearby Peterborough. There from 1839 to 1846 they were increasingly beset by problems. Traill had four healthy children and then lost two children at birth; her health was weakened and her energy much diminished. For his part Thomas found himself facing stressful financial pressures, and, by 1845, their situation had become grave. Lack of income, ill health, and a failed investment left the family in desperate circumstances.[38] Given his age and sedentary inclinations, Thomas had little capacity to meet his family's financial needs. But their circumstances changed for the better in 1846 when the Traills were offered free use of Wolf Tower, a curious house on the western shore of Rice Lake. Their benefactor was the Rev. George Bridges, an Englishman who had come to Upper Canada from Jamaica after reading Traill's "Backwoods" book. Admiring her as a courageous pioneer and an evocative writer, he had sought her out in Peterborough. His kindly offer saved the Traills from serious financial embarrassment – perhaps bankruptcy.[39] While the move to Wolf Tower relieved their immediate financial anxieties, it also helped Catharine to recover her energies; she spent much of her time exploring their new environment of hills and ravines while leading her children on flower-gathering rambles.

The gentler climate, open fields, and lake breezes of the Rice Lake Plains helped Catharine return to her writing. With editorial and publishing help from her sisters in London, she wrote two new books, *Canadian Crusoes* and *Lady Mary and Her Nurse*. Both were published in London to considerable success and brought much-needed income to the family. She also began writing for a few British, Canadian, and American magazines. In particular, her sketches about the Rice Lake Plains appeared as "Forest Gleanings" in *The Anglo-American Magazine* (Toronto, 1852–53) while she wrote seasonal essays and stories for other outlets, drawing upon her observations, her botanical inventory, and her sensitivity to ecological issues. These essays offered her readers perspectives on the moral teachings of flowers, their merits in educating children, and the long history of female interest in horticultural medicinal purposes. In "The Violet," written for *The Literary Garland*, she called attention to the English tradition of female herbalists working in the "still rooms" of great houses seeking medical treatments for diseases and conditions of various kinds. As well, she familiarized herself with what are now called "the tall grass prairie and savannahs" of the Rice Lake Plains. In her "Forest Gleanings" sketches, she wrote with precision about the rich natural life of that site of original, Ontario tall grass prairie;[40] today ecologists and botanists continue to draw upon her sketches in their studies.

During the 1850s she developed further North American magazine connections. In addition to *The Anglo-American Magazine*, she contacted *The Genesee Farmer*, the *Hamilton Agriculturalist*, and James Vick's *The Horticulturalist* in Rochester, New York. Several of her essays for Vick appeared under her favoured title, "Forest Gleanings." As well, she prepared submissions for the annual Provincial Exhibitions of Canada West, relying on friends to submit her entries. Newspaper reports reveal that she won first prize for "best collection of native plants, dried and named" at the Kingston Fair in 1856 and second prize at Toronto in 1862.[41] However, like some of her friends, she was troubled by the biases of the judges. To Frances Stewart she reported, "I did not get a prize at the Shew for my fine collection of Ferns to the great disgust of Mr. Tully and others who said they were without doubt the best things there of their kind and Mr T. withdrew them and meant to remonstrate against the unfairness of the judges – it has now become so partial a thing the

7.5
Poster of the Rice Lake Plains.

awarding of the prizes that I shall make no further attempts to send any collection to the Provincial Shew."[42]

Other letters indicate her continuing frustrations in seeking out a Canadian publisher for her manuscripts of plants and ferns. In 1863–64, a businessman named Adam Brown offered to publish her plant manuscript through the Hamilton Horticultural Society. A year later, Traill wrote, "It is not using me well to keep me in this suspense after all his fine words … I fear it is doomed to end in nothing."[43] She was right. She also sent manuscripts to John Lowe at the Department of Agriculture with the plea that "Any statistical information that may be thought necessary must I fear be left to my editor as I do not consider myself competent to supply it."[44] She knew that her submissions would likely require scientific editing before they could be published.

Traill also initiated correspondence with leading university professors like Dr William Hincks, the first professor of natural history at University College, Toronto, and Professor George Lawson, professor of natural history and chemistry at Queen's University. Heretofore her only contact in the scientific world of Canada had been John Macoun, a prominent botanist in Canada West, whom she met in the 1850s in Belleville where her sister Susanna Moodie and two of her own children were living.

The Lawson connection emerged in 1861. Recently appointed to Queen's University, Professor George Lawson had founded the Botanical Society of Canada in Kingston. Members of the public were encouraged to join as regular or corresponding members. According to its *Annals*, Traill became a "Corresponding Member" on 15 November 1861.[45] Lawson read her letter at that meeting, though the *Annals* provide no specific detail. When in 1863 Lawson decided to leave Queen's, the society ceased operations. This was yet another disappointment for her.[46]

Traill's letters provide a valuable source of information about her daily plant-related activities, though detailed entries are sporadic. We learn, for example, that she made albums of dried plants and ferns with her daughter Kate as gifts to family and friends. Any payments they received helped them in their struggle to meet domestic expenses. In a letter to Ellen Dunlop she described her flower-pressing activities: "I have been very busy with preparing the sets of dried ferns and flowers for dear Charlie to take home. One is indeed your dear Mothers, and the other two sets are for any of the ladies who wished for them. I hope they will not think the price too great, but they do cost me a good deal of time and close attention to gather, press, and put them down, besides naming them which indeed is sometimes the most laborious part of the work." With characteristic deference, she added, "yet $5 does seem a great deal to give for wild plants, does it not? –"[47]

A few years earlier she had received an encouraging boost from England. The Rev. George Bridges, her former Rice Lake landlord, was now settled in Beachley, England. He brought her botanical studies to the attention of his aristocratic friends there. Lady Charlotte Greville quickly emerged as her champion. She arranged for Traill to receive a grant of 100 pounds from Lord Palmerston in honour of her "literary" work; then in 1859 she sent Traill a handsome maple screw-press (or flower press) to facilitate her scrapbook efforts.

In 1861 the now-widowed Catharine took ownership of a house in Lakefield built for her by her brother Sam Strickland and local friends.[48] There with Kate she continued to create albums of flowers, ferns, and grasses, several of which are now housed in Canadian archives and museums. Over three decades they sold such "scrapbooks" or gifted them to friends. Always practical minded, Catharine reported, "I am so anxious to earn what will pay our bills that I write even when I have no hope of a market."[49]

Despite her many disappointments, she continued to seek out Canadian publishers and booksellers for her manuscripts. In 1865 William Dawson, the editor of *The Canadian Naturalist* in Montreal, sent back her ferns manuscript

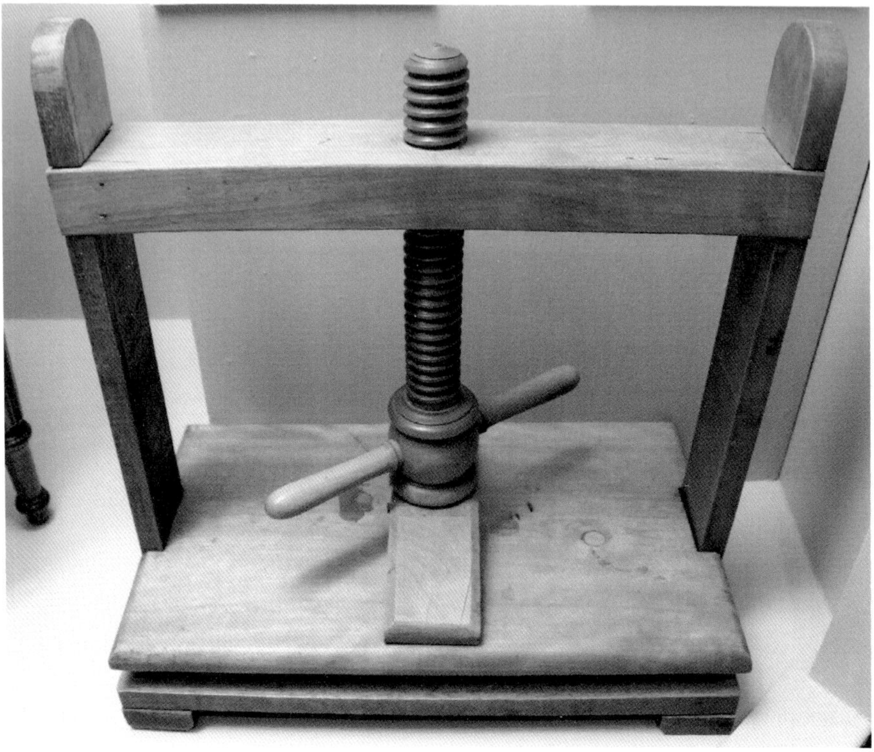

7.6
Catharine Parr Traill's flower press.

with "no remark and no correction."[50] "I do not think he read it," she lamented. "He merely says that they have all their matter for [T]he Naturalist gratis." In her frustration she quoted what the "Great" Sir William Hooker had told John Macoun: "Canada was behind every one of the British colonies and all civilized nations in Scientific literary effort especially in Botany."[51]

A MULTI-TALENTED NIECE MAKES TRAILL'S FLORAL BOOKS HAPPEN

Two of Traill's plant manuscripts, *Canadian Wild Flowers* and *Studies of Plant Life in Canada*, were finally published after Confederation. Neither would have appeared had it not been for the talent, social connections, and industriousness of her niece Agnes Fitzgibbon (1833–1913), then a young widow living in Toronto with a large family and little income.[52] Agnes, well aware of her aunt's "work on Canadian botany,"[53] set out in the mid-1860s to produce

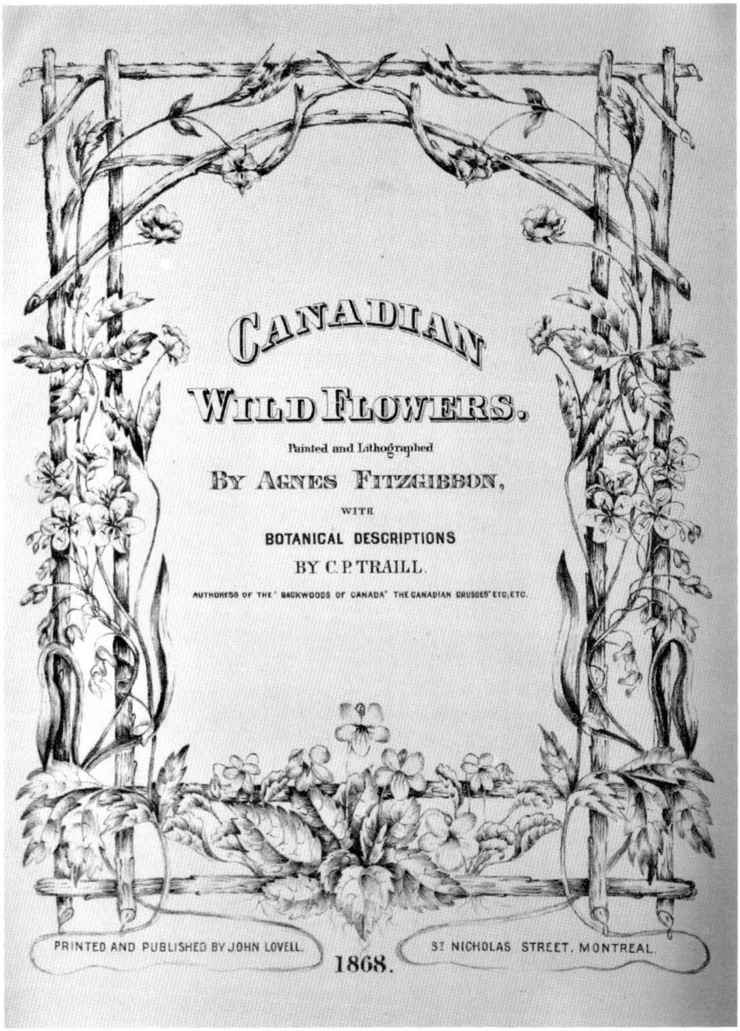

7.7
Title page, *Canadian Wild Flowers* (Montreal: Lovell, 1868).

a large, illustrated book on Canadian wildflowers. Her vision and persistence led to Traill's *Canadian Wild Flowers*, a made-in-Canada, folio-sized book with hand-coloured illustrations. In her youth Agnes had been taught how to paint flowers and birds by her mother, Susanna Moodie. Having won prizes in watercolour competitions, Agnes was now ready for a larger challenge, one that necessitated that she learn lithography, since "there were no professional workers in our country who could do it."[54]

Her efforts included making the drawings, creating the plates, negotiating with a publisher, and organizing a team, including her daughters, to hand-colour the images. Aunt Catharine, now sixty-four, provided the "letter-press" for the floral groupings, drawing material from her manuscript to describe

7.8
Agnes Dunbar Fitzgibbon Chamberlin, 1876.

the flowers in each image. Furthermore, as the volume was to be sold by subscription, buyers had to be signed up and money collected. Again, Agnes was the centre of operations.[55] Socially well positioned, she drew on acquaintances in Toronto, Belleville, Ottawa, and Peterborough, finding enough purchasers, especially among those who valued their British roots, that a second edition was immediately required. In fact, the book went through four editions, cumulatively proving a remarkable success for the Strickland women.[56]

William Dawson's *The Canadian Naturalist* reviewed the book in March 1869, describing it as "a very pretty volume" of "some of our showiest wild flowers."[57] The adjectives "pretty" and "showy" suggest an element of professional disdain because the volume leaned more toward the popular than the scientific. While appending a list of the species illustrated, the reviewer noted

7.9
Ottawa botanists John Macoun (left) and James Fletcher (right) provided scientific guidance and personal encouragement to Catharine Parr Traill for her work on what became *Studies of Plant Life in Canada.*

that Mrs Traill had written "popular descriptions of each plant" drawn from a larger work on the "flowers, shrubs and forest trees of Canada." Congratulating her on providing "excellent English names for the chosen flowers and avoiding such current vulgarities as 'Dutchman's Breeches'" for *Dicentra Cucullaria,*[58] he also made note of the poor quality of the paper and inadequacies in the proofreading. By contrast, *The New Century* magazine in Philadelphia wrote glowingly of Agnes's undertaking, calling it "one of the most remarkable works ever attempted by a woman."[59] In the second edition Agnes sought improvements, though not as fully as her aunt had hoped to see. A third edition followed, catering to a new subscription list. A final edition from William Briggs would follow, adding textual improvements and an "Introduction" by Agnes herself.

Neither would *Studies of Plant Life in Canada* have been published without Agnes's commitment, executive skills, and social connections. A few years after *Canadian Wild Flowers* appeared, she married Col. Brown Chamberlin and moved to Ottawa where she enjoyed with him a prominent place in Ottawa society. As a patron of the Ottawa Field-Naturalists' Club (formed in 1879), she developed close connections with botanists John Macoun and James Fletcher. Both men would provide scientific guidance and personal encouragement to her aunt when, early in 1884, she came to Agnes's New Edinburgh home to complete her manuscript for publication in Ottawa.[60]

Born in Northern Ireland, John Macoun (1831–1920) had served as a member of the Geological Survey of Canada and was the author of numerous botanical treatises. In 1882 he was appointed to the position of "Dominion Botanist." Later he donated his large collection of flora and fauna to the "Dominion Herbarium," an institution that became the National Museum of Natural Sciences. Traill welcomed his suggestion that she use his *Catalogue of Canadian Plants – Part 1, Polypetalae* (a work eventually expanded to seven parts) in preparing her "plant life" text. In her copy of his printed volume, she pencilled notes on her own sightings, the medicinal properties of particular plants, and domestic information she had gathered.[61]

Equally important was the encouragement and editorial advice provided by James Fletcher (1852–1908). When the English-born Fletcher was appointed assistant librarian of Parliament in 1876, he began to share his knowledge of plant and insect life with Ottawa civil servants and politicians. His involvement with the Ottawa Field-Naturalists' Club preceded his appointment in 1884 as Dominion entomologist and botanist. In that role he encouraged a network of Canadian women to submit botanical samples to him. Youthful and messianic about botanical study, he eagerly supported Agnes Chamberlin's plan to bring her aunt's botanical manuscripts together in a single volume. Fletcher began corresponding with Traill in 1883 and a year later in Ottawa he helped her correct and edit her manuscript, undertaking considerable drudgery on her behalf.[62] His letters to her reveal his unfailing courtesy and his admiration for both her fieldwork and personalized botanical descriptions:

> I am charmed with your style and find it so attractive after the irreverent materialistic philosophy, falsely so called, of too many of our modern naturalists. It is very charming to me to see such love for our beneficent creation, & reverence for his perfect works. In all my instructions in botany I have always endeavoured to draw attention to the marvelous & beautiful adaptations, of all objects presented to us in the study of nature, to their required ends and to show how much we have in this lovely world to make us happy.[63]

Traill's work as a natural historian accorded with Fletcher's outlook. Both were anti-materialists who sought to wed accurate botanical observation with an admiration for God's handiwork and the wonders of the natural world. He saw in her literary approach an excellent opportunity to increase the level of interest and respect among Canadian readers for the study of natural history and the productions of "this lovely world."

A decade later – on 26 July 1894 – Fletcher paid her another glowing tribute. Now age ninety-two, Traill had continued to send Fletcher "collection[s] of beautiful specimens," and he wrote to her as follows:

> With regard to your disclaiming the title of botanist, all I can say is, that I wish a fraction of one percent of the students of plants who call themselves botanists, could use their eyes half as well as you have done. I think indeed your work of describing all the wild plants, in your book, so accurately that each one could have the name applied to it without doubt, is one of the greatest botanical triumphs which anyone could achieve, and one which I have frequently spoken of to illustrate how one can develop their powers of observation.[64]

Relatedly, the Trent University Archives holds a scrapbook of grasses that Traill sent to James Fletcher in that same year, with her "kind wishes to her much valued and esteemed friend." The volume contains eighteen pages of grass species.[65]

The conjunction of James Fletcher's editorial help, John Macoun's guidance, Agnes Chamberlin's executive energies, and Traill's unflagging dedication to the collecting and description of Canadian flora led in 1884 to the publication of a book that was more than fifty years in the making. *Studies of Plant Life in Canada* documents her Canadian fieldwork and her zeal as a natural historian. Organized into four sections – "Wild, or Native Flowers," "A Familiar Description of the Flowering Shrubs of Central Canada," "Forest Trees," and "Our Native Canadian Ferns" – this book is a living record of her observational skills, careful collecting, and personalized descriptions.[66] As Fletcher well knew, there were trained botanists now at work in Canada committed to precise scientific language, laboratory methodology, and statistical data. In Traill's book, by contrast, he found a less clinical, natural-history methodology – more traditional, Christian-based, and reader-friendly. Hers was, as she herself insisted, "not a book for the learned."[67] Knowing that his botanical vision accorded with hers, Fletcher provided her with a sympathetic bridge from the old way to the new. Without his informed and sensitive help, Traill, at her advanced age, would likely have been unable to complete her most remarkable book.

Despite many disappointments along the way, Traill's hopes for her floral manuscripts never wavered: she saw them as "an addition to the literature of

STUDIES

OF

PLANT LIFE IN CANADA

OR

GLEANINGS FROM FOREST, LAKE AND PLAIN

BY

MRS. C. P. TRAIL
LAKEFIELD, ONT.

OTTAWA
1885

7.10
Title page, *Studies of Plant Life in Canada* (Ottawa: Woodburn, 1884).

Canada" that "may become a household book, as Gilbert White's *Natural History of Selborne* is to this day among English readers."[68] She aspired to give Canadian readers, particularly women, the cumulative fruits of her research and a record of her "pleasant recreation" as a student of flora in her adopted country. She felt that an informed interest in and respect for the natural world were both very much needed as the nineteenth century drew to its close.[69]

Her success, however, was relatively short-lived. *Studies of Plant Life in Canada* has been largely neglected since the late 1880s. Its second edition in 1906 was copyrighted solely by Agnes Chamberlin and was a different book in several notable ways. It dropped the sections on ferns and trees and tightened up the preceding sections, eliminating some of Traill's entries but respecting her wording for those that were kept. Agnes included twenty plates instead of the original nine and chose a larger typeface for the volume. A full comparison of the two editions is certainly in order.

It is unlikely that Northrop Frye read either edition very carefully as he evaluated Traill's overall contributions to early Canadian literature. Had he examined her sustained interest in the natural world as part of her literary output, he might have seen her work differently. Later commentators like Marion Fowler might then have been less inclined to criticize Traill for her optimism and her botanical enthusiasm: among her mocking comments Fowler wrote, "Mrs. Traill had come, she had named, and she had made herself a household word."[70]

It is a fact of Canadian literary history that few critics have looked closely at the two editions of *Studies of Plant Life in Canada*. That inattention has diminished Traill's overall reputation. The book has been the victim of passing time, changing views of botanical study, and increasing cultural compartmentalization in Canada. Today it is often viewed as of only antiquarian or scientific interest. Nevertheless, *Studies of Plant Life in Canada* (1884) remains a rich volume of nearly 300 pages. Seeing "Every plant, flower and tree" as "a page in the great volume of Nature which lies open before us," Traill emphasized their stories and histories, often appending lines of poetry to enrich her descriptions.

Her "biographies" of individual plants are sometimes a page in length. For example, her text for the Showy Orchis – *Orchis spectabilis*, (L.) begins with four lines from Thomas Gray's "Elegy Written in a Country Church Yard," suggestive of how the beauty and sweetness of flowers are often overlooked: "Full many a gem of purest ray serene, / The dark unfathomed caves of ocean bear; / Full many a flower is born to blush unseen, / And waste its sweetness on the desert air." A full page of text then follows:

> Deep hidden in the damp recesses of the leafy woods, many a rare and precious flower of the Orchis family blooms, flourishes, and decays, unseen by human eye, unsought by human hand, until some curious, flower-loving botanist plunges amid the rank, tangled vegetation, and brings its beauties to the light. One of these lovely natives of our Canadian forests is known as *Orchis spectabilis* – (Beautiful Orchis – or Showy Orchis). This pretty plant is not, indeed, of very rare occurrence; its locality is rich maple and beechen woods in eastern Canada. The colour of the flower is white, shaded, and spotted with pink or purplish lilac; the corolla is what is termed ringent or gaping, the upper petals and sepals arching over the waved lower petal. The scape is smooth and fleshy, terminating in a loosely-flowered and many-bracted spike; the bracts are dark-green, pointed, and leaf-like; the root a bundle of round white fibres; the leaves, two in number, are large, blunt, oblong, shining, smooth, and oily, from three to five inches long, one larger and more

7.11
Agnes Fitzgibbon, Showy Orchis, *Canadian Wild Flowers*, Plate VII.

> pointed than the other. The flowering time of the species is May and June. The exquisite cellular tissues of many of our flowers of this order delight the eye, and give an appearance of great delicacy and grace to the blossoms. In this charming species the contrast between the lilac purple colour of the arching petals and sepals, and the almost pellucid lower lip or somewhat broadly lobed under petal, is very charming. The large shining leaves lie close to the ground when the plant is in flower. Transplanted to gardens the Showy Orchis rarely survives the second season of removal from the forest shade. It will not grow freely exposed to cold wind, or glaring sunlight. It loves moist heat; the conservatory would probably suit it, and it would be worth a trial there, or in the grove or wilderness, or at the root of a large tree near water.[71]

More informal and vivid than standard taxonomy, Traill's descriptions often include literary and nostalgic details that were essential elements of her vision as a natural historian. As James Fletcher observed, she saw deeply and she saw well, providing a model for future botanists about what close and loving observation entails. She offered her readers "Gleanings from Forest, Lake and Plain,"[72] the phrase suggesting the patient gathering undertaken in the wake of harvesters who, having completed their work, left for other employment. She chose to write floral biographies of a familial kind, giving attractive lives to her plants. In her entry "Slender Lady's-tresses," for instance, she offered a sort of *modus operandi* for her approach to floral names:

> The old florists and herbalists of former times were more gallant than our modern botanists, for they gave many pretty names to the flowers instead of the harsh-sounding, unmeaning ones that we find in our scientific manuals of Botany. So we have among our local and familiar names, such prettily sounding ones, as "Ladies' Tresses," "Sweet Cicely," "Sweet Marjoram," or "Marjory," "Mary-gold," "Lady's-slipper," with a number of others that I could name – besides descriptive names, which form a sort of biography of the plant, giving us a correct idea of its characteristics and peculiar uses or habits.[73]

Catharine Parr Traill was a proud and creative heir to the natural-history school. Her arrival in Canada in 1832 triggered her floral study. With belated support from Agnes Chamberlin, James Fletcher, and John Macoun, she was able to see her richly detailed record of Canadian flora published at last in

Canada. As Clara Thomas astutely observed in 1966, "The combination of meticulously observed detail, of practical suggestion, and of affectionate familiarity with all the range of plant life gives a timeless attraction to her work."[74] As a natural historian it was her *magnum opus*. But such have been the changing views of literary genres over the last century that her *Studies of Plant Life in Canada*, a work she prepared "at intervals over a long series of years," has fallen into neglect.

NOTES

1 Cole, "Catharine Parr Traill," 73–9; Collard, "Flowers to Heal and Comfort," 32–6; MacCallum, "Catharine Parr Traill," 39–45.

2 Ainley, "Last in the Field?," 28.

3 Ainley, "Science in Canada's Backwoods," 80.

4 Berger, *Science, God and Nature in Victorian Canada*, 35.

5 Thompson, "With Fire and Axe"; Gerson, "Nobler Savages"; Raglon, "Little Goody Two-Shoes"; Blyth, "History as Storytelling"; Byrne, "My Little Readers"; Bigot, "Did They Go Native?"; Stafford, "How One Subject Springs Out of Another."

6 Ainley, "Science in Canada's Backwoods," 92.

7 Ibid, 85.

8 Elizabeth edited magazines and journals in London beginning in the 1820s. In 1830–31 she served as editor of *The Lady's Magazine* in London where she included work by several of her sisters. As editor she began a series entitled "Biography of Flowers" in the monthly issues – for example, "Primula Family – Cowslips, Auriculas, and Polyanthuses" (June 1831), 341–4.

9 Traill, *I Bless You in My Heart*, 147.

10 See Byrne, "My Little Readers," 88.

11 "Advertisement" in White, *The Natural History of Selborne*, 3.

12 Ibid., viii.

13 Ibid.

14 White corresponded, for example, with the zoologist Thomas Pennant and the explorer-naturalist Daines Barrington. See White, *The Natural History of Shelborne*, xiv.

15 Ibid., 4. White's reference to Providence underlays his Christian view.

16 Traill, *Plant Life in Canada*, 3.

17 See, e.g., *Lady Mary and Her Nurse* (1856) and *Cot and Cradle Stories* (1895).

18 Traill, *The Backwoods of Canada*, 104.

19 Thomas Strickland purchased Reydon Hall near Southwold in Suffolk in 1808. Catharine lived there until her marriage to Thomas Traill in May 1832.

20 For more on the Strickland family at Reydon, see Gray, *Sisters in the Wilderness*; Peterman, *Sisters in Two Worlds*; and Traill, *The Backwoods of Canada* (1997 edition).

21 Traill, *Cot and Cradle Stories*, 61.

22 Thomas Strickland's sudden death in 1818 revised his family's situation. Five of his six daughters began to write for the London literary market in part to alleviate the family's straitened situation. Catharine's engagement to Frank Harral, the son of family friends, ended in 1830.

23 *The Lady's Magazine* (April 1831): 215–16.

24 Sarah did not write for publication. Eliza became an editor in London in the 1820s but did not publish books.

25 Stafford, "How One Subject Springs out of Another," 55.

26 Frye, 245. Frye wrote, "The Wordsworthian sense of nature as a teacher is apparent as early as Mrs. Traill, in whom we note a somewhat selective approach to the subject reminiscent of Miss Muffet."

27 Traill was also influenced by the Reverend William Paley (1743–1805), a highly respected Christian apologist and philosopher. His book, *Natural Theology: or Evidences of the Existence and Attributes of the Deity* (1802) was a bestseller, buttressing the Christian outlook of contemporary natural historians.

28 Susanna Moodie and her family arrived in Upper Canada that same summer; characteristically, she was disturbed by several crude experiences in their new world and felt the loss of her English home and literary connections.

29 Among Traill's early books was *The Young Emigrants* (1826) in which she used letters from Canada to describe how a sister and brother reacted positively to their experiences in Upper Canada. She based her story on letters from family friends and her brother Sam, who immigrated in 1825.

30 Traill, *Backwoods of Canada,* 68. Elizabeth Strickland often wrote about flowers and their "biographies." As editor of *The Lady's Magazine* in London she published biographies of flowers which she signed "E.S." By then Elizabeth had clearly taken the lead among the sisters as a celebrator of flowers.

31 Traill, *Canadian Wild Flowers*, 8.

32 See *The Female Emigrant's Guide* (1856) in which Traill documented domestic and culinary challenges faced by immigrants and provided useful information for female settlers. A new edition edited by Nathalie Cooke and Fiona Lucas was published in 2018

33 Oaklands, the Traill farmhouse near Rice Lake, burned to the ground that August night in 1857.

34 Hull, *British Flora.*

35 See the Traill Family Papers, 3396–7. Both her young son James, then age three, and a family friend were stricken with poison ivy.

36 Frances and Thomas Stewart preceded the Traills in settling near Peterborough in 1822. A literate and educated woman, Frances (1794–1872) became Traill's special friend and mentor and was always eager to help her in her botanical work.

37 Thomas Traill's library, which took many months to reach them in Upper Canada, likely included the Hull and Nuttal books.

38 Thomas's most costly mistake was to lend money to a young Scottish friend to build a mill on the Otonabee River. When the friend drowned, Traill found himself liable for those losses.

39 The Rev'd George Wilson Bridges (1788–1863) was an admirer of Traill and a keen supporter of her work. He immigrated to Upper Canada from Jamaica with his son after the tragic drowning of his four daughters. Later he became a pioneer photographer in Europe and the Middle East. While living at Rice Lake, he occasionally took Anglican services in Peterborough.

40 Today scientists are studying this important example of original and undisturbed tall grass prairie. Traill's descriptions have proved valuable in their pursuit of early records about this landscape.

41 *The Canadian Agriculturalist* vol. 8, no. 11 and vol. 14, no. 24.

42 Traill, *I Bless You in My Heart* [1862], 150.

43 Ibid, 133.

44 Ibid, 196.

45 *The Annals*, 168, 170. On the Botanical Society of Canada, see the chapter by David Galbraith in this volume.

46 Information courtesy of David Galbraith of the Royal Botanical Gardens (Canada). Lawson later named a fern in her honour; it was called "Traill's Shield Fern – A. Marginale."

47 Traill, *I Bless You in My Heart*, 146. The business of accurately naming plants was an ongoing challenge to her.

48 Thomas Traill died in Lakefield in 1859. Catharine named the house built for her in Lakefield Westove after the Traill family estate in the Orkneys.

49 Traill, *I Bless You in My Heart*, 160.

50 Traill also developed a manuscript on native Canadian ferns in these years.

51 Traill, *I Bless You in My Heart*, 159–60.

52 Charles Fitzgibbon was a well-known barrister in Toronto.

53 Dignam, "Women in the Development of Art," 213.

54 Ibid., 212–13. Dignam notes that "The magnitude of the work can be realized from the fact that each plate passed Mrs. Fitzgibbon's hands for the various colours and

details an average of sixteen times. There were fifteen thousand plates in the first three editions." Agnes also invented a useful stencil process that left "little mark of the stencil or brush" on the lithographic plate.

55 For her own comedic view of trying to sell her book on subscription, see Peterman, "Splendid Anachronism," 182. This letter is not included in Traill, *I Bless You in My Heart*.

56 The early editions that appeared in 1868, 1869, and 1870 (the latter under a variant title) were published by John Lovell in Montreal; in 1895 the fourth edition was published by W. Briggs of Toronto. Lovell had published poems and sketches by Susanna Moodie in *The Literary Garland* and novels in serial form. Professor Emeritus Alexander Globe has written a lengthy essay on the four editions and described 135 surviving volumes that he examined in his research. "The Four Editions of *Canadian Wild Flowers* by Catharine Parr Traill and Agnes Fitzgibbon Chamberlin, 1868–1895" remains unpublished to this date.

57 *The Canadian Naturalist*. March 1869, 100.

58 The reviewer preferred Traill's name of "Fly-flower."

59 Cited in Dignam, "Women in the Development of Art," 213.

60 Agnes Chamberlin's papers, including much of her artwork and her subscription lists for the editions of *Canadian Wild Flowers* are in the Thomas Fisher Rare Book Library (University of Toronto).

61 Macoun's text is included in The Traill Family Collection at Library and Archives Canada (8:12, 921–1013).

62 Traill, *Plant Life in Canada*, viii.

63 Fletcher to Traill, November 1883 (Traill Family Collection).

64 Fletcher to Traill, 26 July 1894 (Traill Family Collection).

65 See Blyth, "History as Storytelling."

66 Natural history study did not die at the end of the nineteenth century. The production of popular field books on flowers, birds, and other aspects of the natural world would in some ways replace that old way.

67 Traill, *Plant Life in Canada*, vii.

68 Ibid., xvii.

69 Ibid., xvi.

70 Fowler, *The Embroidered Tent*, 36.

71 Traill, *Plant Life in Canada*, 60. Plate 7 provides an illustration.

72 The phrase was Traill's choice of subtitle for the *Plant Life* volume.

73 Ibid, 85–6.

74 Thomas, "The Strickland Sisters," 67.

BIBLIOGRAPHY

Archives

Fletcher, James. Letters to Catharine Parr Traill (1883–1894) in The Traill Family Collection, Library and Archives Canada, 1: 475–545.

The Traill Family Collection, Library and Archives Canada, MG29.

Books and Articles

Ainley, Marianne Gosztonyi. "Last in the Field? Canadian Women Natural Scientists, 1815–1965." In *Despite the Odds: Essays on Canadian Women and Science*, edited by M.G Ainley, 25–62. Montreal: Véhicule Press, 1990.

– "Science in Canada's Backwoods: Catharine Parr Traill." In *Natural Eloquence: Women Reinscribe Science*, edited by Barbara T. Gates and Ann B. Shteir, 79–97. Madison: University of Wisconsin Press, 1997.

Ballstadt, Carl. "Catharine Parr Traill." In *Canadian Writers and Their Works*, Fiction Series, edited by Robert Lecker, Jack David, and Ellen Quigley. Toronto: ECW Press, 1983.

Berger, Carl. *Science, God and Nature in Victorian Canada*. Toronto: University of Toronto Press, 1983.

Bigot, Corinne, "Did They Go Native? Representations of First Encounters and Personal Interrelations with First Nations Canadians in the Writing of Susanna Moodie and Catharine Parr Traill." *Journal of Commonwealth Literature* 49, no. 1 (2014): 99–111.

Blyth, Molly. "History as Storytelling: Mrs. Traill's Scrapbook of Stoney Lake Grasses." *The Peterborough Review* 1, no. 1 (1994): 97–110.

Byrne, Angela. "My Little Readers: Catharine Parr Traill's Natural Histories for Children." *Journal of Literature and Science* 8 (2015): 86–101.

Chamberlin (née Moodie), Agnes Fitzgibbon. "Introductory Notes." In *Canadian Wild Flowers* by Catharine Parr Traill, 4th edition. Toronto: W. Briggs, 1895.

Cole, Jean. "Catharine Parr Traill – Botanist." In *Portraits: Peterborough Area Women Past and Present*, edited by Gail Corbett, 73–9. Woodview, ON: Portraits Group, 1975.

Collard, Elizabeth, "Flowers to Heal and Comfort: Mrs. Traill's Books for Collectors." *Canadian Collector* 13 (May/June, 1978): 32–6.

Dignam, Mary Ella. "Women in the Development of Art." In *Women in Canada: Their Life and Work*, compiled by the National Council of Women of Canada. Toronto, 1900.

Fowler, Marian. *The Embroidered Tent: Five Gentlewomen in Early Canada*. Toronto: Anansi, 1982.

Frye, Northrop. *The Bush Garden: Essays in the Canadian Imagination*. Toronto: Anansi, 1971.

– "Conclusion." In *The Literary History of Canada*, edited by Carl Klinck. Toronto: University of Toronto Press, 1963.

Gerson, Carole. "Nobler Savages: Representations of Native Women in the Writings of Susanna Moodie and Catharine Parr Traill." *Journal of Canadian Studies* 32, no. 2 (Summer 1997): 5–21.

Gray, Charlotte. *Sisters in the Wilderness: The Lives of Susanna Moodie and Catharine Parr Traill.* Toronto: Viking, 1999.

Hull, John. *British Flora: or a Linnean Arrangement of British Plants.* London: R. Bickerstaff, 1799.

MacCallum, Elizabeth. "Catharine Parr Traill: A Nineteenth-Century Ontario Naturalist." *The Beaver* 306 (Autumn): 39–45.

Macoun, John. *Catalogue of Canadian Plants – Part 1, Polypetalae.* Proceedings of the Canadian Institute, Ottawa no. 42, 23 November 1883.

Nuttal, Thomas. *The Genera of North American Plants; and a Catalogue of the Species, to the year 1817.* Philadelphia: D. Heartt, 1818.

Peterman, Michael A. *Sisters in Two Worlds: A Visual Biography of Susanna Moodie and Catharine Parr Traill.* Toronto: Doubleday, 2008.

– "'Splendid Anachronism': The Record of Catharine Parr Traill's Struggles as an Amateur Botanist in Nineteenth-Century Canada." In *Re(dis)covering Our Foremothers: Nineteenth-Century Canadian Women Writers*, edited by Lorraine McMullen. Ottawa: University of Ottawa Press, 1990.

Peterman, Michael A., and Carl Ballstadt, *Forest and Other Gleanings: The Fugitive Writings of Catharine Parr Traill.* Ottawa: University of Ottawa Press, 1994.

Pursh, Frederick. *Flora Americae Septentriolas; or A Systematic Arrangement and Description of the Plants of North America.* London: White and Cochrane, 1814.

Raglon, Rebecca."Little Goody Two-Shoes: Reassessing the Work of Catharine Parr Traill." In *This Elusive Land: Women and the Canadian Environment*, edited by Melody Hessing, Rebecca Raglon, and Catriona Sandlilands, 4–18. Vancouver: UBC Press, 2005.

Stafford, Jane. "How One Subject Springs Out of Another: The Strickland Family and Early Children's Literature." *Women's Writing* 25, no. 1 (February 2018): 51–6.

Strickland, Elizabeth, ed. *The Lady's Magazine, Or Mirror of the Belles-Lettres, Fine Arts, Music, Drama, Fashions, &c.* London, 1831–32.

Thomas, Clara. "The Strickland Sisters." In *The Clear Spirit*, edited by Mary Quayle Innis. Toronto: University of Toronto Press, 1966.

Thompson, Elizabeth. "'With Axe and Fire': Catharine Parr Traill's Ecological Vision." *Canadian Poetry* 42 (Spring/Summer 1998).

Traill, Catharine Parr. *The Backwoods of Canada.* London: Charles Knight, 1836.

– *The Backwoods of Canada.* Edited by Michael A. Peterman. Ottawa: Carleton University Press, 1997.

– *Canadian Crusoes; or, Lost in the Backwoods.* London: Virtue, 1852.

– *Canadian Wild Flowers*. Montreal: Lovell, 1868 [Rpts. 1869, 1870]. 4th edition, Toronto: W. Briggs, 1895.

– *Catharine Parr Traill's The Female Emigrant's Guide: Cooking with a Canadian Classic*. Edited by Nathalie Cooke, and Fiona Lucas. Montreal and Kingston: McGill-Queen's University Press, 2018.

– *Cot and Cradle Stories*. Toronto: W. Briggs, 1895.

– "Floral Sketches No. 1: The Violet." *The Literary Garland* (February 1843): 87–90.

– "Flowers and Their Moral Teaching." *The British American Magazine* (May 1865): 55.

– "Forest Gleanings." *The Anglo-American Magazine*. Toronto, 1852–53.

– *I Bless You in My Heart: Selected Correspondence of Catharine Parr Traill*. Edited by Carl Ballstadt, Elizabeth Hopkins, and Michael A. Peterman. Toronto: University of Toronto Press, 1996.

– *Lady Mary and Her Nurse; or, A Peep into the Canadian Forest*. London: Virtue, 1856.

– *Sketch Book of a Young Naturalist; or, Hints to the Students of Nature*. London: Harvey and Darton, 1831.

– *Sketches from Nature; or Hints to Juvenile Naturalists*. London: Harvey and Darton, 1830.

– *Studies of Plant Life in Canada, or, Gleanings from Forest, Lake and Plain*. Ottawa: Woodburn, 1884; 2nd edition. Toronto: W. Briggs, 1906.

Walton, Isaak. *The Compleat Angler; or the Contemplative Man's Recreation* (1653). Oxford: Clarendon Press, 1983.

White, Gilbert. *The Natural History of Selborne* (1789). Edited by Richard Mabey. Harmondsworth, UK: Penguin, 1977.

Zeller, Suzanne. *Inventing Canada: Early Victorian Science and the Idea of a Transcontinental Nation*. Toronto: University of Toronto Press, 1987.

8

"Botany … a Prominent Study"

Isabella McIntosh's Ferns and Natural History in 1860s Montreal

KAREN STANWORTH

> Pupils are reminded to continue their practical lessons from school at home during the holidays. With this view *Botany has been made a prominent study*, in the hope that in the further prosecution of that branch of natural science, combining as it does *healthful & spirit stirring recreation with scientific knowledge*, an endless source of pleasure shall be stored up for future years.

With these words, Isabella McIntosh concluded her annual speech to the graduating class of girls attending Bute House School for Young Ladies in Montreal.[1] In describing botany in 1865 as a "healthful & spirit stirring recreation" that results in "scientific knowledge," McIntosh makes it clear that she did not see the study of botany by women as limited to a "polite" pursuit.[2] She emphasized that the pursuit of botany would add "interest to every country walk & in the study of plants unfolding that volume of Creation which reveals the wisdom & goodness of the Creator" (her underline). When McIntosh urges her students to follow their practical or applied lessons of botany, she is urging them to engage in a summer occupation that will elevate their spiritual being and feed their physical and intellectual needs. In this, McIntosh echoes Presbyterian belief that secular life was infused with meaning only in relation to spiritual life. Indeed, Rev. Mathieson, minister of St Andrew's Presbyterian church in Montreal, attended by McIntosh, asserted that it was the duty of "every Christian state" to infuse its civil institutions with Christian morality, which will be made "the foundation of all … public proceedings."[3] This belief system is critical to an understanding of approaches to natural history across the British world at this time. It did not narrow the potential value of natural history studies but rather required the individual to commit more fully to learning how to be in the world. As a consequence, McIntosh's engagement with botany in the British colony of Canada needs to be investigated in relation to her work as an educator and a Presbyterian.

A groundbreaking educator in Montreal during the 1860s and '70s, McIntosh was one of the few working women in Canada East to receive recognition for both her educational innovations and her contributions to pteridology – the study of ferns and related plants. Pursuing the physical, moral, and intellectual development of young women, McIntosh introduced botany, object lessons, and gymnastics to the classical literature and arts curriculum typical of private schools for girls at the time. As there was no public high school for girls in Montreal until the High School for Girls opened in 1875, these private schools merit closer study, if only to gain a better grasp of the early history of education for women in Canada.[4] In particular, I draw attention to the teaching in McIntosh's school of non-traditional subjects – such as botany – alongside the implementation of other innovative educational strategies. An understanding of McIntosh's pedagogical experimentation casts light on the nature and significance of her contribution to the development and spread of interest in botany in midcentury Canada.

An ongoing challenge for historical research is to find evidence of women's professional activities – evidence that seeks to complicate the existing, often myopic, interpretation of history read primarily through the lives of men. Not surprisingly, as with her female contemporaries, there are few traces in print of McIntosh's work and the career she made for herself as a professional educator in nineteenth-century Canada East. Brenda MacKay and Michael Firmin examine how the development of separate school systems after the Education Act of 1841 made this possible.[5] Following on the heels of the Act of Union in 1840, Upper Canada (now Ontario) and Lower Canada (now Quebec) were united as the Province of Canada. However, the newly renamed colonies, Canada West and Canada East respectively, retained control of their own school system. Despite a growing movement toward public education, schooling for girls remained limited and consequently the middling and wealthier classes sought private schooling for their daughters.

New resources about McIntosh that have come to light as a result of this research are largely drawn from two published sources and from archival materials held at the McCord Museum in Montreal. The published sources include advertisements for the school that date from 1863–74, a printed prospectus describing its curriculum,[6] a newspaper article about the school's closing, and two references in the 1864 transactions of the Natural History Society of Montreal (NHSM). The 1864 volume describes the proceedings of the society for the year. Its focus on botany provides insight into how botany was discussed and practised and makes reference to the exhibition of McIntosh's collection of ferns during their October meeting. Fortunately, some ephemera found in the McCord Museum archives help to fill in the picture a bit further. The McCord holds a scrapbook, a photographic album, and eight handwritten drafts of McIntosh's annual end-of-term speeches as head of Bute House School for Young Ladies.[7] Weaving together data gleaned from the archival traces of her life, this chapter presents McIntosh as a well-respected academic whose contributions to science and education were acknowledged as significant by her community.

THE RISE OF NATURAL HISTORY IN 1860S MONTREAL

During the 1860s, Montreal was a city of contrasts and rapid change in urban demographics. Rising from 76,000 in 1844 to more than 100,000 in population in 1861, the city was home to a diverse range of immigrants.[8] From newly arrived, poverty-stricken Irish to long-established French Canadians, the city's populace was far from homogenous. The city was approximately 45 per cent of Anglo descent and governed largely by the rich commercial and industrialist class. The wealthiest lived in what was known as the Golden Square Mile, a one-mile-square area, which is best seen today in the mansions still evident on Sherbrooke, Redpath, and McGill College Streets.[9] The southern slopes of Mont Royal were home to successful merchants, railway men and bankers, such as the Allan, Redpath, Ogilvy, and Stephen families, many of whom were Scots Presbyterians like McIntosh whose families emigrated in the early nineteenth century.[10] McIntosh conducted her boarding (and later, day) school within the limits of the Golden Square Mile, occupying first Burnside House (on upper McGill College Street) until 1864 and then nearby Bute House (on Sherbrooke at the corner of McGill College). In the 1840s, Burnside House was positioned in an expansive rural setting that reached south to what is now known as Boulevard René-Lévesque. By the 1860s, Bute House School was relocated to the south side of the rapidly expanding grounds of the University of McGill College. As we shall see, the locations are significant since Mont Royal was recommended as a local hunting ground for botanists.

What might botany have meant as what McIntosh termed a "spirit stirring" subject and a physical practice for her and her contemporaries in an increasingly industrial urban environment when the pursuit of specimens for natural

8.1
"Burnside," residence of the Late James McGill, Montreal, QC, c. 1842, engraving by John H. McNaughton.

science was evolving as a scientific pursuit? In early eighteenth-century European thought, "science" was considered part of natural philosophy, a term that was understood to include topics related to natural history, as well as mathematics, astronomy, and other earth studies. However, by the middle of the nineteenth century, the notion of a "scientific community" was developing, natural history increasingly was defined separately from natural philosophy, and the disciplines we know today were soon to be seen as distinct and separate sciences.[11] In many minds "the nineteenth century" and "science" became synonymous with "progress."[12] Nevertheless, Carl Berger's classic study *Science, God, and Nature in Victorian Canada* reminds us that a significant sense of reverence continued to permeate the study of the natural world well after the initial appearance of Darwinian thought – both in Canada and the rest of the Western world.[13] In other words, the study of natural history was seen as entirely consistent with a deep faith in a Christian God as the creator of nature. At the same time, botany was perceived as one of several sciences that could also serve both British imperial and local needs.[14] In her examination of Victorian science in Canada, Suzanne Zeller argues plant studies grew more important in Canada when it became apparent midcentury that agricultural practices needed reforming.[15] Potato blight had ruined crops in Canada West

8.2
Bute House, 1873, page 2 in Rynas Scrapbook.

(1830s to 60s), farmers felt that the seeds had "run out" (due to repetitive planting practices), and the British government repealed the Corn Laws (1847) which had favoured corn imports from Canada.[16] Thus, toward the end of the century, there was increased emphasis on a practical education in botany and chemistry as a means of ensuring Canada's agricultural future.

The emergence of Montreal as a scientific community supported by a knowledgeable public in the first half of the nineteenth century is a history of individual passions, social tensions, and colonial rule. Below I briefly consider the development of the scientific community and then how the citizenry was nurtured so as to establish the requisite financial backing and intellectual milieu necessary to support science. This provides context for how the category of natural history, with its separate disciplines of geology, botany, and zoology, emerged at several institutions in Montreal by the end of the nineteenth century. This evolving scientific community supported McIntosh's introduction of botany to the Bute House curriculum as well as her botanical pursuits.

With the influx of British immigrants to Montreal came aspiring professionals, including many Scottish physicians and educators. Richard Vaudry traces the influence of Alexander Skakel, a schoolteacher trained at King's

College, Aberdeen, who migrated to Montreal in 1798.[17] As a "science popularizer," Skakel gave public lectures on natural philosophy in the 1820s and 30s and arguably was "crucial to the creation of both pure and applied science in Montreal."[18] One of Skakel's protégés, physician Andrew Holmes also offered lectures on experimental chemistry – lectures that were offered at Skakel's house.[19] A few years later, Holmes was one of the founders of the Montreal General Hospital (1819) and its affiliated teaching college, the Medical Institute of Montreal (1821). Unable to grant degrees, the Medical Institute established a strategic merger with the newly chartered, but inactive, University of McGill College in 1828, thus forming the Faculty of Medicine.[20] For the institute, this move allowed them to receive royal assent to give medical degrees and provided a functioning faculty for the floundering University of McGill College. However, it was to take nearly two decades of court cases over the land and property (which included Indigenous and Black enslaved persons), near bankruptcy, inactive principals, civic unrest, and a duel before the university opened a faculty on campus.[21]

Before 1843,[22] all courses were part of the Medical Institute program. William Lyons taught *materia medica* (pharmacy and medical plants) and dietetics; Holmes taught chemistry and pharmacy. There is no record of botany being taught except for the mention at the bottom of an 1823 advertisement for the schedule of medical lectures, which advised that Holmes would teach botany in summer.[23] Holmes, as well as his fellow physician professor, John Stephenson, were trained at the University of Edinburgh where they had opportunity to study botany as well as *materia medica*. Holmes was a keen botanist, prepared a large herbarium of Canadian plants (not just medicinal plants), and was a major participant in the activities of the NHSM. He was the president from 1836 to 1841 and took responsibility for building a network of correspondents throughout the Atlantic world.[24] In 1856, he donated his herbarium to the university, where it now forms the core of the McGill Herbarium. When Holmes assumed senior roles at the Medical Institute, Dr Archibald Hall took over *materia medica* in 1835 and chemistry in 1842. Lactance Papineau is recorded as lecturing on botany at McGill, again in the summer session like Holmes. However, the record indicates that Papineau's various illnesses prevented him from fulfilling his teaching duties.[25] In 1855–56, botany was finally taught in the Faculty of Arts and Science at McGill when James Barnston, a physician and botanist, delivered a series of lectures on structural botany and was appointed the first professor of botany at McGill in 1857.[26] Tragically, he too died of an unspecified illness in 1858, but nevertheless, he managed to make a significant contribution to the promotion of botany when he organized the Botanical Society of Montreal, which found its home in the NHSM. He also was the society's librarian and was the curator for the museum, which was open to the public. Additionally, he published several articles on botany in the society's transactions of 1857 and 1859.[27]

Arguably, published articles, open museums, and public lectures stimulated the growth of a public sphere that encouraged the advancement of science.[28] It is unclear whether Holmes's summer lectures served to continue the Skakel tradition of a public, money-making venture or were part of the Medical Institute. However, given that botany was not a required course for a medical degree and it was not required for the Faculty of Arts at the university until 1845, it seems likely that Holmes's botany lectures were driven by a combination of passion and poverty, which functioned to public benefit.[29] Holmes's lectures, open to a paying public, helped to create a demand for what might be called scientific "edutainment." His public lectures in chemistry continued until at least 1836, when John Samuel McCord mentioned the scientific equipment used by Holmes as a means of embellishing his presentation.[30] Both men were members of the NHSM and seemed to share an interest in practical sciences. The popularization of natural sciences was enhanced further by the donation of Presbyterian minister Rev. James Somerville, who left £1000 upon his death in 1833 to support the Annual Somerville Science Lectures.[31] Beginning in 1840, the NHSM offered the Somerville annual lectures, which also were open to the public.

The public forum for botany that was nurtured by the NHSM temporarily declined in the 1840s during the ongoing political issues arising from civic unrest, the Rebellion Losses Bill, and eventually the burning of the parliament in 1849.[32] One near casualty was the NHSM, which declined due to a lack of membership. The society's revitalization in the mid-1850s is credited to John Dawson. His appointment in 1855 as principal of McGill and professor of chemistry, agriculture, and natural history[33] led to an expansion of the campus and his involvement in the society. Dawson's course on natural history included field trips on Mont Royal as the practical side of studies in zoology, botany, and geology. Botany is described in the McGill College prospectus as "including Anatomy and Physiology of Plants, and Systematic botany, with notices of the Flora of Canada, and of the principal plants applied to medicinal and other useful purposes."[34] Natural history was a required fourth-year course for all students in the Faculty of Arts,[35] and optional in earlier years. It also became an optional course in the Faculty of Medicine, where students were only required to study *materia medica*.[36] McGill College had also introduced a "High School Department," which offered a "Liberal Education." This included classics and natural philosophy, but natural history was "left to the pupil's future opportunities."[37]

By 1850, Montreal was rapidly recovering from the tensions of the 1840s and the economic depression of the late 1840s; industry and manufacturing increased, immigration continued, and the Golden Square Mile was established.[38] As the hub for Presbyterian thought in Canada at the time, Montreal provided a fertile ground for pedagogical and professional developments in science, industry, and culture. John Dawson, substantially self-trained, was committed to modern scientific practices such as field-based research, plant systematics, taphonomy (a branch of paleontology dealing with fossilization), and ecosystem reconstruction.[39] Dawson quickly became involved with the NHSM and was named its president in 1856. Mainly consisting of Anglophone businessmen and industrialists, membership was augmented significantly during the late 1850s and 1860s as a result of Dawson's leadership and through the participation of professional scientists who were involved with the Geological Survey of the province. Such was the case, for example, of Joseph Frederick Whiteaves, a paleontologist with the Geological Survey, who served as the curator of the NHSM's museum from 1864 to 1876.[40]

In 1863, an annual conversazione was added to the NHSM's public offerings. It was open to both men and women who entered through separate doors, women presenting their pink tickets and men their blue; this suggests that single women such as Isabella McIntosh would have been able to enjoy what Stanley Frost calls "a regular feature of Montreal life."[41] Another initiative by NHSM that would have encouraged women botanists was the 1867 launch of an annual event called "exploring field day."[42] Additionally, the category of "associate member" for women was introduced in 1868 to increase income as well as "attendance at the meetings, the visits to the Museum, and the general interest felt in the concerns of the society."[43]

Under Dawson's stewardship, the NHSM began publishing transactions on a bi-monthly basis and participated more extensively in the exchange of scientific literature produced in the United States, England, Scotland, and some European countries (mainly Germany and France).[44] The society had its greatest number of members in 1864, when the transactions were revamped and volume 1 in the new series appeared under the title *The Canadian Naturalist and Geologist*.[45] While the transactions are not the only source of information on the state of natural history in the 1860s – there were other societies that produced scientific literature (such as the Literary and Historical Society of Quebec [LHSQ]) – the NHSM publication provides significant evidence of the ways in which new knowledge was disseminated. In addition to printing papers of interest, the journal included details of the public conversazione, monthly meetings, donations to the library and museum, and reprints of pertinent articles from other publications.

Volume 1 of the new series of NHSM transactions focused on botany, with 148 out of 460 pages dealing with flora and fauna. Geology was the second largest area of focus, with other topics such as zoology having a fairly minor presence. The first article, "Botanical Science – Record of Progress," was by George Lawson, professor of chemistry and natural history at Queen's University. As the founder of the short-lived Botanical Society of Canada, located at Queen's, Lawson set the tone for the volume's focus on botany.[46] Under the title "Flora of Canada," his article records a request from British botanist William Jackson Hooker for "information for the occurrence in Canada not hitherto recorded as Canadian, when accompanied by authenticated specimens."[47] The rest of the article consists of twenty-five observations on various aspects of "progress" in the understanding of Canadian flora, including a seemingly random listing of observations about, for example, Canadian nuts, Indian bamboo, and nettle fibre.

The same volume records speeches made and the entertainments featured at the second annual conversazione of the NHSM held on 2 February 1864. William Sheppard, an elder statesman of the society, spoke about natural history and its progress in Canada. Sheppard briefly mentioned the work done by men such as André Michaux on Canadian flora, Francis Masson, a collector for the Royal Garden at Kew, and the botanical work of Frederick Pursh. He specifically addressed "the ladies" in attendance at the conversazione, offering a word of encouragement, and cited the herbarium of Countess Dalhousie as proof of what can be accomplished "by those having the leisure to study."[48] Sheppard discussed at some length in his speech the significance of having a museum for the advancement of knowledge and its valuable collection, which had been augmented by the merger of the NHSM with the Geological Survey. Finally, Sheppard asserted the importance of the society's facilities for the study of natural history at colleges and universities in Canada East and Canada West.

At that same gathering, a Professor Wiles of Bishop's College, Lennoxville, spoke on the importance of the NHSM as one of the "chief ornaments of the city, and one of the most efficient promoters of progress."[49] Wiles also drew attention to the museum "*collection of specimens, illustrative of facts and phenomena of natural history*"[50] (original italics), and noted that "the impressions producible by verbal description, even when accompanied by good drawings, is neither so vivid nor so permanent as that which is created by the sight and handling of the objects."[51] In particular he drew attention to "specimens appropriately arranged." Specimens, he explained, "as perfect as possible of their several kinds, not neglecting artistic display, but at the same time sacrificing even that (when necessary) to the conditions of order in a series, position, and other requirements for rendering illustrative objects of natural history really useful … It is not so much the extent of a museum that renders it useful in the cause of science, as attention to unity of purpose, and to natural conditions."[52]

ISABELLA MCINTOSH AND FERN COLLECTING

The emphasis in the speeches at the conversazione on the value of collections of specimens highlights the significance of herbaria, as these provided the opportunity for students of nature in Canada to make their observations and comparisons in person, if not in situ. The difficulty of travel via river and rough roads had improved slightly in midcentury Canada with the advent of rail travel, and new recordings of plants outside the cities took on even greater significance, especially in terms of how plants resembled or differed from the more familiar European specimens. Significantly in this regard, Isabella McIntosh is cited twice in the 1864 volume of the NHSM for her botanical work. Her name appears first in "Notes on the Habitats and Varieties of Some Canadian Ferns" by David Ross McCord (October 1864), in which forty-five specimens of ferns are listed by scientific name, location of verified occurrence, and the name of the botanical authority.[53] McCord's commentary is sparse and generally limited to comments on difficulties or differences that he encountered as a collector, but his last entry is more fulsome. He records the following: "*Ophioglossum Vulgatum*. Melbourne, C.E., where exceedingly fine specimens are to be found, Miss Isabella McIntosh, Burnside House, Montreal. This fern, with the *Botrychium Lunaria* mentioned above, are now for the first time recorded as being found in Canada proper."[54]

The second reference to McIntosh occurs later in the volume under the synopsis of the first monthly meeting of the 1864–65 season, where it is noted that McCord also "exhibited a collection of native ferns, collected and prepared by Miss Isabella McIntosh (of Burnside House), among which were three species of peculiar interest."[55] The author of the report, probably McCord himself as he was the recording secretary of the NHSM, provides further details about McIntosh's three species "of peculiar interest": "The first was the 'green spleenwort' (*Asplenium viride*, Hudson), a small species occurring somewhat rarely in mountainous districts in England and in various localities in Europe. It had been previously detected in Gaspé, in the summer of 1863, by John Bell and this was the only station in which it was previously known to occur in Canada. The other species are the 'Adder's Tongue fern,' (*Ophioglossum vulgatum*), of which fine specimens were collected at Melbourne, in the Eastern Townships; and the 'Moonwort' (*Botrychium Lunaria*), two species well known to inhabit Europe, but *now for the first time recorded* as occurring in Canada."[56]

McCord's paper on the habitats and varieties of Canadian ferns highlights the significance of a first recording and of corrections to the existing record.[57] McCord describes McIntosh as not only having "fine specimens" in her collection but also indicates that she was the first to record the *Ophioglossum vulgatum* in Melbourne. Melbourne is a small community in the Eastern Townships of Quebec, southeast of Montreal, consisting of rolling hills, grassy meadows, and woodland. The location of Melbourne is noteworthy because McCord frequently referred to ferns found there. It appears to have been a newly discovered fern habitat in the mid-1860s, with seventeen ferns found there by John A. Bothwell in addition to the ferns found and recorded by McIntosh.[58] In comparison to most of rural Canada East, Melbourne had become relatively accessible as the Grand Trunk Railway ran from Montreal to Richmond, which is about five kilometres from Melbourne. Nevertheless, the trip would have taken most of a day and even five kilometres over rough roads would be time consuming. Given that it is 120 kilometres from Montreal, McIntosh must have taken significant journeys in the course of her botanical collecting. Her excursions likely would have been confined to the summer season when the school was closed during July and August. Not a lady "of leisure" as Sheppard imagined in his advice to the botanical ladies, McIntosh was a working woman who did her botanical research in her summer holidays.

The NHSM report about the native ferns "collected and prepared" by McIntosh, also refers to an exhibition of "a series of ferns collected in Canada West, by Mrs. Traill, the well-known authoress."[59] A comment followed by Dawson that "the study of non-flowering plants of Canada was as yet but in its infancy" but that the studies were "an investigation full of interest and promise."[60] Presumably, the exhibition of these ferns in 1864 was meant to illustrate and educate the largely well-informed members of the NHSM about those non-flowering plants. The physical display of McIntosh's fern collection alongside that of Catharine Parr Traill suggests that the two collections were regarded as compatible and complementary. Beyond fulfilling the need to observe specimens, the fern collections seem to have functioned as a form of Flora for the native ferns found in the united province of Canada, with McIntosh's representing Canada East and Traill's representing Canada West. Traill's herbaria, well known today,[61] follow the standard protocol for preparation as described by her contemporaries (see discussion of Barnston's "Hints to the Young Botanist," below). Her Maidenhair fern, for example, shows Traill's attention to preparation, including common and Latin names, date, location, and short description of the habitat as "woody and shade rich vegetable mould." She also created a number of scrapbooks of plants that were meant as gifts, so her work varies from a more standardized display to decorative treatments. Even the albums intended as gifts usually contain the common and Latin names and location.[62] Marianne Ainley has made a strong argument for reconsidering the amateur status typically ascribed to Traill, especially given James Fletcher's insistence in 1894 that her work "is one of the greatest botanical achievements that anyone could achieve."[63] As Fletcher became the first permanent Dominion entomologist and botanist in 1887, his judgment indicates that Traill met his standards for botanical work.

Since the collections of ferns by Traill and McIntosh were exhibited together at the NHSM, it is likely that McIntosh's sheets were "prepared and described" much like Traill's. Arguably, the use of McIntosh's collection in this context establishes her status as both a collector and an authority. Certainly McCord and, by extension, the NHSM seem to consider the ferns collected by McIntosh as authoritative. Functioning to identify new and previously unknown species in Lower Canada, McIntosh's collection served as definitive a purpose as the samples in the collection created by Traill. With McCord's act of recording and the act of exhibition of the sheets, McIntosh's samples did not need further authentication, at least not for the members of the NHSM.

TEACHING FROM NATURE

The education of young people in natural history was an ongoing concern of the NHSM. In 1857, Elkanah Billings prefaced the first volume of the old series of *The Canadian Naturalist and Geologist* by advising the public that the NHSM would publish articles on geology and natural history and would record discoveries in the two domains.[64] Billings, a paleontologist of the Geological Survey of Canada and a member of the NHSM, is credited with launching the bi-monthly publication, which was to be sold by subscription and could be bound into annual volumes of 480 pages, complete with illustrations and an index. As such, it would be useful for educators and students alike.

The educational thrust was made explicit in James Barnston's "Hints to the Young Botanist, regarding the collection, naming and preservation of Plants,"

8.3
Catharine Parr Traill, pressed specimen of Maiden Hair Fern, c. 1888.

published in the second volume of the same series, in 1857.[65] Barnston describes how to collect and carry plants for further study, and the tools required for this task, including a tin carry case known as a "vasculum," in which to place the plants between sheets of paper interspersed with one-inch wood board. The young botanist is advised to "confine himself to a certain well-defined district and, to collect all the plants within it," such as Montreal's Mont Royal. All the species of a genus should be placed together and he recommends "using the Natural Orders" and arranging the samples according to "an approved system." He describes in detail how to prepare and attach the samples to the sheets and the use of "stiff portfolios or volumes" to form herbaria. He notes that each sample should be identified by name, locality, date of collecting, "and other particulars if noteworthy."[66] It is notable that Barnston refers to the "young" botanist, not the amateur. This discrimination does not appear to be in use, at least not in Barnston's writing. Equally, in terms of language, Barnston refers to such collections as herbaria, thus assuming that the act of preparation is for research purposes.

Sometime during the school year of 1856–57, Dawson presented a lecture on the educational aspects of natural history as part of a "popular course of the Natural History Society of Montreal."[67] The "popular course" seems to have been both an effort to educate the general public and a means of introducing the subject to young scholars, such as high school students intending to become teachers. According to the printed excerpts, the lecture was one of a series of popular talks aimed at "the cultivation among its members and the public in general, [of] a taste for those useful and ennobling pursuits to which it is devoted."[68] Relatively little of the extract actually speaks to the presumed objects of natural science, rather it seems to function as a defence of the subject in the face of doubt from some "modern educators" as to whether any training is required in the observation of nature.[69]

Dawson opines on modes of instruction when he argues that "much of the teaching of natural history by object lessons and popular text-books is too inaccurate and superficial, and too deficient in enlargement of view, to be really useful. In our colleges and scientific institutions, nature should be illustrated by the best possible collections of typical forms, explained and examined under the guidance of the best scientific skill."[70] While Dawson is critiquing how commercial object lessons were used, he is not dismissing the use of collections for teaching with objects. His objection is to the relatively elementary object lessons that were printed on cards sold by the Normal School's Book Depository.[71] The educational theory behind object lessons was that young children would learn concepts better through accurate observation and questioning of local and foreign objects (ranging from cones and spheres to cards with samples of raw resources such as a piece of raw cotton stuck to a printed card that explained how cotton grew, its uses, etc.).[72] Dawson argued for a kind of guided learning, in which students describe and sense the qualities of the natural object. This is similar to the guided questions used with commercial object lessons but the questions and observations are applied to actual objects. In this pedagogy, classrooms in elementary schools often held small cabinets of samples of local minerals, seeds, dried plants, etc. for this purpose.[73]

Dawson advocated a modern approach to teaching the observation and understanding of natural history, starting with elementary object lessons about nature, then moving to natural history studies in advanced schools, and culminating in natural science as taught in colleges. He wrote, "Observation in external things begins in earliest infancy, and is essential to the happiness, the utilities, and even the safety of after life … The prevailing deficiency in accurate observation … can in part be remedied by the object lessons in our primary schools, by the drawing, the music, the natural history of our more advanced schools; by the natural science of our colleges … [And so] we shall have more independent observation and originality."[74]

McIntosh was clearly interested in similar modern education theory and practice. She experimented with an early introduction of kindergarten and tried a variety of innovations in the curriculum. An extant published prospectus for Bute House provides an overview of the curriculum, fee schedules for both resident and day pupils, and costs of optional expenses.[75] Courses include English, French, Latin, and music, similar to those in the classics and arts found at other seminaries for girls.[76] Other offerings are atypical, notably the "Infant Class," which is described as "perhaps, the most interesting part of the establishment."[77] A newly built addition was "fitted up with a gallery and has all the appliances for carrying out what is known as the 'Training System,' which is based on a regard to a child's threefold nature, physical, intellectual and moral."[78] In referring to the "Training System," McIntosh assumes that her audience will know that she is using the well-known Scottish Normal School approach to education. The 1859 edition of *The Training System of Education* by David Stow lays out – in its 588 pages – details for developing the threefold nature of the child.[79] Also known as the "moral training system," Stow's program sought to integrate the moral and physical development of the child with intellectual growth.

McIntosh continued her emphasis on the threefold development across the curriculum with the introduction of classes in gymnastics and French. The

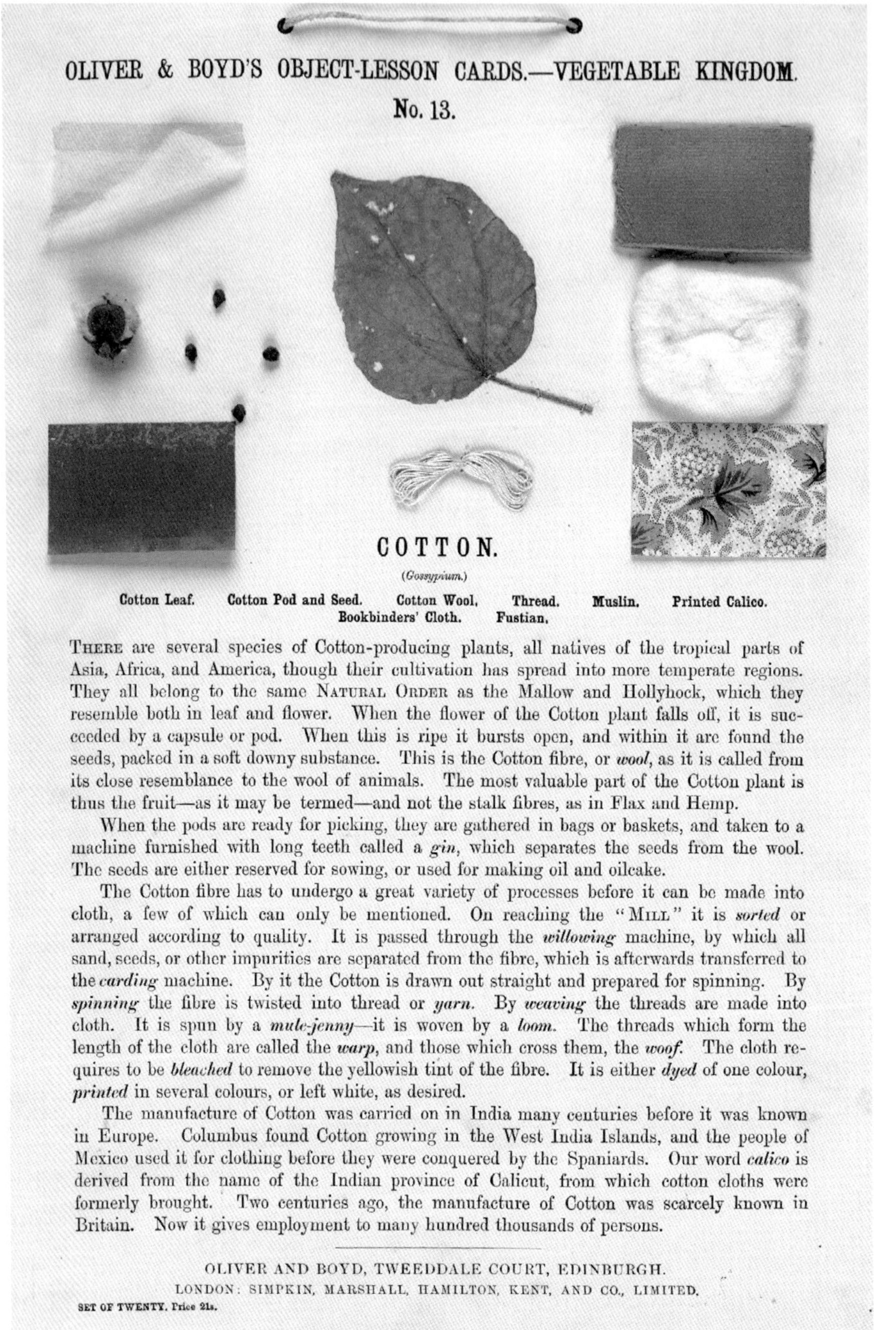

OLIVER & BOYD'S OBJECT-LESSON CARDS.—VEGETABLE KINGDOM.

No. 13.

COTTON.

(*Gossypium.*)

Cotton Leaf. Cotton Pod and Seed. Cotton Wool. Thread. Muslin. Printed Calico. Bookbinders' Cloth. Fustian.

THERE are several species of Cotton-producing plants, all natives of the tropical parts of Asia, Africa, and America, though their cultivation has spread into more temperate regions. They all belong to the same NATURAL ORDER as the Mallow and Hollyhock, which they resemble both in leaf and flower. When the flower of the Cotton plant falls off, it is succeeded by a capsule or pod. When this is ripe it bursts open, and within it are found the seeds, packed in a soft downy substance. This is the Cotton fibre, or *wool*, as it is called from its close resemblance to the wool of animals. The most valuable part of the Cotton plant is thus the fruit—as it may be termed—and not the stalk fibres, as in Flax and Hemp.

When the pods are ready for picking, they are gathered in bags or baskets, and taken to a machine furnished with long teeth called a *gin*, which separates the seeds from the wool. The seeds are either reserved for sowing, or used for making oil and oilcake.

The Cotton fibre has to undergo a great variety of processes before it can be made into cloth, a few of which can only be mentioned. On reaching the "MILL" it is *sorted* or arranged according to quality. It is passed through the *willowing* machine, by which all sand, seeds, or other impurities are separated from the fibre, which is afterwards transferred to the *carding* machine. By it the Cotton is drawn out straight and prepared for spinning. By *spinning* the fibre is twisted into thread or *yarn*. By *weaving* the threads are made into cloth. It is spun by a *mule-jenny*—it is woven by a *loom*. The threads which form the length of the cloth are called the *warp*, and those which cross them, the *woof*. The cloth requires to be *bleached* to remove the yellowish tint of the fibre. It is either *dyed* of one colour, *printed* in several colours, or left white, as desired.

The manufacture of Cotton was carried on in India many centuries before it was known in Europe. Columbus found Cotton growing in the West India Islands, and the people of Mexico used it for clothing before they were conquered by the Spaniards. Our word *calico* is derived from the name of the Indian province of Calicut, from which cotton cloths were formerly brought. Two centuries ago, the manufacture of Cotton was scarcely known in Britain. Now it gives employment to many hundred thousands of persons.

OLIVER AND BOYD, TWEEDDALE COURT, EDINBURGH.

LONDON: SIMPKIN, MARSHALL, HAMILTON, KENT, AND CO., LIMITED.

SET OF TWENTY. Price 21s.

8.4

Cotton, *Oliver and Boyd's Object-Lesson Cards*, Vegetable Kingdom, No. 13 (London: Simpkin and Co., 1880).

report for 1866–67 describes how the new room served both the infant school and the gymnastics program. It allowed McIntosh "to carry out what has always been a favourite scheme, that is making physical education under the direction of a competent instructor a part of the regular training of the establishment." She ends her report that year by asking the parents to support this "experiment," which they obviously agreed to, as the following year she reported, "our scholars had two lessons weekly."[80] Sometimes experiments failed, such as the temporary addition of a French instructor for the younger students. While French was well integrated into the senior curriculum, in 1867–68, she reports that the parents were not prepared to pay extra for access by the younger students.[81] However, the younger students did enjoy the variety and innovative use of gallery seating so that all the students could see the teacher's demonstrations with object lessons. The object lessons then would be used much as a chalkboard is today to point out various visual characteristics of the object of study. As McIntosh noted, these lessons were "nicely interwoven" into reading and writing lessons for junior students.[82]

Senior pupils benefited from "the privilege of attending a course of lectures on geology, given weekly by Principal Dawson at the Normal School." They also enjoyed another course on astronomy prepared by Dr Edwards expressly for schools which were illustrated by a valuable series of diagrams that used a "Lime Light," an early form of magic lantern that used a combination of oxygen and hydrogen to project a bright image.[83] Thus, in addition to regular classes in natural history at Bute, the older girls were attending at least some lectures given at the Normal School, which was administered by the University of McGill College. The lectures at the teaching college may have encouraged some of the girls to take training there. As Dawson reports in his autobiography, "educationally the [normal] school has thus been a professional college for women providing a thorough course extending over three years … which has opened a field of usefulness for educated women."[84] He also refers to how in addition to his duties for the university, he accepted the principalship of the Normal School and taught weekly lectures in natural history.[85] Presumably, some of these were the geology lectures that the Bute girls were attending. In the fifteen years prior to the opening of the Montreal High School for Girls, McIntosh carried out a program of rigorous training, with specialized lectures in geology, astronomy, and botany augmented by curricular innovations and experimentation.

McIntosh's emphasis on botany as a "healthful & spirit stirring recreation with scientific knowledge" that "reveals the wisdom & goodness of the Creator" echoes the views of Dawson, who was well known to McIntosh. She shared his firm convictions about the power of the study of nature. When McIntosh urges her students to follow their practical lessons in botany, she is counselling them toward a summer occupation that will elevate their spiritual being and feed their physical and intellectual needs. Indeed, she echoes Barnston's belief that the study of creation would raise their souls. In this, McIntosh, Dawson, and Barnston echo the Presbyterian belief that their secular life is infused with meaning through its relation to their spiritual life.

Ann Shteir has argued that British botanist John Lindley's intent in the 1830s was to "separate botany from the domains of politeness and accomplishment that for decades had linked botany, in his view, to women and to gentility" in order to shape a new kind of botanist who would be a scientific expert.[86] This is the effort in which I believe McIntosh was participating. The brief comment made by McIntosh in her 1865 end of year speech signals her approach to nature. She does not just encourage her students to get out into the country for health or for drawing plants as a polite accomplishment. She urges her students to use their practical lessons as "healthful & spirit stirring recreation with scientific knowledge." In this short statement, McIntosh reveals how she sees botany and where it fits in a curriculum for young women. A subject worthy of "prominent study" in the curriculum, it is not a minor area for leisure activity or an addition to art class. As in all areas of education for a Presbyterian, study of the world brings the student joy in the knowledge of "Creation." Education historian Roderick McLeod has argued that, whereas Anglican education focused on social reproduction, Presbyterian education was driven by the notion of personal salvation. In this regard, McIntosh insists, girls may have finished their schooling with her but they are not finished building their knowledge of the spiritual in creation. Rather, her graduates have "passed the portal to that higher education which every one may give to herself."[87]

BIOGRAPHY MATTERS

Research on Isabella McIntosh (1828–c. 1915) and her background gives glimpses into how family circumstances and social opportunity fed into her life decisions and achievements and led to her interests in education and botanical science. Biographical detail is significant here. Although many details are not clear, her family moved to Montreal from Martintown, Ontario, and she grew up in direct contact with Montreal's cultural and scientific community. The rental of part of the old McGill farmstead, Burnside House, and an unspecified acreage in the early 1860s as their first home in Montreal led to the development of important social connections that later contributed to the success of McIntosh's approach to teaching and the establishment of her school.[88] It also matters that her father died when she was in her early thirties, suddenly thrusting her and her sisters into the need to survive by working.[89] They must have had sufficient schooling as they turned to education as a profession.

Further archival evidence about McIntosh is found in the pages of *The Presbyterian*, the monthly publication of the Presbyterian church of Canada. This journal focused primarily on religious activity in the approved churches but also followed the social circles of significant figures. Isabella McIntosh and her school are frequently mentioned. Constrained by formalities and a faith that forbad aggrandizement, *The Presbyterian* "humbly invoked" the attention of its readers to the speech given by Rev. Gavin Lang, minister of St Andrew's Presbyterian, in which he characterized the seminary at Bute House "as an educational Marvel … A thousand young ladies had been trained under that roof … who had permanently and richly benefited by the instruction and good principles" of Isabella McIntosh.[90] Further revealing McIntosh's connections and influence is the advertisement for the school, which appeared regularly in the *Presbyterian*.[91] The advertisement entitled "Board and Education" informs "their friends" that they are prepared to receive a few additional boarders. Most importantly, the ad includes a list of referees willing to answer enquiries from interested parents. The names include Rev. Dr Mathieson and Rev. W.M. Inglis, both of St Andrew's Presbyterian; Rev. Dr Wilkes, head of the Congregationalist Church in Montreal; Alex. Morris, Esq., MPP, interested in academic matters; the Hon. L.H. Holton, MPP and later MP, who defined himself as a progressive Liberal and supported free public education;[92] and Rev. W. Leitch, principal of Queen's College. Also a Presbyterian minister, Leitch had a strong interest in botany, having worked with Dr George Lawson,[93] professor of chemistry at Queen's, where they founded the short-lived Botanical Society of Canada in Kingston in 1860.[94] Thus, McIntosh was very familiar with some of the primary male figures in Montreal's religious, economic, scientific, and educational communities. These familial, social, and cultural connections allowed McIntosh to build a professional profile in Montreal at a time when women of middling classes had little or no opportunity for access to scientific community or even the expectation of proximity to its activities.

A scrapbook held at the McCord Archives in Montreal and known as the Rynas Scrapbook provides insight into McIntosh's web of social and scientific connections. Internal evidence suggests that McIntosh filled the scrapbook over at least two decades from the 1860s to 1880s, pasting into its pages a variety

8.5
Pressed leaves in Rynas Scrapbook, possibly Osier Willows.

of articles cut from printed materials and watercolour images of botanical specimens (painted by her sister); some pressed leaves are inserted at various intervals in the scrapbook as well. Obituaries in the scrapbook, for example, suggest McIntosh's social and educational connections. The only obituary for a non-family member, other than that of Rev. Alexander Mathieson, the Presbyterian minister of St Andrew's, Montreal, was for Hannah Lyman, a principal of Vassar College.[95] As the obituary points out, Lyman was a renowned educationalist who had opened Cote House, a fashionable girls' school in Montreal, in 1839. McIntosh was evidently following Lyman's career, and it is possible that Lyman either taught McIntosh or provided some mentorship.

The Rynas Scrapbook also functioned as a vasculum for McIntosh. Between its pages are pressed a dozen or more plant leaves. Of varying lengths and widths, the leaves are possibly from Osier Willows. They are distributed, seemingly randomly, throughout the scrapbook and are neither identified nor labelled. Evidence of an ongoing interest in plants, they leave me with an unfulfilled desire. Why did McIntosh press them? Were they all from one tree or a cluster of trees? Did they demonstrate anomalies that she wanted to consider? Perhaps we cannot know why, but it is clear to me that McIntosh was engaged with plants for many years. She taught botany, collected samples, and contributed to public discussion about Ontario flora. I have found no evidence that McIntosh assembled a herbarium other than the reference in the NHSM transactions, but it is possible that some of her vouchers or specimen sheets are in the McGill University Herbarium, which received the collection of the NHSM in 1927.[96] In the 1880s, publications such as John Macoun's *Catalogue of Canadian Plants* (1883–1902) and George Lawson's *The School Fern-Flora of Canada* (1889) (meant as an appendix to Asa Gray's *Botany for Young People & Common Schools*, 1872), served to consolidate earlier research and popularize botany as a school subject. While the complete extent of McIntosh's contribution to botany may not be fully known, further examination of fern collections in the last decades of the century may cast a clearer light on the specific work of midcentury plant collectors such as Isabella McIntosh to the development of botany as a distinct science.

NOTES

1 Isabella McIntosh, untitled (Annual Report for Bute School, 1865), unpaginated.

2 Ann Shteir provides a detailed overview of the development of this polite culture of botany, particularly in the eighteenth century, in chapter 2 of *Cultivating Women, Cultivating Science*; Brittany Blachford surveys the contributions of nineteenth-century Canadian and other colonial women to botany in "Exploring the History of Women in Botany." See also Gates, *Kindred Nature*, 132; and Shirk, "Contributions to Botany," 297.

3 McKim, *Boundless Dominion*, 154.

4 Research on schooling for girls in nineteenth-century Canada is limited. For a general overview of early education in Canada, see Axelrod, *The Promise of Schooling*, and Wilson and Audet, *Canadian Education*, 214–40.

5 MacKay and Firmin, "Development of Private Education in Canada," 62–3.

6 MG 2020, Acc. No. 077, Ref. 1, McGill University Archives.

7 McIntosh, "End of Year Reports," McCord Archives, 1864–72.

8 According to Statistics Canada, the population of Montreal grew from 76,000 in 1844 to 100,830 in 1861. See https://www150.statcan.gc.ca/n1/pub/11-627-m/11-627-m2017020-eng.htm.

9 The area was roughly the south slope of Mount Royal, bounded by Boulevard Robert-Bourassa, Boulevard René-Lévesque, and Rue Guy. For more on Montreal at this time, see Linteau, *Brève histoire de Montréal*, 75–87.

10 This immigration followed the defeat of the French by the British in 1760. Scottish merchants created the North West Company to compete with the English-run Hudson's Bay Company, creating an opportunity for more Scots to find work in the Canadas. Particularly relevant is Lucille Campey's *An Unstoppable Force* in which she notes the Scots valuing of culture and religion and that many emigrated by choice, paying their own way. See also J.M. Bumstead's classic, *The Scots in Canada*, and E. Cowen's "The Scots' Imagining of Canada."

11 Cahan, "Looking at Nineteenth-Century Science," 3–15.

12 Ibid., 5. Cahan reviews the contributions of three significant twentieth-century writers who attempted to capture the shifts of nineteenth-century thought: John T. Merz, John D. Bernal, and Joseph Ben-David. On the professionalization of science, see also: Shortt, "Physicians, Science, and Status," 51–68.

13 Berger, *Science, God, and Nature.*

14 For a comprehensive overview of science and culture in nineteenth-century Canada, see Berger, *Science, God, and Nature*; Zeller, *Inventing Canada*. For a detailed description of nineteenth-century science and culture in Montreal, also see Richard Vaudry, *Andrew Fernando Holmes.*

15 Zeller, *Inventing Canada*, 181–269, esp. 200–30.

16 Ibid., 200–1.

17 On Skakel, see Vaudry, *Andrew Fernando Holmes*, 38–44.

18 Vaudry covers in detail the career of Skakel using local advertisements to support his argument that Skakel was able to amass a collection of scientific instruments through the popularization of science. Skakel even used a subscription by elite citizens, including James McGill, to pay for the instruments. Vaudry, *Andrew Fernando Holmes*, 43–4.

19 Vaudry, *Andrew Fernando Holmes*, 42.

20 The Faculty of Arts was formed in name only when the university got its charter in 1821; the first four appointed professors were not paid, never taught, and did not live in Montreal. Hannaway and Cruess, *McGill Medicine*, 4.

21 Most of the intrigue is covered by McMillan, or Hannaway and Cruess, and more recently by Vaudry. Reference to the enslaved people held by James McGill and later by Desrivieres, his son-in-law, who claimed the land and property, can be found in Nelson, *Slavery, Geography, and Empire* or Mackey, *Done with Slavery*. The possession of enslaved people during the court cases is seen in Else and McCrossan, "Biographies of People Enslaved by James McGill," 62–7.

22 McMillan covers various aspects of the opening. McMillan, *McGill and Its Story*, 140.

23 The newspaper advertisement for the summer lectures is reproduced in Hannaway and Cruess, *McGill Medicine*, 13.

24 Vaudry, *Andrew Fernando Holmes*, 144–6.

25 Papineau died in early 1851. Papineau's appointment to the faculty and his failure to fulfill his teaching duties is noted in Vaudry, *Andrew Fernando Holmes*, 206–7.

26 Barnston, "Introductory lecture to the course on botany, delivered before the students of arts and medicine, McGill College, session, 1857," 334–45. Also see Zeller and Noble, "James Barnston."

27 The second article appeared posthumously. Zeller and Noble, "James Barnston."

28 Vaudry discusses this "public sphere of science" in *Andrew Fernando Holmes*, 43.

29 It is not clear if Holmes actually did the summer lectures every year as claimed in the secondary literature. More research on the advertisements would be required. Botany was housed in and a compulsory course of the Faculty of Arts and was optional for medical students. Hannaway and Cruess, *McGill Medicine*, 63.

30 Vaudry, *Andrew Fernando Holmes*, 102.

31 Frost, "Science Education in the Nineteenth Century," 36.

32 Conflict over the Rebellion Losses Bill along with growing violence on the streets deeply affected the public sphere at the time. See Horner, *Taking to the Streets*, esp. ch. 6 for insight into how the divisiveness affected daily life.

33 Eakins and Eakins, "Dawson, Sir John William."

34 McGill College, *Prospectus*, 11. For more on the High School of Montreal and its Presbyterian roots, see Rider and McNabb, *Kingdom of the Mind*, 273.

35 McGill College, *Prospectus*, 7.

36 Taught by Prof. W. Wright, M.D., McGill College, *Prospectus*, 13, https://archive.org/details/prospectusofmcgioomcgi/page/n1/mode/2up.

37 Ibid., 21.

38 What is now Sherbrooke Street was built across the lower portion of James McGill's original forty-seven acre donation to the college; as a result many bankers, industrialists, railway presidents, etc. built their grand mansions in the newly available land to the north of the city proper. See Linteau, *Brève histoire de Montréal*, esp. ch. 7. For an illustrated, somewhat updated version of this history, see the McCord Museum's "The Industrial City" webpage, http://collections.musee-mccord.qc.ca/scripts/printtour.php?tourID=VQ_P2_14_EN.

39 Sheets-Pyenson, *John William Dawson*; Falcon-Lang and Calder, "Sir William Dawson."

40 Gagnon, "Natural History Society of Montreal's Museum."

41 Ibid.

42 Frost, "Science Education," 33.

43 NHSM, *The Canadian Naturalist and Geologist* 3 (New Series), 408.

44 For more on learned societies in Canada and their social, pedagogical, and professional concerns, see Stanworth, "The Politics of Display," 120–41; McTavish, "Learning to See in New Brunswick, 1862–1929," 553–81; Murray, *Come Bright Improvement!*; virtualmuseum.ca, "Athena's Heirs: Science Education in 19th Century Quebec," http://www.virtualmuseum.ca/edu/ViewLoitLo.do;jsessionid=592207A721B58D6CD9CED4E410D09513?method=preview&lang=EN&id=4032.

45 The first issue of what was to be known as the "old series" or "first series" appeared in 1857, shortly after Dawson became the president of the NHSM. The new series commenced in 1864.

46 On George Lawson and the Botanical Society of Canada, see the chapter by David Galbraith in this volume.

47 Lawson, "Botanical Science – Record of Progress," 1.

48 Anon, "Annual Conversazione," *The Canadian Naturalist and Geologist*, New Series 1, (1864): 53–7. The use of quotation marks suggests that the author of the report on the conversazione is recording Wiles's entire speech.

49 Anon, "Annual Conversazione," 59.

50 Ibid., 60.

51 Ibid., 61.

52 Ibid., 63.

53 On McCord, see Miller, "David Ross McCord," and Harvey, "Location, Location, Location," 57–82.

54 McCord, "Notes on the Habitats," 361.

55 Anon, "Natural History Society – Report on first monthly meeting for 1864–65," *The Canadian Naturalist and Geologist* 2, New Series, (1864): 371–2.

56 My emphasis. Ibid., 372.

57 McCord, "Notes on the Habitats," 361.

58 Bothwell earned his MA from McGill and practised as a lawyer in Drummondville, near Melbourne.

59 See the chapter on Catharine Parr Traill by Michael Peterman in this volume.

60 Recording Secretary (McCord), "Proceedings," 372.

61 The largest collection of Traill's herbaria is in the National Herbarium of Canada at the Canadian Museum of Nature.

62 See. e.g., Evergreen Wood fern CAN 588 171.2; Botrichium sampler, CAN 588169.3. National Herbarium of Canada.

63 Ainley, "Science in Canada's Backwoods," 93. On Fletcher, see Riegert, "James Fletcher."

64 Billings, *The Canadian Naturalist and Geologist*, 1. For more on Billings, see Désilets and Pageau, "Billings, Elkanah."

65 Barnston, "Hints to the Young Botanist," 127–35; 335–45. Barnston, a medical doctor,

was the son of George Barnston, a well-known Canadian botanist who collected specimens with Professor John Macoun (author of the *Catalogue of Canadian Plants*, 1883–1902).

66 Barnston, "Hints to the Young Botanist," 127–35.

67 Dawson, "Natural History in its Educational Aspects." The nine-page document does not have publication details, but below the title is "reprinted from Barnard's American Journal of Education for June, 1857." The lecture may have been intended for McGill's Normal School; the article seems to have been reprinted for use in other schools. On the title page below the name of the author is added, "(Extracts from the Introductory Course of the Natural History Society of Montreal, Winter of 1856–57.)"

68 Dawson, "Natural History in its Educational Aspects," 428.

69 Ibid., 429.

70 Ibid., 430.

71 Stanworth, *Visibly Canadian*, chap. 2, esp. 87–92.

72 Hodgins, *Hints and Suggestions*, 92. Recording secretary and deputy superintendent of schools in Ontario, George Hodgins noted that by 1886, "Object lessons are popular as June picnics. The lyceum has caught the fever. With the speaker comes the stereopticon. The lecture must be a picture gallery. Some are introducing the blackboard into the pulpit. The chalk competes with the pen. Ideas must be pictures, they tell us."

73 Stanworth, *Visibly Canadian*, chap. 2.

74 Dawson, "Natural History in its Educational Aspect," 429–30.

75 McIntosh, "Educational Circular of the Misses Neil McIntosh, 844 Sherbrooke Street," MG 2020 (975, item 2). This "Educational Circular," an undated, folded sheet of A4 paper, was probably printed around 1866, when "Bute House" replaced the name of "Bute Place" in the city's directory listings. The name change probably occurred when the school moved to the larger quarters provided at Bute. City directories, such as Lovell's *Montreal Street Directory*, are useful resources for this kind of research because they provide a glimpse into the relationship between similar industries, as advertisements reveal details about goods and services in relation to each other.

76 A competitor, Mrs Simpson's Establishment for the Board and Education of Young Ladies, offered classes in arithmetic, English grammar, history, geography, religious instruction, French, and Latin. Equally, at the Select Grammar School for male youth, classes were in English, mathematics, and classics.

77 McIntosh, Report of 1866–67.

78 A "gallery" consisted of a raised platform with rows of elevated seating. McIntosh, Report of 1866–67.

79 Stow. *The Training System*, 307. First published in 1836, the training system developed from Stow's interest in providing Sunday school for poor youth in Glasgow.

80 McIntosh, Report of 1866–67.

81 McIntosh, Report of 1862–63. McIntosh adds, in an "effort to have all students in Junior and Senior study French," the extra fees were "reduced by 50% and [we] engaged a Mr. Dondrick [Dondick?] however the parents of the junior school have removed all but 8 of 25 from French lessons."

82 McIntosh, Report of 1867–68.

83 McIntosh, Report of 1868–69.

84 Dawson, *Fifty Years of Work in Canada*, 119.

85 Ibid., 121.

86 Shteir, "Botany and 'Modern' Science," 33. See also Kennedy, "Science Culture in the Nineteenth Century," 29.

87 MacLeod, "In the Hallowed Name of Religion," 233. Indeed these tensions were dominant in the 1850s, as the Presbyterians feared that McGill would be an Anglican university. Thus the hiring of Dawson further signalled the success of the Presbyterian social-scientific community.

88 Stanworth, *Visibly Canadian*, 280. Note: my original understanding was that McIntosh rented all forty-seven acres. That is unlikely as the Faculty of Arts was active by that time. Presumably, by 1860, McIntosh's grocer father rented a much smaller acreage.

89 In 1860–61, the women were listed in *Mackay's Directory* as running a preparatory school at 13 Phillip's Square, where their brother was also listed under "carpenter." Presumably it was their father's death in 1861 which forced them to turn their home, Burnside House, into a boarding school in order to earn more income. For more details on the father, Neil, see Stanworth, *Visibly Canadian*, 280, n32 and n33.

90 *The Presbyterian*, June 1873, 154.

91 McIntosh, "Board and Education," 1862. The earliest advertisement dates from when the school was at Burnside House on McGill College Avenue.

92 Klassen, "Luther Hamilton Holton."

93 Connor, "To Promote the Cause of Science," 3–33.

94 Neatby, "William Leitch."

95 Rynas Scrapbook, McCord Archives.

96 McGill Herbarium curator Dr Frieda Beauregard confirms that there are "a number" of sheets from unknown collectors in the NHSM collection, pers. email, 4 April 2021.

BIBLIOGRAPHY

Archives

McCord Museum Archives, *Rynas Scrapbook*, C258/B1.1.

McGill University Archives MG 2020, Acc. No. 077, Ref. 1 McGill University Archives.

The Canadian Naturalist and Geologist 1, 1856–1857; New Series 1–3, 1864–68.

Catharine Parr Traill Collection, The National Herbarium of Canada, Canadian Museum of Nature, CAN 588 165–171; CAN 588 169, https://nature.ca/en/research-collections/collections/plants-algae.

Lovell's Montreal Street Directory, 1842–1977.

McIntosh, Isabella. "Educational Circular of the Misses Neil McIntosh, 844 Sherbrooke Street," McGill Archives, MG 2020 Acc. No. 975, item 2.

– "Untitled Annual Reports for Bute School, Montreal, 1862–1873." Some years missing. Envelope inserted in scrapbook. McCord Archives, C258/B1.1

Books and Articles

Ainley, Marianne Gosztonyi. "Science in Canada's Backwoods: Catharine Parr Traill." In *Natural Eloquence: Women Reinscribe Science*, edited by Barbara T. Gates and Ann B. Shteir, 79–97. Madison: University of Wisconsin Press, 1997.

Anderson, R.D. *Education and Opportunity in Victorian Scotland: School and Universities*. Oxford: Oxford University Press, 1983.

Axelrod, Paul. *The Promise of Schooling: Education in Canada, 1800–1914*. Scarborough, ON: Prentice-Hall of Canada, 1997.

Barnston, James. "General Remarks on the Study of Nature, with Special Reference to Botany." *The Canadian Naturalist and Geologist* 2 (1857): 34–40.

– "Hints to the Young Botanist, regarding the collection, naming and preserving of plants." *The Canadian Naturalist and Geologist* 2 (1857): 127–35.

– "Introductory Lecture to the Course on Botany, delivered before the students in Arts and Medicine, McGill College, Session 1857." *The Canadian Naturalist and Geologist* 2 (1857): 335–45.

Berger, Carl. *Science, God, and Nature in Victorian Canada*. Toronto, University of Toronto Press, 1983.

Billings, E. *The Canadian Naturalist and Geologist* 1 (1856): 1.

Blachford, Brittany. "Exploring the History of Women in Botany: Tracing Seven Female Contributors of the UBC Herbarium." University of British Columbia, 2013, https://open.library.ubc.ca/cIRcle/collections/undergraduateresearch/52966/items/1.0075696.

Bumstead, J.M. *The Scots in Canada*. Ottawa: Canadian Historical Association, 1982.

Cahan, David. "Looking at Nineteenth-Century Science: An Introduction." In *From Natural Philosophy to the Sciences: Writing the History of Nineteenth-Century Science*, edited by D. Cahan, 3–15. Chicago: University of Chicago Press, 2003.

Campey, Lucille H. *An Unstoppable Force: The Scottish Exodus to Canada*. Toronto: Dundurn Press, 2008.

Cowan, E.J. "The Scots' Imagining of Canada." In *A Kingdom of the Mind*, edited by Peter E. Heather McNabb, 3–21. Montreal and Kingston: McGill-Queen's University Press, 2006.

Connor, J.T.H. "To Promote the Cause of Science: George Lawson and the Botanical Society of Canada, 1860–1863." *Scientia Canadensis* 10, no. 1 (1986): 3–33

Dalzel, Andrew. *History of the University of Edinburgh from Its Foundation: By Andrew Dalzel. With a Memoir of the Author. II.* United Kingdom: Edmonston & Douglas, 1862.

Davie, George E. *The Democratic Intellect: Scotland and Her Universities in the Nineteenth Century*. Edinburgh: Edinburgh University Press, 1961.

Dawson, William. *Fifty Years of Work in Canada, Scientific and Educational*. London & Edinburgh: Balantyne, Hanson & Co, 1901.

– "Introductory Lecture to the Course on Botany, Delivered before the Students of Arts and Medicine, McGill College, Session, 1857." 334–45.

– "Natural History in Its Educational Aspects (extracts from a lecture)." 1856–57. Reprinted from *Barnard's American Journal of Education*, June 1857: 428–36.

Désilets, Andrée, and Yvon Pageau. "Billings, Elkanah." In *Dictionary of Canadian Biography* 10. Toronto and Laval: University of Toronto/Université Laval, 2003, http://www.biographi.ca/en/bio/billings_elkanah_10E.html.

Eakins, Peter, and Jean Sinnamon Eakins. "Dawson, Sir John William." In *Dictionary of Canadian Biography* 12, University of Toronto/Université Laval, 2003.

Else, Gemma, and Colin McCrossan. "Biographies of People Enslaved by James McGill." In *Slavery, Geography and Empire in Nineteenth-Century Marine Landscapes of Montreal and Jamaica*, 62–7. London: Routledge, Taylor & Francis, 2016.

Falcon-Lang, Howard J., and John H. Calder. "Sir William Dawson (1820–1899): A Very Modern Paleobotanist." *Atlantic Geology* 41, no. 2–3, online, https://journals.lib.unb.ca/index.php/ag/article/view/181.

Frost, Stanley B. "Science Education in the Nineteeth Century: The Natural History Society of Montreal, 1828–1925." *The McGill Journal of Education* 28, no. 1 (1981): 31–43, https://mje.mcgill.ca/article/view/7440/5370.

Gagnon, Hervé. "The Natural History Society of Montreal's Museum and the Socio-Economic Significance of Museums in 19th-Century Canada." *Scientia Canadensis* 18, no.2 (1994): 103–35.

Gates, Barbara T. *Kindred Nature: Victorian and Edwardian Women Embrace the Living World*. Chicago: The University of Chicago Press, 1998.

Hannaway, Joseph, and Richard L. Cruess. *McGill Medicine. First Half-Century, 1829–1885*. Montreal and Kingston: McGill-Queen's University Press, 1996.

Harvey, Kathryn. "Location, Location, Location: David Ross McCord and the Makings of Canadian History." *Journal of the Canadian Historical Association* 19, no. 1 (2008): 57–82.

Hodgins, George. *Hints and Suggestions on School Architecture and Hygiene*. Education Department, Toronto: Department of Public Instruction for Upper Canada, 1887.

Horner, Dan. *Taking to the Streets: Crowds, Politics, and the Urban Experience in Mid-Nineteenth-Century Montreal*. Montreal and Kingston: McGill-Queen's University Press, 2020.

Kennedy, Kerrie. "Science Culture in the Nineteenth Century: Women and the Botanical Society of Canada." *Resources for Feminist Research* 33, no. 3–4 (2010): 47–70.

Klassen, H.C. "Holton, Luther Hamilton." In *Dictionary of Canadian Biography* 10, University of Toronto/Université Laval, 2003, http://www.biographi.ca/en/bio/holton_luther_hamilton_10E.html.

Lawson, George. "Botanical Science – Record of Progress," *The Canadian Naturalist and Geologist* 1, New Series (1864): 1–13.

– *The School Fern-Flora of Canada: Comprising descriptions of all ferns known to inhabit the Dominion, together with an account of their geographical range or prevalence in the several provinces, and special localities for the rarer species*. Halifax, NS: A.&W. Mackinlay, 1889.

– "Synopsis of Canadian Ferns and Filicoid Plants." *The Canadian Naturalist and Geologist* 1, New Series (1864): 262–300.

Lenoir, Timothy. *Instituting Science: The Cultural Production of Scientific Disciplines*. Stanford: Stanford University Press, 1997.

Linteau, Paul-André. *Brève histoire de Montréal*. Montreal: Boréal, 1992.

– "The Industrial City." The McCord Museum website, http://collections.muse-emccord.qc.ca/scripts/printtour.php?tourID=VQ_P2_14_EN&Lang=2.

Lloydlangston, A. "Women in Botany and the Canadian Federal Department of Agriculture, 1887–1919." *Scientia Canadensis* 29, no. 2 (2006): 99–130.

MacKay, B., and M. Firmin. "The Historical Development of Private Education in Canada." *Educational Research & Perspectives* 35 (2008): 57–72.

Masters, D.C. "The Scottish Tradition in Higher Education." In *The Scottish Tradition in Canada*, edited by W. Stanford Reid, 248–72. Toronto: University of Toronto Press, 1976.

McCord, David Ross. "Notes on the Habitats and Some Varieties of Canadian Ferns." *The Canadian Naturalist and Geologist*, New Series 1 (1864): 354–62.

McGill College. *Prospectus of McGill College, Montreal, 1856–57*. Montreal: McGill College, 1857.

McIntosh, Isabella. "Board and Education." *The Presbyterian*, February 1862.

McKim, Dennis. *Boundless Dominion: Providence, Politics, and the Early Canadian Presbyterian Worldview*. Montreal and Kingston: McGill-Queen's University Press, 2017.

– "God's Garden: Nature, Order, and the Presbyterian Conception of the British North American 'Wilderness.'" *Journal of Canadian Studies/Revue d'études canadiennes* 51, no. 2 (2017): 398–433.

McLeod, Roderick. "In the Hallowed Name of Religion. Scots and Public Education in Nineteenth Century Montreal." In *Kingdom of the Mind*, edited by Peter E. Rider and Heather McNabb, 227–41. Montreal and Kingston: McGill-Queen's University Press, 2006.

McMillan, Cyrus. *McGill and Its Story, 1821–1921*. London; New York : John Lane Company, 1921, 2009.

McTavish, Lianne. "Learning to See in New Brunswick, 1862–1929." *The Canadian Historical Review* 87, no. 4 (2006): 553–81.

Miller, Pamela. "David Ross McCord." In *Dictionary of Canadian Biography* 15, University of Toronto/Université Laval, 2003, http://www.biographi.ca/en/bio/mccord_david_ross_15E.html.

Murray, H. *Come Bright Improvement! The Literary Societies of Ontario*. Toronto: University of Toronto, 2002.

Neatby, Hilda. "William Leitch." In *Dictionary of Canadian Biography* 9, University of Toronto/Université Laval, 2003, http://www.biographi.ca/en/bio/leitch_william_9E.html.

Nelson, Charmaine A. *Slavery, Geography and Empire in Nineteenth-Century Marine Landscapes of Montreal and Jamaica*. London: Routledge, Taylor & Francis, 2016.

Nelson, Charmaine A., and Student Authors. *Slavery and McGill University: Bicentenary Recommendations*. Report to McGill University, 2020, https://www.blackcanadianstudies.com/Recommendations_and_Report.pdf.

Pickstone, John. *Ways of Knowing: A New History of Science, Technology and Medicine*. Chicago: University of Chicago Press, 2000.

Recording Secretary (McCord). "Proceedings for Oct 1864." *The Canadian Naturalist and Geologist*, New Series 1 (1864): 370–2

Riegert, Paul W. "James Fletcher." In *Dictionary of Canadian Biography* 13, University of Toronto/Université Laval, 2003, http://www.biographi.ca/en/bio/fletcher_james_13E.html.

Rider, Peter E., and Heather McNabb. *Kingdom of the Mind: How the Scots Helped Make Canada*. Montreal and Kingston: McGill-Queen's University Press, 2006.

Sheets-Pyenson, Susan. *John William Dawson, Faith, Hope and Science*. Montreal and Kingston: McGill-Queen's University Press, 1995.

Shirk, Henrietta. "Contributions to Botany, the Female Science, by Two Eighteenth-Century Women Technical Communicators." UBC Herbarium Course Project, https://open.library.ubc.ca/media/download/pdf/52966/1.0075696/3.

Shortt, S.E. "Physicians, Science, and Status: Issues in the Professionalization of Anglo-American Medicine in the Nineteenth Century." *Medical History* 27, no. 1 (1983): 51–68.

Shteir, Ann B. "Botany and 'Modern' Science in Victorian England." *Women, Gender, and Science: New Directions, Osiris* 12 (1997): 29–38.

– *Cultivating Women, Cultivating Science: Flora's Daughters and Botany in England, 1760 to 1860.* Baltimore, MD: John Hopkin's University Press, 1996.

Sigrist, René, and Eric D. Widmer. "Training Links and Transmission of Knowledge in 18th Century Botany: a social network analysis." REDES – *Revista hispana para el análisis de redes sociales* 21, no. 7 (2011). http://revista-redes.rediris.es.

Stanworth, Karen. "The Politics of Display: A 'Literary and Historical' Definition of Quebec in 1830s British North America." In *Art Apart. Artefacts, Institutions and Ideology in Britain and North America*, edited by Marcia Pointon, 120–41. Manchester: Manchester University Press, 1994.

– "In Sight of Visual Culture. Pedagogy and the Discipline of Art History." *Symploke Journal of Comparative Literature and Theory*, special issue: "Sites of Pedagogy" 10, no. 1 (2003): 106–17.

– *Visibly Canadian: Imaging Collective Identities in the Canadas, 1820–1910.* Montreal and Kingston: McGill-Queen's University Press, 2014.

Statistics Canada. "Montreal, 375 Years: Celebrate Our History." https://www150.statcan.gc.ca/n1/pub/11-627-m/11-627-m2017020-eng.htm.

Stoll, Mark. *Inherit the Holy Mountain.* Oxford: Oxford University Press, 2015.

Stow, David. *The Training System of Education Including Moral School Training for Large Towns, and Normal Seminary, for training teachers to conduct the system.* London: Longman Green, 1859.

Sylvain, Philippe. "Henry Wilkes." In *Dictionary of Canadian Biography* 11, University of Toronto/Université Laval, 2003, http://www.biographi.ca/en/bio/wilkes_henry_11E.html.

Triggar, R. "Protestant Restructuring in the Canadian City: Church and Mission in the Industrial Working-Class District of Griffintown, Montreal." *Urban History Review* 31, no. 1 (2002): 5–18.

Vaudry, Richard W. *Andrew Fernando Holmes: Protestantism, Medicine, and Science in Nineteenth-Century Montreal.* Toronto: University of Toronto Press, 2020.

Wilson, J. Donald, Robert M. Stamp, and L. Audet, eds. *Canadian Education: A History.* Scarborough, ON: Prentice-Hall of Canada, 1970.

Zeller, Suzanne E. *Inventing Canada. Early Victorian Science and the Idea of a Transcontinental Nation.* Montreal and Kingston: McGill-Queen's University Press, 2009.

– "Sir William Logan and Sir J.W. Dawson: Victorian Geology as Scottish Science in a New World Environment." In *Kingdom of the Mind*, edited by Peter E. Rider and Heather McNabb, 167–82. Montreal and Kingston: McGill-Queen's University Press, 2006.

Zeller, Suzanne E., and John H. Noble. "James Barnston." In *Dictionary of Canadian Biography*, 8, University of Toronto/Université Laval, 2003.

PART FOUR
Seeing and Making

9

Botanical Albums as Theoretical Objects

Sophie Pemberton and the Logic of Identity

KRISTINA HUNEAULT

Leaf through one of the two albums of British Columbia wildflower drawings created by the Canadian artist Sophie Pemberton in 1895 and you will come up against this supremely Victorian assertion: "Men are God's trees, and women are God's flowers."[1] The snippet of verse, taken from the poet laureate Tennyson, is offset against its page by artfully cropped sprigs of a flowering plant, closely and accurately rendered. Turn the watercolour over and you will learn that the plant is a member of the *Caprifoliaceae* family: *Lonicera involucrata* or Bush Honeysuckle. The Latin name has been carefully inscribed – family, genus, and species – in keeping with botany's conventions for scientific classification. The page comes near the end of the album and there is a sense in which it knits the project together. Whereas in the album's earlier pages the rigours of binomial nomenclature vie somewhat uneasily with the lyricism of the accompanying verses, here, with the line from Tennyson, poetic and botanical knowledge systems fall into alignment as parallel articulations of a worldview in which entities may be clearly defined and categorized. The naturalism of the accompanying image only reinforces this impression of a readily known and knowable world. For a moment, indeed, the whole album slips into place – verse, image, and scientific nomenclature all working together in the service of a particular kind of conceptual certainty: to each thing its identity, the page proclaims, well delineated and safely beyond doubt.

Such certitude is one of the hallmarks of botanical drawing. Like conventional portraiture, the botanical genre is supremely an art of identity, focussed on exactitude of likeness and oriented toward viewer recognition. While portraiture traditionally allies identity and interiority, however, botanical art focuses on surface morphology, and whereas the portraitist commonly seeks to

9.1
A page from one of Sophie Pemberton's two 1895 albums of wildflower drawings. She gave this album, entitled *Wild Flowers BC*, as a gift to her brother Frederick and his wife Mary Ann. Scientific nomenclature was included on the backs of pages in both albums. The combination of images, Latin nomenclature, and snippets of poetry was familiar from floral dictionaries of the day. (See figure 9.4.)

capture the uniqueness of an individual, the botanical artist conventionally depicts those qualities most typical of a species group. In nineteenth-century Canada, the primary producers of such images were women, and, much like the specimens they drew, the female artists who created collections of plant drawings found themselves defined in similar ways. As women living in the Victorian era, their physical appearances were closely scrutinized, their physiologies were widely taken as determining characteristics, and their identities were clearly delineated according to the idealizing norms of a generalized femininity that their culture endeavoured to secure beyond doubt or confusion.

This chapter looks closely at Sophie Pemberton's two albums of British Columbia wildflowers in order to better understand how knowledge about plants in nineteenth-century Canada was bound up with larger patterns of social thinking about identity. Taking Pemberton's albums as a representative case study of a much larger popular practice, it asks how women who made collections of wildflower drawings were positioned by those larger patterns of thought, and how they responded to them. Above all, it explores how the art they created enables us to see that logic at work. Today, women's wildflower albums are preserved as exponents of local or regional cultural heritage and are most often approached as material traces of the past. But the significance of that heritage is also forward-facing. The Victorians, we know, used nature as a resource to bolster social hierarchies and the stereotypes that supported them,[2] and albums such as Pemberton's enable us to *visualize* a conceptual architecture that simultaneously sustained and opposed that tendency. As such, they continue to offer insights for our contemporary cultural context, still so much preoccupied by thorny questions of identity.

In exploring those insights, my analysis keeps company with that vein of contemporary feminist theory which looks for ways to "render more mobile, fluid, and transformable" the means by which female subjects are produced.[3] However unexpectedly, women's wildflower drawings offer rich scope for understanding how the visual can militate both for and against such transformative practice. Approached in this way, these albums claim our attention as something more than just a popular expression of the bygone arts of polite feminine accomplishment. Indeed, I will argue that they can also be considered as "theoretical objects" in their own right – that is to say, as objects whose particular combination of pictorial and material qualities endows them with the capacity to engage and even embody thought.[4] Knowledge of plants has long been inseparable from the pictorial imagery that both creates and disseminates it, and knowledge of identities is no less thoroughly entwined with the visible. The theoretical purchase of Pemberton's wildflower albums lies in their rehearsal of these conjunctions. Carefully attended to, her albums, like those of so many other nineteenth-century Canadian women, actuate complex perspectives on an idea – identity – that continues to shape our world.

SOPHIE PEMBERTON, BOTANICAL ART, AND FLORAL CULTURE IN NINETEENTH-CENTURY CANADA

The Victoria-born artist Sophie Pemberton (1869–1959) drew and painted flowers all her life: in oil on canvas for the Royal Academy, on lacquered trays to raise funds for the Red Cross, on decorative household screens and notecards to friends, even on bathroom walls.[5] Her two albums of British Columbia wildflower drawings are early articulations of this pictorial interest, created when she was twenty-six years old and on the cusp of a short but distinguished career. The watercolours fit comfortably within a tradition of women's botanical album making that spanned much of the nineteenth century, brought from England to Canada in the travel trunks of immigrants and on the pages of metropolitan printers. As early as the 1820s, collections of pressed plants, known as herbaria, were being created by the wives of British colonial officials, who used them as a means to participate in European and American networks of scientific exchange.[6] By the 1830s women were also creating collections of drawings that would enable their viewers to appreciate both the colour and the structure of North American plants.

Making a collection of botanical watercolours had various attractions for women. It enabled them to develop their skills of observation and documentation, engaged their creativity, and provided a congenial framework for the accumulation of knowledge. The prolonged attentiveness that drawing from nature occasioned was, for many, an opportunity to rejoice in God's creation, and Pemberton was typical of nineteenth-century Canadian album makers in being an active church goer. Collecting parties to gather specimens offered such women respectable opportunities for a sociability that would be further reinforced when the finished albums were displayed and discussed in the family sitting room. There, they served as proof of feminine application and refinement, and so had a role to play in the nineteenth-century marriage market.[7] Albums also served needs that were particular to their makers. Pemberton's, for instance, were created at the time of her art training in England, and their carefully dated folios reflect the rigours of the young artist's South Kensington education, with its injunction to draw every day.[8] More than just exercises in technical skill, however, the drawings tied Pemberton to her loved ones. William and Frederick, the artist's brothers, had an abiding interest in the natural world, and were members of the province's natural history society, first organized in 1890 by – among others – the artist's future husband.[9] As much as her watercolours testified to shared family interests, however, they were also memorials to family feeling, sentimental records of the spring and summer days she spent back home in Victoria, BC. Once returned to London, the artist would gather her folios together and bring them to a bookseller in the Kensington High Street to be bound into volumes that she subsequently gave to her elder siblings as "affectionate souvenirs" of their too-fleeting time together.[10]

The two gifts differ from each other. One, destined for Frederick and his new wife Mary Ann, pairs snippets of floral-themed verse with appropriately

9.2
Millicent Mary Chaplin, *Rudbeckia*, 1840–42. The 1830s and '40s saw the emergence of women's botanical albums in Canada, with early examples by Maria Morris Miller in Halifax and Millicent Mary Chaplin and Fanny Bayfield in the city of Quebec. Albums were especially popular among English-speaking women associated, whether through commercial or familial ties, with British military officers trained in natural history.

9.3
Sophie Pemberton, *Lathyris maritimus*, 6 June 1895, from the album for her sister Ada in the BC Archives. Compared with the album for Frederick and Mary Ann, the folios in this volume are more straightforwardly botanical in format; poetry is confined to the frontispiece.

chosen plants, cropped and positioned as frames for the text. This is the volume with the lines from Tennyson, and it is now in the collection of the Art Gallery of Greater Victoria. The other album is less sentimental in tone. Made for her sister Ada and currently housed in the BC Archives, it confines poetry to the frontispiece and adopts a more conventionally botanical stance. Specimens are skilfully but straightforwardly rendered in thirty-one folios, each of which isolates the plant from its natural environment, centres it on the page, and gives species and genus names on the back. A hastily scribbled reference to botanist Frederick Pursh, cited as the authority for one identification, completes the album's botanical credentials.[11]

Taken together, the two albums – one poetic in nature, the other more scientific in tone – reflect the diverse contexts within which botanical knowledge circulated in nineteenth-century Canada. Like most women, Pemberton engaged botany and its conventions as part of a rich panoply of floral representations, ranging from drawings to découpage, pressed plants to seed catalogues.[12] The fashion for combining botanical nomenclature and snippets of poetry, for example, can be traced back to the Victorian craze for the "language of flowers" and to books such as Elizabeth Wirt's wildly popular *Flora's Dictionary*, which paired flowers with specific emotions to enable timid or circumspect lovers to speak their heart's desires.[13] Such books were not simple lexicons of floral feeling, however; they also trained readers to recognize plants and contained a surprising amount of botanical history and information, rendered more palatable to a general audience by the addition of short passages of verse.

Scholarly approaches to botanical art have not consistently recognized this broader floral culture, preferring instead to divide plant imagery into distinct representational categories. Of these, "the botanical" is the most rarified and exclusive. Curator Gill Saunders, for example, reserves the designation for the copperplate engravings published in scientific texts, asserting that "botanically-correct studies of plants painted for their own sake rather than for publication in a botanical journal or monograph" are, strictly speaking, "flower painting."[14] In this way, floral representations have been subjected to a classification structure not dissimilar from the taxonomic impulse that eighteenth- and nineteenth-century botanists brought to the task of understanding plant life itself.

At one level, Canadian women's wildflower albums do appear to take their distance from the quintessentially "botanical" images found in periodicals such as *Curtis's Botanical Magazine*. The drawings that Pemberton created for Ada's album, for example, lack the highly specialized pictorial techniques developed by the elite corps of European and British artists who were employed and closely trained by botanists to illustrate these publications. Exploded details, line drawings, and cross-sections enabled such artists to better convey the taxonomic information prized by the scientists who supervised their output. By contrast, botanical art by nineteenth-century Canadian women is largely the product of artists like Pemberton, working to please themselves or those nearest to them. On the few occasions when female artists partnered

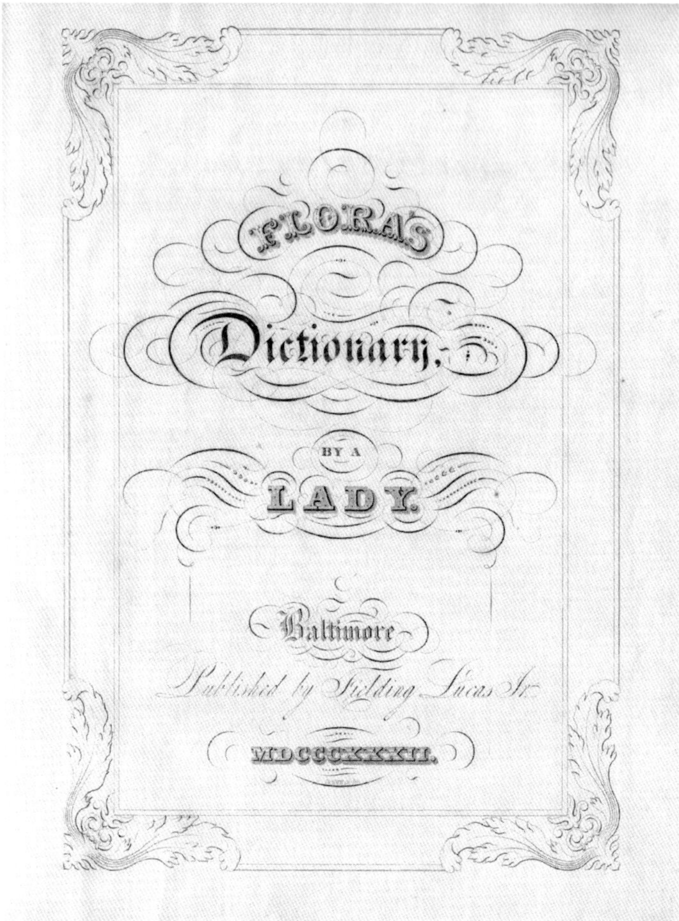
FLORA'S
Dictionary,
BY A
LADY.
Baltimore
Published by Fielding Lucas Jr.
MDCCCXXXII.

with botanists, it was the women who instigated and even controlled the collaborations, using scientific text to lend legitimacy to their art.[15] Because they did not have the same training as European illustrators and were not working to the same ends, these women did not widely employ the same techniques, and as a result their drawings are less overtly scientific in appearance.

In other ways, however, the albums that Canadian women produced are deeply enmeshed in a pictorial genealogy that is just as surely botanical as that reference to Frederick Pursh hastily scribbled on the back of Pemberton's drawing. Pursh's *Flora Americae Septentrionalis* (1814) was a long-standing authority for Canadian women botanizers from Lady Dalhousie (1786–1839) to Catharine Parr Traill (1802–1899), and the basic format adopted by many album makers clearly resembles the formula favoured by Pursh in his own watercolours.[16] Like him, album makers isolated their specimens against plain

9.4 *Opposite*
The title page of *Flora's Dictionary,* first published "by a lady" (Elizabeth Wirt) in 1829. The book, which combined extensive botanical information with poetry and a romantic lexicon of floral meanings, was part of the broader floral culture for women's botanical albums. In 1831, Montreal's Anne Ross, an amateur botanical artist, received a copy from her fiancé, John Samuel McCord.

9.5 *Left*
Curtis's Botanical Magazine was a leading English forum for botanical illustrations. While the artists who worked closely with botanists to pioneer scientific illustration in the eighteenth century were mostly European men, the English periodical employed several women in the nineteenth century, including Anne Henslow Barnard, whose *Aster Townshendii* of 1879 is pictured here.

backgrounds, adopted neutral lighting, and often employed a tightly controlled watercolour technique. Stems, leaves, buds, and blossoms are reliably centred on their pages. The visual idiom is naturalistic and the perspective is frontal, with the flower placed at eye level and no horizon line given. Plants are isolated, usually with one specimen per page. Beyond this visual similarity, women's albums of botanical drawings also shared other qualities with botanical publications. As with the scientific *floras* that exhaustively catalogued the plants native to a certain region, women's albums focused on a given geographical area, and while their collected drawings possessed neither the taxonomic detail nor the exhaustive scope of a *flora*, the experience of creating them alerted women to the need for a new kind of botanical publication geared to botanizers in the field, seeking to name the flowers they saw. When the first botanical field guides hit the market in the 1890s, their authors were

9.6
Frederick Traugott Pursh's original watercolour of *Digitalis purpurea*, pictured here, does not appear in his classic volume *Flora Americae Septentrionalis*, but an engraved version is preserved in the library of the American Philosophical Society. The engraved version contains additional anatomical information, suggesting that the scientific cutaways and magnified details typically absent from Canadian women's albums were sometimes added by engravers at a later stage.

female and their illustrations were a clear continuation of the wildflower albums that their mothers and grandmothers had been creating for the better part of a century.[17] Such manifest connections between women's albums and the major traditions of botanical publishing caution us against an overly narrow definition of the genre that unnecessarily debars women's amateur activity, precludes much colonial output, and obscures the connections between botanical forms of visual knowledge and the wider floral culture of which they were a part.

LEARNING TO KNOW THEMSELVES

A favoured trope of this broader floral culture was its insistent correlation of flowers and femininity. The flowers that marked the most important rites of passage in a young woman's life were used as enhancements to female attractiveness and metaphors for feminine character. From the poetry they recited to the sheet music they played from, nineteenth-century women and girls were conditioned to think of themselves as delicate and decorative blossoms and botanical education was marshalled to reinforce the connection. "Remember," one advice book for young women urged, "that in learning to know nature you are learning to know yourselves."[18]

The artistic and archival record of Sophie Pemberton's life suggests that she participated fully in this identification. Her mature oils, beginning in the years immediately following the 1895 albums, repeatedly foreground the link between flowers and feminine interiority. Snapshots in family photograph albums often portray her amid flowering plants, and the quickly jotted entries of her diary bear out their importance to her: "April 10 1906 To Cordova Bay a picnic alone to get ferns, gorse, arbutus trees and wild currant … April 11 1906 Effie and I luncheon at the Arden. The daffodils and lilies in the garden … June 1 1907 Painted the Broom at Beacon Hill. The winding road & lupins … June 28 1907 Rose show Victoria."[19] In this way, Pemberton used plants to mark the passage of time, the pulse of social life, the landscape that surrounded her, and the moments of her creative engagement with it. Her life, her connections to family and friends, her position as a social subject in a natural world – all these she mediated through flowers.

Art-historical analysis of women's botanical imagery has attended to such auto-performative aspects. In her work on the history of drawing in eighteenth- and nineteenth-century England, for example, Ann Bermingham highlights the potential for female agency that was lodged deep within floral subject matter. For Bermingham, the beauty of the botanical genre is that it allowed women to use their overdetermined relationship to flowers in order to subvert the stereotypes of decorative passivity on which that relationship was based.[20] Sheltered by the associations of improving activity that hung thick in the air around botanical study, women took to the drawing of flowers as a socially acceptable means of developing professional careers, whether as artists, teachers, authors, or exhibitors. Pemberton is among the Canadian

9.7
Pemberton's large canvas, *Spring*, was exhibited at the Royal Academy in 1903. It continued the thematic pairing of women and flowers which had also marked *Daffodils*, her inaugural contribution to the Academy exhibition in 1897.

counterparts to Bermingham's English examples, and although the audience for her albums of BC wildflowers was strictly private, the conjunction of women and flowers would be professionally serviceable to her in other contexts. Two years later, her inaugural submission to the Royal Academy, *Daffodils*, was a highly polished articulation of the link between flowers and feminine grace, and she was rewarded by seeing it hung on the line. Later paintings continued the association.

Socially informed analysis of this connection has done much to salvage women's flower drawing from the general art-historical neglect that had long accrued to it as a genteel feminine pastime. Key aspects of the genre's significance remain to be explored, however. While the overarching history of botanical art has been charted and specific artists have been studied in depth, there has been comparatively little sustained visual analysis of botanical artworks themselves or the effects of the genre's pictorial conventions.[21] This is especially true of nineteenth-century women's work, which has been largely addressed through a contextual lens. And yet the distinctive visual appearance of such art is its most engaging quality. Don a pair of cotton gloves and unwrap the acid-free paper that now protects them, and the pictorial appeal of wildflower albums is immediate. In their pages, the great wealth of the natural world is arrayed for our delectation – its profusion of colour, multiplicity of structure, and almost infinite variety of form, distilled into a consistent, and consistently compelling, visual formula.

In asserting the significance of this pictorial appearance, my impulse is not simply or even primarily aestheticizing. Rather, my goal is to map the *look* of women's botanical albums back on to their social function by calling attention to the ways in which the genre's distinctive visual codes feed into a performance of female identity. In so doing, it becomes possible to draw out not just the social performativity but also the theoretical agency that resides within botanical art and cements its importance both to its own time and to ours.

THEORETICAL OBJECTS

The notion that pictorial form can embody a theoretical disposition or orientation toward the self is now well-established.[22] Take, for example, the quattrocento pictorial construct of linear perspective, which casts the individual as the irreducible centre around which objects dispose themselves, thus visually instantiating the developing logic of humanism.[23] In shaping our vantage point on the world, technologies of visual representation also become techniques of the self. Botanical albums, too, are possessed of an intellectual history that interacts with their pictorial form. In her book *Draughtsmen, Botanists and Nature*, the historian of science Karin Nickelsen parses what she calls the "visual language" of eighteenth-century botanical engravings, drawing out the syntax and semantics of their formal conventions and assessing the cognitive value these patterns had for the European men who produced and consumed them.[24] No female artists are considered, however, and questions of subjectivity and identity do not arise. To move in that direction, a closer analysis of the visual experience the albums create is in order.

To look with care at a skilfully made wildflower album – we might take either of Pemberton's two versions for now – is to be drawn into the work in a very particular way. Each of the albums is roughly fourteen by ten inches in size. Owners would originally have held them comfortably in their laps, and that sense of physical connection is maintained even in today's more controlled archival settings, as viewers incline their heads toward the thick, creamy pages to get a better look. The physical requirement of turning those pages furthers the sense of connection. That gesture, performed upwards of fifty times in the passage from the front cover to the back of Frederick's album, also emphasizes that its images are not experienced in isolation but rather in

9.8 *Above and opposite*
Visual similarities and differences repeat like a refrain across the various folios of Pemberton's 1895 albums. Pictured here are her *Aster douglasii* (left) and *Sidalcea hendersonii* (right), both from Ada's album in the BC Archives.

sequence. As the drawings begin to pile up, their repetition of form creates a pattern of visual expectation that is never fulfilled in exactly the same way twice. The succession of drawings both whets our appetite for what is to follow and piques our curiosity about how it will be special and different from what has come before.

We can get a better sense of this pictorial effect by looking closely at a selection of pages from Ada's album, noticing the play of similarity and difference that they establish. Here, on a succession of thirty-one carefully trimmed sheets

of watercolour paper, Pemberton presents her plants to the viewer with evident care for the coherence of their cumulative impact. Beyond the obvious similarities of size, medium, and material, the artist's compositional and colouristic choices prioritize regularity; hues are of uniform intensity and brushwork is consistently handled. Stems, leaves, buds, and blossoms are reliably centred on the pages and set off from them by background washes. Because these washes are too generalized to read clearly as cast shadow, they contribute to a sense of spatial indeterminacy that contrasts with the highly specified particularities of the plants. Some aspects of this visual appearance speak to Pemberton's unique hand. Her skilled contrasting of hue and value, for example, together with the sophisticated use of the white of her support (visible in her *Lathyris maritimus*, fig. 9.3), enabled the Kensington-trained artist to create pictorial precision without relying on the independent line so common in albums by earlier Canadian artists. Nevertheless, Ada's album clearly conforms to the same general mode of pictorial organization employed by earlier women such as Millicent Mary Chaplin (1790–1858) in the city of Quebec and Mary Rebecca Wilkinson (1808–1874) in St John.[25] The central placement of the plants, their isolation on the page, the neutral lighting and frontal positioning: these are the structural conventions of botanical art, and they persist over time, despite changes in artistic style.

And yet, to insist too forcefully on a dynamic of sameness in botanical albums would be quite spectacularly to miss what is most significant about them. To the women who made dozens, sometimes hundreds of drawings, each plant, each leaf, each petal was uniquely important. Indeed, the whole point of the genre is that it records diversity. Such diversity is most obviously registered in the distinct morphologies of individual plants: the edges of their leaves, serrated or smooth, their petals long and thin or short and rounded. Different plants placed distinct pictorial demands upon artists and elicited differing aesthetic responses. Compare the airy openness of Pemberton's *Aster douglasii* to the weight and solidity of her *Sidalcea hendersonii* (figs. 9.8a and 9.8b). Faced with the massed geometric volumes of the larger blossom, the artist takes a much greater care to establish depth of field, fully capturing the spatial projection and recession of the crimson-pink florets that cluster so closely together on the head of the foremost composite blossom. Volume is achieved additively here, through the careful delineation of each petal and its placement in a relation of structural contiguity to its neighbours. By contrast, in her handling of the *Spirea discolor* plant a volumetric effect is attained by letting barely-suggested petals blur and blend with each other, producing the fuzzy appearance and softer occupation of space so familiar to gardeners of the western coastal regions. In still other folios – those of orchids and lilies most especially – Pemberton replaces the concern for volume with an interest in sinuous movement. Focussing on the curving edges of supple stalks and swelling blossoms, she creates lithe and graceful decorative effects. Such difference repeats like a refrain across the pictorial and conceptual fields of botanical albums, offering a formal counterpoint to the similarities that so visibly link individual pages, individual albums, and individual artists to each other.

Difference and sameness, then. This is the pictorial hook on which the visual fascination of botanical albums depends. From it unfolds the balance of austerity and abundance that captivates our senses, and to it accrues an interest that belies the works' apparent simplicity. Against a familiar framework

9.9
St John artist Mary Rebecca Wilkinson included this *Lady's slipper, Cypripedium acaule,* in her 1868 album, *Wildflowers of New Brunswick*, a gift commissioned from the artist for Margaret Medley, a pioneering nurse and the wife of the local bishop. Despite differences in technique and style that develop over time, this album of wildflower watercolours shares a basic pictorial organization with those by Sophie Pemberton and Millicent Mary Chaplin.

9.10
Sophie Pemberton, *Spirea discolor*, 1895, from the album for Ada in the BC Archives.

of rigorously secured parameters, the drawings' provision of specificity keeps us engaged. But there is more than just aesthetic pleasure at stake here, and if we wish to understand the epistemology of identity embodied in these objects, we will do well to attend to this dual dynamic, for sameness and difference are not only visual phenomena, they are also the logical principles that determine our understandings of identity. As such they become the fulcrum around which the conceptual power of the genre mobilizes, drawing the albums into focus as theoretical engines for self-perception and revealing a divided approach to identity at their core.

THE PRINCIPLE OF SAMENESS: BOTANICAL ART AND IDENTITY

We have seen that, in key respects, all botanical images look somewhat alike, for they all conform to the same basic set of pictorial strictures. If this is the most obvious way in which the genre registers sameness, however, it is not the most significant. With its clear lighting, controlled outlines, and pronounced distinction between figure and ground, the pictorial logic of the botanical formula is in fact much less about sameness *between* examples than it is about rigorously establishing the parameters for species *self-sameness* – in other words, "identity" in the sense that philosophers mean it: an entity that is demonstrably the same as itself and not something other than itself. Thus, to the extent that botanical illustrations are intended as rigorous and exact delineations of the essential attributes of a given plant – unambiguously capturing and setting forth their distinctive appearance – they also stand as one of visual culture's clearest pictorial articulations of two of the logical foundations on which Western philosophy has grounded itself: the "law of identity" (A is A) and the "principle of non-contradiction" (A is not not-A). To make a botanical drawing, then, is to participate in the logic of self-sameness that determines, and so affirms, identity.

The botanical drawings made by women in colonial contexts were, as it happens, especially concerned with identity and the corollary practice of identification. For most of the nineteenth century, Canadian women lacked opportunities for the training in plant anatomy that advanced scientific study provided, and as a result they oriented themselves and their drawing to the task of plant identification in preference to advanced taxonomy or the functioning of the organism. This orientation was reinforced by the rise of economic botany, with its focus on the usefulness of plants to people. Since the first step to more fully exploiting plants is to know what they are and what they are good for, identification was the starting point for the vast majority of botanical instruction in Canada and the other British colonies, where it conjoined with a nascent national pride and a growing impetus to celebrate empire's regional distinctiveness through knowledge of the plants indigenous to its varied locales. Whether collecting specimens in the fields or assembling drawings in an album, the goal of most botanical pursuits undertaken by British North American women in the nineteenth century was thus to name, and so proclaim, identity.

Yet even the ostensibly simple task of botanical naming placed Canadian women at a disadvantage, immersing them in a world of specialized Latin terminology to which their education gave them little access. Most women struggled to arrive at correct identifications, and Pemberton was no exception. In Frederick's album the nomenclature is elaborately and consistently formatted in the same ink as that of the verses, suggesting that its presence was planned by the artist as an integral part of the project.[26] Yet the artist was not always certain of herself and pencilled corrections have been added later, in a different hand, likely that of her brother, who was an avid horticulturalist and botanical amateur.[27] It is a familiar enough phenomenon in drawings by women, who frequently turned to the men of their acquaintance to make or correct their identifications for them. Indeed, Sophie Pemberton's decision to relegate the scientific names to the backs of her folios raises the question of how much she really cared about plant identification, and it is true that beyond these two albums there is no evidence that she was actively interested in scientific pursuits. Her diaries, despite their abundant references to plants and flowers, bear no trace of botanical language; nor did science writing figure among the many books she noted in their margins. Still, the scientific names are there on the drawings, as if it did not seem quite proper to leave them out. Scientific nomenclature brought a useful, perhaps even reassuring, stability to one's knowledge of the world. With it came the assurance of one authoritative designation – *Spirea discolor*, say – for an entity that otherwise has a potentially confusing variety of names: Ironwood, Creambush, Oceanspray. *Spirea discolor* may be demanding, even unfriendly, but it offers certainty in return for its exigency and so facilitates communication.

That value of that certainty was not only pragmatic; it was also existential, and the emergence of taxonomy in the eighteenth century had been profoundly shaped by the Enlightenment project of grasping reality by the horns. As botanists laboured to fashion truly modern classification systems, they held out the promise that these could be grounded on something firmer than fancy and less arbitrary than convention. Thus, in the Linnaean system, floral designations were determined by the visible structure of the plant itself. In this way, botany aspired to deliver what philosopher Michel Foucault describes as a "well-made language" – that is, a system of representation so grounded in reality that it offered conclusive knowledge of it. This, Foucault explains, was taxonomy's early project and its allure: that through its articulation of the visible it would permit "an absolutely certain knowledge of identities."[28]

Prominent among these certainties were those attached to sexual identity. By rigidly insisting on the incommensurability of those parts of the plant that it designated "male" and "female," Linnaeus's "sexual system" of plant identification naturalized social attitudes toward sex and gender.[29] This was especially so in England and its colonies, where the particular longevity of Linnaean conventions meant that they overlapped with the full span of the industrial revolution. As distinctions between public and private spheres intensified, the sexual system of classification was drawn into the hardening of gendered identity that characterized the Victorian era.[30] In this world, where men were God's trees and women God's flowers, correct delineations of floral identity were also enlisted in the service of well-regulated female identity. Pemberton's embrace of Tennyson suggests that she both understood the invitation and was inclined to accept it.

But if Tennyson's trees and flowers were a touchstone for the prospect of womanly women and manly men, they are also a reminder that self-sameness is not, on its own, sufficient to determine identity. For while the conventional understanding of gender enshrines conformity to sameness (*women are like this*), it can only do so by virtue of a countering principle of opposition through difference (*because they're not like that*). And as with men and women, so too with botanical entities: they cannot be defined in a vacuum. For this reason, the naming practices of scientific identification are inseparable from the taxonomic practice of scientific classification. Botany, in other words, does not just ask its adherents to define entities but to do so by placing them in relation to other entities within a larger system of relations.

Crucially, the principle of sameness would reign supreme within those botanical systems. Linnaeus's sexual system of botanical classification offers a clear example. The system began by picking out one physical characteristic (say, number of stamens), then delineating all the differing variables of species that shared that one feature, thus privileging relations of sameness. The system was clear and easy to learn, but botanists were troubled by its transparently arbitrary character. The fact that it imposed its organizational structure on nature, rather than reflecting an order that resided within it, limited the credibility of its claim to represent the world *as it really was* and not simply as people had agreed to speak about it.

The "natural method" that succeeded the sexual system was an attempt to rectify this shortcoming. By the 1890s, the natural method of classification espoused by Asa Gray, John Torrey, Joseph Hooker, and others, was predominant, and at first glance it appeared to prioritize difference. Instead of beginning from one unifying quality (like number of stamens) practitioners of the method began by considering the totality of a plant's different characteristics. When, for example, Sophie Pemberton carefully indicated that her Shooting

Star was of the "natural order" *Primulaceae*, that act of identification proceeded by categorizing it according to a wide variety of characteristics that, taken together, distinguish members of the primrose family. The ability to determine what might or might not belong to a given family required a broad knowledge, based either on considerable experience or, more commonly, on recourse to a recognized authority.

Despite this apparent inversion, however, philosophical commentators on natural history have concluded that both the Linnaean system and the natural method ultimately privileged a principle of sameness. Thus, while recognizing that classification was "clearly always a problem of ordering differences," philosopher Gilles Deleuze makes the larger point that such considerations always unfolded within a network of continuities based on resemblance. Difference may appear to run riot through botany's panoply of individuated species, but its celebration of species diversity is carefully contained within a bigger envelope of generic similarity. The very idea of "specific difference," Deleuze notes, "presupposes an identical concept in the form of a genus."[31] To the extent that they participate in the histories of plant identification and plant classification, then, botanical albums are not just *pictorially* geared toward a repetition of the same elements; they are also *conceptually* grounded in a principle of sameness that was structurally rooted within the larger botanical project.

This privileging of sameness within botanical taxonomy is not an isolated practice, and its repetition across a wide range of Western intellectual practices has had far-reaching consequences for how our society is able to conceptualize difference, and how it relates to those – like women – who are marked by it.[32] Contemporary philosophers of difference such as Deleuze and Luce Irigaray note that "we tend to subordinate difference to identity in order to think it" and lament the social impact of that conceptual arrangement.[33] Since Aristotle, Western thought has privileged sameness as the basis of its ontology. But when a thing can only be A, or not-A, it is hard to find room for B, much less X. The privileged term is established as the norm and soon nothing else quite measures up. In this vein, for instance, we might note that botany's founding classification system did not just identify the male and female characteristics of the plant but also ranked them, assigning the taxonomic unit based on the male attribute (the class) a higher position than that based on the female part (the order).[34] The natural method was less sexually overdetermined, but it too continued to entrench its practitioners in a hierarchically organized world of kingdoms, phylums, classes, and orders. Indeed, the closer adherents of the natural method felt they had come to discerning a classification system that reflected the ordering principles of a benevolent creator, the more they were tempted to project those principles back onto society.[35]

In this way, the "absolutely certain knowledge of identities" that botany sought to secure was part of a much larger approach to knowledge that classed and categorized the entities of the world, inserting them into well-defined schemas and normative standards. With sexual anatomy as its taxonomic touchstone, botany's correct delineation of floral identity was readily turned to the social regulation of gender. Under the logical regime of sameness, however, the constraints that gender imposes on men and women alike are not experienced equally, for one term must always be privileged as the standard from which alterity is measured and by which it is defined. In a nineteenth-century Canada where women could not vote and were not even recognized as persons under the law, their relegation to the subordinate position was unmistakable. The straightjacket of femininity has always been the tighter and more confining one.

Was there, then, a bitter irony in the fact that women like Pemberton so enthusiastically took up practices of botanical identification that were fundamentally grounded in a logic that reinforced their subordinate status? By 1895, the discoveries of Charles Darwin had profoundly unsettled scientific faith in the divinely ordered regularity of nature, but popular science continued to be shaped by the older rhetoric. On the frontispiece of Ada's volume, words from Robert Browning testify to Pemberton's understanding of flowers as the visible manifestation of a divine plan that also guided human action. "O world / as God has made it / all is Beauty, / And knowing this is love / Love is Duty / What further may be / sought for / or declared." Browning walks arm in arm with Tennyson here, whistling softly in the meadows of the Victorian specimen collector. But did their complementary melodies of clearly determined and divinely ordered identities constitute the only song to be heard there? Or was there also a contrapuntal refrain that carried women's botanical imagery in another, more liberating, direction?

INTERLUDE: ARBOREAL WOMEN AND INDETERMINATE HUES

An image from an entirely different Pemberton album offers a beguiling rejoinder to this question. The image is part of an unstudied collection of family photographs taken on a trip to Italy. In it, a well-dressed Pemberton poses next to an intricate stand of trees. The photograph has been carefully constructed to emphasize the similarity of its two pictorial subjects. The artist's

9.11
This photograph of Sophie Pemberton was taken in 1911 while the artist was on holiday in Italy. The photographer has skilfully constructed the composition to highlight formal similarities between the woman and the tree.

stylish clothing, for instance, is answered by the elegant sweep of the trees; her waist and hips align with the uppermost trunk and are roughly equal to it in girth; the contrasting tiers of her hat, neckpiece, and arm, echo in miniature the sloping rhythms of the three trunks on the right of the composition. Both the woman and the tree have *character*. The vigour of her stance and intensity of her gaze proclaim her individuality as loudly as the resplendent bend of the wood does that of the tree. Whatever our normal expectations about women and flowers may be, *this* woman is clearly arboreal. And trees, the image also tells us, can be elegant and graceful just as well as strong and tall.

Evidence for this more open kind of thinking about identity is scattered throughout Pemberton's botanical albums and merits our attention in the contemporary moment when paradigms of identity (some more rigidly conceived than others) are so much to the fore. Consider the album for Frederick – the one with the poems and flowers combined. Heavy-handed as the didacticism of Tennyson may be, it is not evenly sustained throughout the volume's pages. On 13 May, for example, a very different sentiment from the decadent poet Swinburne holds sway: "Even to the upland verge / Whence the woods gathering watch new cliffs emerge / Higher than their highest of crowns that sea-winds fret / Hold fast, for all that night or wind can say / Some pale pure colour yet / Too dim for green, too luminous for grey."[36] In place of the clear identities and identifications propounded by Tennyson, Swinburne is evocative, celebrating ambiguity in nature, attending to the subtleties of colouration, and affirming a capacity for differentiation that exceeds pre-established categories. Where the poet laureate used plants to carve out socially sanctioned roles for women and men, the very possibility of such determinate meaning is upset by Swinburne, a queer intellectual who venerated Sappho and whose dazzling complexity of language is notoriously hard to pin down. In the passage chosen by Pemberton, plants inhabit a world of nuanced indeterminacy and the ineffable richness of the natural world surpasses our vocabulary for it. With Swinburne, then, Frederick's album shifts from the Tennysonian project of *identification*, into the significantly different realm of *individuation*. If we look closely, something quite similar is to be seen in Ada's album as well, and it brings us to the second organizing principle at work in women's botanical watercolours: that of difference.

THE PRINCIPLE OF DIFFERENCE: BOTANICAL ART AND INDIVIDUATION

The norms and conventions of modern botanical illustration were calculated to convey an air of rigorous objectivity. In actuality, however, most botanical illustrations contain numerous departures from strict representational fidelity.[37] Colours are heightened, proportions are altered, and orientations are changed. In many cases, sequential phases of plant development are shown as occurring simultaneously. Once alerted to them, botanical imagery's deviations from nature are everywhere visible. Leaves that consistently face the viewer on paper, for example, are quite differently oriented in nature, where they more reliably point upwards to the sky.

In all of this, the aim of classical botanical imagery is to render a type: the ideal of a plant arrayed for our consumption rather than the individuality of any given specimen.[38] Because such specimens were subject to the vagaries of soil and climate, they were widely considered inadequate to the task of fully exemplifying the species to which they belonged. Illustrators, therefore, regularly smoothed over imperfections and banished uniqueness in order to capture a species' most characteristic elements. Whereas taxonomy itself honours singularity by means of the *holotype* – a particular specimen selected to serve as a reference point when a species is first named – the typification most operative in botanical drawing takes its cue from the quite different idea of the *typological species concept*: a group of organisms whose members conform to certain fixed properties. Interestingly enough, a holotype need not faithfully "typify" the most distinguishing characteristic of a species; classical botanical illustration, by contrast, operates under exactly that compunction. Thus, the pages of nineteenth-century publications like *Curtis's Botanical Magazine* are filled with paradigmatic and impossibly perfect plants.

This was not Pemberton's approach, however. Or at least it was not so consistently. True, the largest leaf of her *Stachys ciliata* is suspiciously aligned with the picture plane, but that same orientation also highlights a morphological idiosyncrasy that no professional botanical illustrator would have included: there, just to the left of the leaf's central vein, an insect hole, shaped like a tiny Christmas tree. Other images contain similar irregularities. The lower leaves of her *Spirea* plant, for example (fig. 9.10), are riddled with holes and their edges are turning brown, as is the foliage in her *Lonicera involucrata* and *Grindelia integrifolia.* Such "imperfections" are not exclusive to women but they are especially common to them, from the seventeenth-century works of German-Dutch artist Maria Sibylla Merian (1647–1717) to those made by Canadian women more than two hundred years later. Brown and misshapen leaves appear repeatedly in drawings by Elizabeth Hazen (1840–1935) and Mary Rebecca Wilkinson (1808–1874) in New Brunswick, and Agnes Fitzgibbon (1833–1913) in Ontario. Insect damage makes a surprising appearance in the otherwise picture-perfect work of Anne Ross McCord (1807–1870) in Quebec, and even Alberta's Mary Schäffer (1861–1939), who was otherwise quite insistent that the botanical game should be played in strict accordance with the rules, individualized her specimens in this way.[39]

More often than not, there is no good botanical justification for such inclusions. True, some plants, like Morning Glory, are so commonly host to hole-eating organisms that damage to their leaves might almost be considered a characteristic of the species, but neither *Spirea* nor *Stachys* is unusually susceptible to infestation. When insects do strike *Spirea*, they are most likely to be aphids, which do not have the chewing mouth parts necessary to create large holes like the ones Pemberton gives us. In place of a concern for species,

9.12
Like her *Spirea discolor*, Pemberton's *Stachys ciliata*, 1895, shows insect holes and dying leaves. This emphasis on the particularities of individual specimens is common to many Canadian women album makers and contrasts with the idealized floral types so characteristic of scientific publications.

then, the drawings of *Stachys* and *Spirea* demonstrate a preference for the particularity of specimens, highlighting the sensibility of an artist who was fully attuned to the individuality of each and every flower that she drew. Images that foreground morphological particularities are sprinkled throughout Ada's album, following no apparent botanical pattern or purpose beyond that of individuation.

That concern is also a cumulative characteristic of the album as a whole in which each successive iteration of the botanical formula engages singularity and produces it. The serial nature of botanical albums prompts us to recognize each of its folios as importantly different from the one which preceded it or the one that was to follow. In this way, repetition is the fountainhead of the difference that is so foundational to the pictorial sensibility of the botanical genre. Crucially, this serial principle of difference functions as a means of destabilizing the logic of identity inherent in the genre's other wellspring – the principle of sameness. For if, as we have seen, sameness is tied to the realm of idealized and normative identities, difference leads altogether elsewhere: to fields in which no flower is quite like another, where the richness and variety of existence are there to be reckoned with.

The value of such reckoning may well have been brought home to Pemberton through the writing of John Ruskin, whose views on beauty in nature as the gift of God closely paralleled Pemberton's own. Ruskin's writings were given as prize-books to students in Victoria while Pemberton was growing up,[40] and given his status as the most important English-language art theorist of the nineteenth century she is sure to have encountered his advice to young artists, either directly or indirectly. At the heart of this advice was the exhortation to look as closely as possible at nature and then to represent it *exactly* as it appeared. In this, Ruskin, who was an avid but irreverent botanist, pushed back against then-dominant aesthetic and botanical paradigms that eschewed overly detailed descriptions of plants and recommended their distillation into generalized components.[41]

A similar disdain for knowledge rooted in generalizing convention was voiced by another, and quite different, aesthetic theorist of the nineteenth-century: Friedrich Nieztsche. "No leaf ever wholly equals another," Nieztsche observed, "and the concept 'leaf' is formed through an arbitrary abstraction from these individual differences, through forgetting the distinctions."[42] Like Ruskin, Nietzsche advocated for a richer encounter with the world than our convention-bound conceptual armatures are able to furnish, exhorting readers to be open to the unique and wholly individualized nature of experience. While the German philosopher was almost certainly unknown to Pemberton (and while his atheism would have been anathema to her if he had been), Nietzsche's sensitivity to the workings of sameness and difference offers contemporary interpreters of her art additional insight into the dual principles that underpin it. Crucially, what Nietzsche brings to the table that Ruskin, a far more normative thinker, does not, is an awareness of the extent to which generalizations not only miss the full beauty of the world but also inflict a violence upon it, forcing inherently different entities into the same conceptual box. "We separate things according to gender (*Geschlechtern*)," Nietzsche writes, "designating the tree as masculine and the plant as feminine. What arbitrary assignments!"[43] Tellingly, the German term *Geschlechtern*, which signifies gender in both the grammatical and the sexual senses, also means "genus" or "type" – tying observations about language and gender back into the typifying practice of botanical illustration. Nietzsche's passing references to leaves and trees thus have as their greater concern a desire to reveal the restrictive impetus that all too often accompanies the generalized circumscription of entities through categories; as demarcations between beings are promoted to immutable truths, he argues, the identities thus created come to be ranked in a pyramidal order of privilege and subordination that is inherent in the principle of sameness.

How fascinating, then, that botanical art's lessons in the conventions of typification were ones that female artists so commonly ignored. The American artist and author Mary Vaux Walcott (1860–1940) was explicit on this point in the introduction to her *North American Wildflowers* (1925), where she stated that her aim was "to depict the natural grace and beauty of the plant without conventional design."[44] In Canada, too, women departed from botanical conventions in a variety of ways. Over a span of six decades, the Nova Scotia artist Annie Prat (1861–1960) moved from carefully inscribing scientific nomenclature to frequently abandoning it altogether. On other occasions she took it upon herself to invent names of her own – sometimes when no reliable authority was available, and sometimes just to please herself. "I've called [it] golden glow," she noted of the fungus *Polyporus sulphurous*, "as it looks like nothing else."[45] Perhaps she took her lead from Ruskin in this, or perhaps it came from the Canadian example of Catharine Parr Traill, who had unapologetically done the same decades earlier.[46] Either way, women repeatedly drew inspiration from their own unmediated encounters with plants, and their drawings call attention to this fact.

In this way, botanical albums by women in Canada capture their makers' satisfaction in moments of *encounter* with plants that are scrutinized attentively, repeatedly, and to the exclusion of all else. The women who created these albums took evident pleasure in the things that appeared in the world, and the drawings they produced extoll the value of seeing a thing plainly, for what it is, unfettered by external expectations and pre-existing categories. Untrained in the priorities of advanced botanical drawing, these women were also comparatively free from its occupational hazard of shoehorning reality into idealized and easily classifiable types. In this, whether they were aware of it or not, they shared something with newer directions in botany itself, which by the 1890s had expanded well beyond taxonomic practice into plant anatomy and cellular biology. Between evolutionary theory and the exponential growth in the discovery of new species, the new botany was centrally concerned with diversity. And it is precisely such diversity that Pemberton and other Canadian women courted in their drawings. They may be characteristically colonial in their failure to attain the heights of technical sophistication achieved by British and European practitioners of the art, but their failure in this regard was also an opportunity to rejoice in a world of infinite variety where things were, quite marvellously, what they were. In place of the pictorial conventions developed in order to meet the normative needs of taxonomic practice, their drawings record particularity, making room for individuation and difference in a way that effectively offset the generalizing thrust of botanical art, countered the taxonomic drive to confine beings within their appropriate categories, and contested its pictorial perpetuation of impossible standards of normative perfection, so idealized that few if any actual individuals could hope to live up to the models so widely circulated through representation.

CONCLUSION: IDENTITY, APPEARANCE, EXISTENCE

There are, then, two sides to the logic of identity that underwrites the botanical albums that women created in Canada during the nineteenth century. At a time when women were so often not seen for what they were but rather told what they should be, their albums reproduced the framework of normative identification within which they operated on a day-to-day basis. Yet these same albums also worked against such typification, offering opportunities for close looking that produced a vivid appreciation of individuation and difference, and challenging botanical illustration's orientation toward a logic of identity predicated on relations of sameness. To the extent that women shied away from botany's grand and somewhat aggrandizing project of idealizing reality into systematized categories, it may be that they were able to more fully appreciate at least one important facet of that reality: its infinite diversity. This is the dual epistemology of identity set forth in the pictorial and material structure of the botanical album considered as theoretical object. Once recognized, the albums' double logic of sameness and difference serves as a conceptual catalyst. It enables us to take account of the distinctive appearance of the genre and, through the intermediary of the broader floral culture, to arrive at an appreciation of how that pictorial formula presented women with two quite disparate ways of understanding their own identities.

Regrettably, the women who made botanical art in Canada almost never wrote about what their practice meant to them, and in the absence of such documentary evidence we can do little more than speculate about their perspectives on the genre and its conventions. It is extremely unlikely, of course, that Pemberton or the other women I have named in this chapter would have articulated their thought processes in anything like the terms I have employed here. Indeed, it is one of the prerogatives of art that it speaks differently to one generation than it does to another. Still, it may be possible to transpose my analysis into a language that Pemberton would have appreciated and might even have used herself. Consider, for example, the important notion of *appearance*. Close observation of the world's appearance was precisely what the later nineteenth century equipped its artists to do best, awakening practioners to the possibilities of painting the world as they saw it, rather than as they had learned to believe it to be. By 1895, Naturalism's precept that it was the business of artists to capture appearances was widely established, and Pemberton's albums partake in such prioritization of seeing. From a comfortable viewing distance (held in the lap, say, or placed on a table), the folios are highly convincing records of floral entities, but get right up to the leaves and petals and they seem to dissolve as objects and resolve themselves as paint, reflecting the artist's understanding that our eyes do not simultaneously capture the whole of an entity and each of its details. Appearance is a multivalent term, however, and as a woman raised to assume her place amongst the elite of colonial Victoria, Pemberton was trained to observe appearances of another sort as well: not those associated with perception and representation but those tied to the maintenance of social norms and standards. The evidence that remains suggests that she adhered to these quite stringently. She was a devotee of fashion and a devout Anglican with a highly developed sense of duty and propriety. Judging from her diary, her life was a daily routine of highly orchestrated social intercourse: dinners, plays, lectures, European travel, and hotels. Such privileged observances exact their own toll, however, and in the face of them I cannot help but wonder about the enjoyment – perhaps even the relief – that the artist may have felt in those moments when the act of representing the natural world enabled her to trade in one kind of appearance for the other. In opening a view onto a world in which experience does at least occasionally exceed our pre-existing conceptual frameworks for it, perhaps her botanical drawing also insulated her from the expectations that were dominant and prescribed in her time.

If so, the effect was no more than temporary. The archival record of Pemberton's personal engagement with flowers and plants reveals their doubled function in her life. One day she is lunching in society, garden blooms demarcating the well-tended spaces of social convention, the next she is picnicking alone, the gathering of wild plants offering an occasion for solitude and independence. Family snapshots repeat this twofold reality: the excessively civilized dresses and hats of Victorian and Edwardian decorum coexist with the wild, windswept shoreline of Vancouver Island, often in the same images. For Pemberton, it was never an even contest. Like the poets who sang the praises of uncultivated flowers in highly cultivated language, her view of plants was never free from gendered overdetermination; even in her final years, her correspondence made overt comparisons between the female friends she loved and the lilies that she drew.[47] By means of her engagement with flowers, Pemberton embraced Victorian articulations of femininity and bound herself to their constraints. She lived her life in conformity to a normative model of female identity that was founded, like all such identities are, on a principle of generic similarity no different from that which echoes through botanical thought. We cannot just ignore the Browning nor the Tennyson either.

But though I clearly do not seek to reclaim Pemberton for feminism, I am quite interested in exploring how the albums that she and so many other women created might contribute to our ways of thinking about gendered identity today. If philosopher Elizabeth Grosz is right in her assertion, quoted at the outset of this chapter, that the work of feminism is to render more mobile the alignment of forces that constitute women's identity, then there is great value in attending to the patterns of thinking that sustain identity and those that loosen its stranglehold on us. To give women's botanical albums their due as theoretical objects is, I would argue, one way of sensitizing ourselves to these patterns, for while the rhetoric of Victorian floral culture may have changed, the rhetoric of identity that underpinned it has not gone away; the challenge of navigating contemporary existence is still very much a matter of finding one's way through its bewildering maze of constraints and possibilities for being.

Botanical art by women in nineteenth-century Canada is both an inscription of identity and a paean to diversity, subject to and expressive of a double logic of similarity and difference whose impact is visible throughout Pemberton's albums and, beyond that, on the genre as a whole. In their most positive aspect the albums function as engaged records of the sheer fact of a flower's existence, directing our focus, with great sensitivity, to the full richness and variety of what *is*. They recognize it. They acknowledge it. They even celebrate it. Wonderfully, however, such acts of recognition do not necessarily – or at least not exclusively – lead us to the idealizing belief in types and essences that underpins the logic of identity but also enable us, through attentiveness, to move beyond it.

ACKNOWLEDGMENTS

My thanks go to Ann Shteir, Jacques Cayouette, and the anonymous readers of this manuscript for their invaluable input – and to Larry and Nathan Deck, Brian Foss, and Charlie Hill for the time and space in which to write. For more information about women's botanical art in Canada see chapter 4 of my book *I'm Not Myself at All: Women, Art, and Subjectivity in Canada*. My aim in this chapter is to focalize, through the lens of Pemberton's albums, an argument which I make at greater length but more diffusely there.

NOTES

1 Tennyson, *Becket*, 12. The line was a popular late-Victorian bromide, quoted on greeting cards such as one preserved in the Baylor Library's Texas Collection.

2 Gates, "Ordering Nature," 180.

3 Grosz, *Time Travels*, 193.

4 On works of art as theoretical objects, see writing by Mieke Bal, for example, "Narrative Inside Out."

5 The main sources on Pemberton currently remain Tuele's MA thesis (1980) and the 1978 exhibition and catalogue from which it stemmed. A new study by Kathryn Bridge is forthcoming.

6 See Shteir and Cayouette, "Collecting with 'botanical friends,'" as well as Shteir's chapter on Mary Brenton in this volume.

7 Huneault, *I'm Not Myself*, 169–70.

8 The South Kensington system was a state-run national art training and examination program, centrally administered by the South Kensington School of Art (later the Royal College of Art). A network of local schools and private art academies prepared students for its examinations. On 12 June 1889 Pemberton entered such an academy in nearby Pelham Street, run by Arthur S. Cope.

9 Sophia Pemberton and Arthur John Beanlands were married in 1905. William, a mining engineer, and Frederick, a surveyor and horticulturalist, appear on the published membership lists of the Natural History Society of British Columbia in 1901 and 1913 respectively; Beanlands is consistently present as a "foundation member." See: the Natural History Society of British Columbia, *Revised Constitution and List of Members of the Natural History Society of British Columbia* (Victoria, BC: The Colonist Presses, 1901), and the Natural History Society of British Columbia, *Revised Constitution and List of Members of the Natural History Society of British Columbia* (Victoria, BC: The Colonist Presses, 1913).

10 The dedication page of Frederick's album in the Art Gallery of Greater Victoria reads "To F.A.P. and M.A.P an affectionate souvenir from S.T. Pemberton." The

cover is embossed with the title "Wild Flowers BC," and a sticker indicates that the binding was done by Young's Library, 36 Kensington High Street. Newspapers confirm the store's location there in 1895. Ada's album, also professionally bound, was then wrapped in a hand-made cover of silk brocade.

11 The annotation is not obviously in Pemberton's distinctive hand and may have been added by a family member, perhaps the album's recipient, Ada.

12 See Scourse, *The Victorians and Their Flowers*.

13 See Seaton, *The Language of Flowers*.

14 Saunders, *Picturing Plants*, 96.

15 Maria Morris Miller, for example, solicited texts by Titus Smith (1839), Alexander Forrester (1853), and George Lawson (1866 and 1867) for the different parts of her *Wild Flowers of Nova Scotia*. See Huneault, *I'm Not Myself*, 161–3, n41, 328.

16 Original watercolours by Pursh, preserved in the American Philosophical Library, expand our knowledge of his artistry beyond the plates for *Flora Americae Septentrionalis*. For his influence on Dalhousie and her circle see Shteir and Cayouette, "Collecting with 'botanical friends'; on Traill see Peterman's chapter in this volume.

17 See, for example, Frances Theodora Parsons, *How to Know the Wild Flowers* and Ellen Miller and Margaret Christine Whiting, *Wild Flowers of the North-Eastern United States*. For more on botanical books see the chapter by Dawn Bazely and Kate McPherson in this volume.

18 Annie Ryder, *Hold Up Your Heads, Girls!* (Boston: Lothrop, 1886), 44, quoted in Seaton, *Language of Flowers*, 22.

19 BCA, Pemberton Family Fonds, MS 1295.1, Sophie Pemberton, diary and daily scripture readings. The photograph albums are housed at the Art Gallery of Greater Victoria.

20 Bermingham, *Learning to Draw*, 215–18.

21 The classic text is Blunt's *The Art of Botanical Illustration* (1950); rev. ed. with botanist W.T. Stearn, appeared as recently as 2015. On female botanical illustrators in the Victorian era see also Kramer, *Women of Flowers*.

22 Daston and Galison, *Objectivity*. See also Crary, *Techniques of the Observer.*

23 Damisch, *Origin of Perspective*.

24 Nickelsen, *Draughtsmen, Botanists and Nature*, 149–60.

25 Chaplin, along with her friend Fanny Bayfield, was connected to an enthusiastic circle of amateur naturalists in the city of Quebec, many of them British military officers or their family members, as was the case with the two women; their fine botanical albums are preserved in Library and Archives Canada. Mary Rebecca Wilkinson, the wife of a military cartographer, was commissioned by the women of St John to create an album of wildflower drawings as a gift for the local bishop's wife. It is now in the University of New Brunswick Archives. On the wider field of

women's botanical art in nineteenth-century Canada, see Huneault, *I'm Not Myself*, 152–72.

26 This impression is strengthened by the fact that one inscription has been partially cut off by the trimming process, indicating that the names were there before the volume was bound.

27 The uncertainty is even more pronounced in Ada's album, where the folios have lightly pencilled attempts at binomial nomenclature, often erased and sometimes scratched out, before a final identification appears in a neater hand.

28 Foucault, *The Order of Things*, 61.

29 Shteir, *Cultivating Women*, 16–17; the classic exploration is Schiebinger, "Private Lives of Plants."

30 Gates and Shteir, *Natural Eloquence*, 7.

31 Deleuze, *Difference and Repetition*, xv and 247–8.

32 See Irigaray, *An Ethics of Sexual Difference*.

33 Deleuze, *Difference and Repetition*, xv. Irigaray, *This Sex Which Is Not One*, 74.

34 Shteir, *Cultivating Women*, 16.

35 For the Canadian context see Berger, *Science, God, and Nature*, 32. For the English precedent see Gates, "Ordering Nature," esp. 179–80.

36 The passage is from the 1879 poem "On the Cliffs," published the following year in *Songs of the Springtides*.

37 Nickelsen, *Draughtsmen, Botanists and Nature*, 4–9. See also Blunt, *Art of Botanical Illustration*, 256.

38 Saunders, *Picturing Plants*, 22.

39 Some examples of drawings with damaged foliage are Hazen's *Cornus canadensis*, and *Fragaria vesca* (University of New Brunswick Archives and Special Collections), Wilkinson's *Convolvulus sepium*, *Epigaea repens*, and *Pyrola elliptica* (also at UNB), Anne Ross McCord's *Moss Rose*, *Sarracenia purpurea*, and *Convallaria Bifolia* (McCord Museum), Fitzgibbon's *Rubus odoratus* (Thomas Fisher Rare Books Library), and Schäffer's *Lonicera involucrata* and *Rubus parviflorus* (Whyte Museum of the Canadian Rockies). All have been digitized and may be consulted online. The women named here represent a cross-section of Canadian women's nineteenth-century botanical drawing practice, ranging from the arts of accomplishment (McCord) to professional publishing ventures (Fitzgibbon and Schäffer), and offering their practitioners opportunities for civic engagement (Wilkinson) and public exhibition (Hazen).

40 Finlay, *A Woman's Place*, 72. Ruskin's ideas were sufficiently influential in British Columbia that the settlement of Ruskin, BC, was founded in 1895 as an ideal workers' community.

41 Ruskin, *Modern Painters*, 3: 151 and 295. For Ruskin's botanical irreverence, see *Prosperpina*; on his anti-generalizing stance see Kobayashi, "John Ruskin on Natural Shape," 92–101.

42 Nietzsche, *Portable Nietzsche*, 46.

43 Nietzsche, "On Truth and Lies," 82.

44 Walcott, foreword to *North American Wildflowers*.

45 Nova Scotia Archives, Prat, Starr Family Fonds, 1979–147.594.61. Prat, who studied art at the School of the Art Institute of Chicago in her youth, was later the first dean of women at King's College, Windsor, NS.

46 Traill's *Backwoods of Canada* was first published in 1836; Ruskin's botanical book *Prosperpina* appeared in 1873.

47 In a sympathy card from Pemberton to her friend Flora Burns in 1940, Pemberton compares her friend's mother to a lily: "I always thought your dear Mother looked like a lily in her grace, stately beauty" (BC Archives, Flora Alfreda Burns Family Fonds, MS-2786.3.23.).

BIBLIOGRAPHY

Archives

British Columbia Archives (BCA), Victoria, BC

Pemberton Family Fonds, MS 1295.1.

Flora Alfreda Burns Family Fonds, MS-2786.3.23

Public Archives of Nova Scotia, Halifax, NS

Prat, Starr Family Fonds, 1979–147.594.61.

Books and Articles

Bal, Mieke. "Narrative Inside Out: Louise Bourgeois' Spider as Theoretical Object." *Oxford Art Journal* 22, no. 2 (1999): 103–26.

Berger, Carl. *Science, God and Nature in Victorian Canada*. Toronto: University of Toronto Press, 1983.

Bermingham, Ann. *Learning to Draw: Studies in the Cultural History of a Polite and Useful Art*. New Haven: Yale University Press, 2000.

Blunt, Wilfrid, and W.T. Stearn. *The Art of Botanical Illustration*. Woodbridge: Antique Collectors' Club in association with the Royal Botanic Gardens, Kew, 1994.

Burton, Samantha. "Canadian Girls in London: Negotiating Home and Away in the British World at the Turn of the Twentieth Century." PhD diss., McGill University, 2011.

Crary, Jonathan. *Techniques of the Observer: On Vision and Modernity in the Nineteenth Century*. Cambridge, MA: MIT Press, 1990.

Damisch, Hubert. *The Origin of Perspective*. Translated by John Goodman. Cambridge, MA: MIT Press, 1994.

Daston, Lorraine and Peter Galison. *Objectivity*. Cambridge, MA: MIT Press, 2007.

Deleuze, Gilles. *Difference and Repetition*. Translated by Paul Patton. New York: Columbia University Press, 1994.

Dickenson, Victoria. *First Impressions: European Views of the Natural History of Canada from the 16th to the 19th Century*. Kingston: Agnes Etherington Art Centre, 1992.

Finlay, Karen A., ed. *A Woman's Place: Art and the Role of Women in the Cultural Formation of Victoria, B.C., 1850s–1920s*. Victoria, BC: University of Victoria, 2004.

Foucault, Michel. *The Order of Things*. Translated anonymously. London: Tavistock, 1970.

Gates, Barbara T. "Ordering Nature: Revisioning Victorian Science Culture." In *Victorian Science in Context*, edited by Bernard Lightman, 179–86. Chicago: University of Chicago Press, 1997.

– "Retelling the Story of Science." *Victorian Literature and Culture* 21 (1993): 289–306.

Gates, Barbara T., and Ann B. Shteir, eds. *Natural Eloquence: Women Reinscribe Science*. Madison: University of Wisconsin Press, 1997.

George, Sam. *Botany, Sexuality and Women's Writing, 1760–1830: From Modest Shoot to Forward Plant*. Manchester: Manchester University Press, 2007.

Globe, Alexander. "The Story of Canadian Wild Flowers." *Historical Perspectives on Canadian Publishing* website. McMaster University, 2009. http://hpcanpub.mcmaster.ca/case-study/story-canadian-wild-flowers.

Grosz, Elizabeth. *Time Travels: Feminism, Nature, Power*. Durham: Duke University Press, 2005.

Guildford, Janet. "Maria Morris Miller: The Many Functions of Her Art." *Atlantis* 20, no. 1 (1995): 113–24.

Huneault, Kristina. *I'm Not Myself at All: Women, Art, and Subjectivity in Canada*. Montreal: McGill-Queen's University Press, 2018.

Irigaray, Luce. *An Ethics of Sexual Difference*. Translated by Carolyn Burke and Gillian C. Gill. Ithaca: Cornell University Press, 1993.

– *This Sex Which Is Not One*. Translated by Catherine Porter. Ithaca: Cornell University Press, 1985.

Keeney, Elizabeth. *The Botanizers: Amateur Scientists in Nineteenth-Century America*. Chapel Hill: University of North Carolina Press, 1992.

Kobayashi, Chinatsu. "John Ruskin on Natural Shape and Ornamentation and the Birth of Art Nouveau." PhD thesis, Université du Québec à Montréal, 2019.

Kramer, Jack. *Women of Flowers: A Tribute to Victorian Women Illustrators*. New York: Stewart, Tabori & Chang, 1996.

Miller, Ellen and Margaret Christine Whiting. *Wild Flowers of the North-Eastern United States*. New York: G.P. Putnam's Sons, 1895.

Nickelsen, Kärin. *Draughtsmen, Botanists and Nature: The Construction of Eighteenth-Century Botanical Illustrations*. Dordrecht: Springer, 2006.

Nietzsche, Friedrich. "On Truth and Lies in a Nonmoral Sense." In *Philosophy and Truth: Selections from Nietzsche's Notebooks of the Early 1870s*, edited and translated by Daniel Brazeale, 79–97. Atlantic Highlands, NJ: Humanities Press, 1979.

– *The Portable Nietzsche*. Edited and translated by Walter Kaufmann. New York: Viking Press, 1954.

Parsons, Frances Theodora. *How to Know the Wild Flowers*. New York: C. Scribner's sons, 1893.

Ruskin, John. *The Complete Works of John Ruskin* (Library Edition). E.T. Cook and A. Wedderburn (eds.), 39 vols. London: George Allen, 1903-12.

Saunders, Gill. *Picturing Plants: An Analytical History of Botanical Illustration*. Berkeley: Zwemmer and University of California Press in association with the Victoria and Albert Museum, 1995.

Schiebinger, Londa. "The Private Lives of Plants." In *Nature's Body: Gender in the Making of Modern Science*, 11–39. Boston: Beacon Press, 1993.

Scourse, Nicolette. *The Victorians and Their Flowers*. London: Croom Helm, 1983.

Seaton, Beverly. *The Language of Flowers: A History*. Charlottesville: University of Virginia Press, 1995.

Secord, Anne. "Botany on a Plate: Pleasure and the Power of Pictures in Promoting Early Nineteenth-Century Scientific Knowledge." *Isis* 93, no. 1 (March 2002): 28–57.

Shteir, Ann B. "Botany in the Breakfast Room: Women and Early Nineteenth-Century British Plant Society." In *Uneasy Careers and Intimate Lives: Women in Science, 1789–1979*, edited by Pnina Abir-am and Dorinda Outram, 31–43. New Brunswick, NJ: Rutgers University Press, 1982.

– *Cultivating Women, Cultivating Science: Flora's Daughters and Botany in England, 1760 to 1860*. Baltimore: Johns Hopkins University Press, 1996.

– "Elegant Recreations? Configuring Science Writing for Women." In *Victorian Science in Context*, edited by Bernard Lightman, 236–55. Chicago: University of Chicago Press, 1997.

– "Let Us Examine the Flower: Botany in Women's Magazines, 1800–1830." In *Science Serialized: Representations of the Sciences in Nineteenth-Century Periodicals*, edited by Geoffrey Cantor and Sally Shuttleworth, 17–36. Cambridge: MIT Press, 2004.

Shteir, Ann B. and Jacques Cayouette. "Collecting with 'botanical friends': Four Women in Colonial Quebec and Newfoundland." *Scientia Canadensis* 41, no. 1 (2019): 1–30.

Swinburne, Algernon Charles. "On the Cliffs." In *Songs of the Springtides*. London: Chatto and Windus, 1880.

Tennyson, Alfred. *Becket*. London: Macmillan, 1884.

Tuele, Nicholas. *Sophia Theresa Pemberton (1869–1959)*. Victoria: Art Gallery of Greater Victoria, 1978.

– "Sophia Theresa Pemberton: Her Life and Art." MA thesis. University of British Columbia, 1980.

Walcott, Mary Vaux. *North American Wildflowers*. Washington: Smithsonian Institution, 1925.

10

Slips and Seeds

Botany and Horticulture in Two Nineteenth-Century Canadian Quilts

VANESSA NICHOLAS

One of the earliest quilts in the Heritage Quilt Collection at the Agnes Etherington Art Centre at Queen's University in Kingston, Ontario, is a linen patchwork quilt embroidered with various floral designs known as the Fallowfield quilt that has recently been attributed to Elizabeth Bell (1824–1919).[1] Bell was an Irish settler and she most likely made the quilt in the late 1840s before she married her neighbour in the village of Fallowfield, which is now considered part of the City of Ottawa in Ontario. Another quilt adorned with floral embroidery can be found at the Canadian Museum of History in Gatineau, Quebec. This black wool quilt, which features a brightly coloured garden scene, was also made in southeastern Ontario, by Margaret Ann McCrum (1847–1888) sometime before her marriage in 1881.[2] Both of these embroidered quilts were preserved by family members, indicating that they have private significance as heirlooms.[3] They also, however, carry public significance as representations of the decorative traditions and botanical pursuits that enriched the lives of Canadian women throughout the nineteenth century. Elizabeth Bell's quilt is a tribute to historical British embroidery and textile conventions and exhibits her interest in Canadian wildflowers.[4] Margaret McCrum's quilt, on the other hand, borrows formal elements from American quilt conventions and Woodland Indigenous beadwork to depict numerous identifiable garden flowers. Together, these quilts show Victorian Canadian women using homecraft and flowers to engage with questions of place and cultivate cultural identities.

Little attention has been paid to historical Canadian quilts, and the rarity of embroidered examples has led to their being almost overlooked by scholars. Mary Conroy's *300 Years of Canada's Quilts*, for example, includes only two

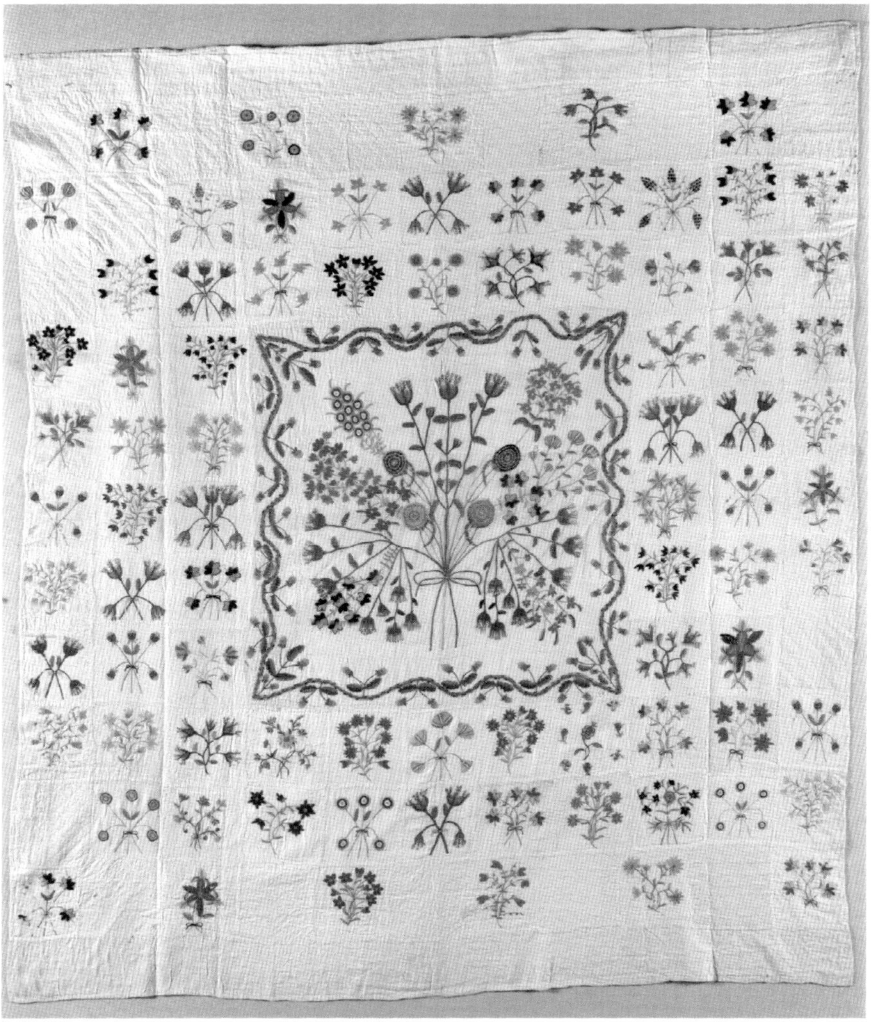

10.1
Fallowfield quilt, attributed to Elizabeth Bell, 1840s.

embroidered quilts among more than one hundred illustrations, and only nine embroidered quilts appear among more than four hundred quilts illustrated in Ruth McKendry's *Quilts and Other Bed Coverings in the Canadian Tradition*. Neither Elizabeth Bell's quilt nor Margaret McCrum's quilt appears in these texts, and Margaret McCrum's quilt has no known bibliography. Ruth McKendry discusses the Fallowfield quilt in a number of publications, including a short article on "The Use of Embroidery on Quilts in Canada," where she attributes the rarity of embroidered quilts to the degree of skill that they demand: "Embroidered bedcovers were never common because women

10.2
Coverlet, Margaret Ann McCrum, 1860–81.

making such quilts had to be artistic as well as skilful with their needles in order to plan a design that would hold together over such a large area, and, at the same time, be pleasing and harmonious in colour."[5] Embroidered details, she explains, were fashionable amongst quilt makers from the end of the nineteenth century through to the 1930s, but these often derived from stamped patterns and lacked the originality of the earlier all-over embroidery styles exemplified by Elizabeth Bell and Margaret McCrum's quilts.

Elizabeth Bell and Margaret McCrum's quilts are part of a British tradition of domestic embroidery and needlework that dates back to the Elizabethan period.[6] Initially a professional male occupation, embroidery was later feminized in part because it enabled women to demonstrate their taste as well as their practical skills, traits that were considered desirable amongst an emergent class of mercantile men.[7] As a result, the Elizabethan period saw the development of samplers, plain grounds embroidered with motifs, patterns, and stitches, that serve as exemplars for future needlework. Many that survive in museum collections are embroidered with the names and ages of the young women who made or "worked" them, to use the historical term that continues to mean being engaged with needle and thread.[8] The term "embroidery" once referred to the embellishment of costly ground fabric, like silk or velvet, with silk floss and metal threads, while "needlework" described the humbler pursuit of stitching wool yarn or silk floss decorations onto plain-weave linen fabric.[9] These terms became interchangeable by the eighteenth century.

In keeping with the convention of British domestic embroidery, Elizabeth Bell and Margaret McCrum embellished their quilt tops with floral designs. Flowers have long been a defining feature of British design and embroidery. The earliest British samplers are worked with designs based on flowers, and English publishers in the seventeenth century began producing special features on flowers for designers and needleworkers.[10] Floral embroidery styles were in part a response to Indian chintz, a cotton dress and furnishing fabric that was made in India between the seventeenth and eighteenth centuries. Chintz fabrics were adorned with scrolling all-over vegetal and floral designs produced using a mordant- and resist-dying process that resulted in bright, fast colours.[11] Chintz delighted British eyes upon its introduction to Western markets, and British embroiderers were quick to emulate its effect. The predominance of floral designs in domestic embroideries produced by British women at that time has also been attributed to the introduction of new plant species to the British Isles by way of global exploration and trade, which generated interest in gardening.[12] British women expressed their general interest in flowers in their needle arts, endeavouring to embroider truthful representations of real plants.[13]

The embroidered quilts made by Elizabeth Bell and Margaret McCrum in mid- to late-nineteenth-century Canada were likewise informed by the plant sciences of their time. Starting in the eighteenth century, British women had been encouraged to study plants and flowers within the bounds of the household. Ann Shteir has shown that eighteenth-century British literature wielded illustration and narrative to foster a female niche in modern science culture. Frontispieces in botany books, for example, pictured the Goddess Flora as a maternal figure that signified order and propriety.[14] Additionally, there emerged a genre of popular science books written predominately by British women that chronicled young girls learning Linnaean botany from their maternal elders. For example, Priscilla Wakefield's pioneering text *An Introduction to Botany, in a Series of Familiar Letters* (1796) takes the form of a correspondence between sisters, one of whom is learning Linnaean botany from her governess.[15] Elizabeth Bell and Margaret McCrum's quilts are visual and material representations of this popular and gendered form of botanical science that emerged in eighteenth-century British culture.

More specifically, their quilts are the products of the British female botanical tradition that, as Kristina Huneault has shown, flourished in nineteenth-century Canada.[16] According to Huneault, Victorian Canadian women were interested in familiarizing themselves with plants and practising plant identification; however, they did not take a strong interest in studying the inner workings of plants. As a result, the botanical art that they produced tends to record the particular details of an individual plant, rather than articulating type specimens. This approach was arguably informed by their perception that a Canadian identity was forged in direct, physical experiences with nature.[17] This somewhat subjective approach to botany is exemplified by naturalist Catharine Parr Traill. Writing in southeastern Ontario in the 1830s, Traill describes her interest in regional plants in decidedly creative and personal terms: "I take the liberty of bestowing names upon them according to inclination or fancy."[18] In keeping with this approach, Victorian Canadian women pursued generalized botanical pastimes, like gardening, and they also developed a visual culture of floral imagery and decoration through flower painting, découpage, and embroidery that had little to do with scientific study.[19] Elizabeth Bell and Margaret McCrum's quilts belong within this culture of floral imagery, but they also incorporate plant identification and natural observation. As such, these objects integrate the botanical pastimes and decorative traditions that were typical of the female botanical tradition in Canada.

THE FALLOWFIELD QUILT, BRITISH HERBAL DESIGN TRADITIONS, AND CANADIAN BOTANY

Elizabeth Bell's quilt, which is named for the village where she married and lived, is a medallion quilt. The term "medallion" describes quilts that have a distinctive centrepiece bordered by smaller quilt blocks or pieces. The medallion at the centre of the Fallowfield quilt is approximately seventy-five centimetres square, and the smaller quilt blocks that surround it measure approximately fifteen centimetres square. All of the pieces in this quilt were cut from lengths of plain-woven linen cloth and hand sewn together along their edges. The quilt's medallion and eighty-one of its small squares are embroidered with naturally dyed wool thread worked using a combination of tent, tambour, and satin stitches. All of the embroidered designs are floral. The medallion features a spray of eight flower varieties tied together with a string looped into a bow, and it is bordered with a festoon of strawberry plants. With two exceptions, the embroidered sprays on the quilt's smaller blocks do not contain multiple flower varieties. Rather, they are composed of single flower varieties and at least ten unique designs can be found across the quilt's surface.

Elizabeth Bell emigrated with her family from Ireland in 1837 when she was a child of twelve or thirteen.[20] Her mother, Anne Little Bell, presumably died in Ireland because she did not accompany her husband and children to Upper Canada.[21] Design historian Lisa Binkley has concluded that twenty-eight of the quilt's embroidered blocks, as well as its central medallion, were worked in Ireland by Anne Little Bell, and that Elizabeth Bell embroidered the remaining fifty-three blocks during the year of her engagement in the late 1840s, attaching all the squares together to make her wedding quilt. This conclusion is informed by two distinct styles of embroidery on the quilt, suggesting that two different women worked its squares. Nine of the thirteen sprays of rosebuds that figure on the quilt's squares and in its central medallion, for example, are very schematic and would be unrecognizable as rosebuds were it not for their overwrought sepals.[22] By contrast, no two of the seven other rosebud sprays are alike, and their variation reveals a needleworker with an objective eye and naturalistic style. It is possible that Anne Little Bell made a portion of the quilt's embroidered blocks and that her family brought these to Canada in remembrance of her. Visual evidence suggests that Elizabeth Bell produced the quilt blocks worked with naturalistic embroideries.

The botanical aspect of the Fallowfield quilt begins with its design principles, which almost certainly reference early modern British "slips." In the Elizabethan and early Stuart periods, British needleworkers produced small squares of linen or canvas embroidered with isolated floral sprigs resembling gardeners' cuttings. These embroidered squares were called "slips." Slips were made individually and then applied in rows or patterns to finer fabric grounds, like velvet.[23] Slips featured common flowers such as columbines, cornflowers, honeysuckles, and daffodils.[24] Slip designs were often copied from antique and medieval herbals whose texts and illustrations provided information about plants before the advent of modern botanical science.[25] The earliest printed herbals date to the fifteenth century and feature plain woodcut illustrations of single plant varieties.[26] As Albert Frank Kendrick observes, these illustrations made for excellent embroidery patterns because "the essential character of the plants was interpreted with the greatest economy of line."[27] The synergy between early natural science and the needle arts is seen in *La Clef des Champs* (1586), a natural history book by Jacques Le Moyne de Morgues. Printed in London, this text contains illustrations of wild and garden flowers, animals, insects, birds, and fruit and is dedicated to embroiderers, tapestry weavers, and other craftsmen.

The illustrations in some surviving copies show signs of having been pricked with pins and pounced with chalk in order to transfer their designs to another surface.[28] By 1559, printmakers had attempted to capitalize on the decorative appeal of herbals by issuing embroidery pattern books.[29]

The embroidered blocks that make up the Fallowfield quilt look remarkably like the early modern slips that derived from illustrated herbals and other early natural history texts. Take for example, a set of slips in the collection of the Victoria and Albert Museum, which date to the early seventeenth century. Each of these slips is only just slightly larger than the blocks in the Fallowfield quilt. Additionally, both the seventeenth-century slips and the embroidered blocks in the Fallowfield quilt can be described as squares of plain-woven linen grounds worked with polychrome embroideries. Most significantly, the seventeenth-century slips and the quilt's blocks both feature single flower varieties. Katharine Brett identifies the following flowering plants in a set of thirteen early modern slips in the collection of the Royal Ontario Museum: red and white strawberries, roses, carnations, marigolds, periwinkles and pansies, lilies, columbines, and mallows.[30] Many of these plants also figure in the embroideries on the Fallowfield quilt. However, whereas the flowers embroidered onto the slips in the ROM's collection have roots, this scientific detail is not part of the embroideries on the Fallowfield quilt. Instead, the flowers in the quilt's medallion and a number of its squares are tied with a bow and read as floral arrangements rather than as specimens. When compared to historical slips, it seems evident that the embroidered squares of the Fallowfield quilt were at least partly inspired by this decorative art.

For British settlers in nineteenth-century Canada, New Zealand, and Australia, flowers and floral imagery invoked the British countryside and thereby became emblematic of nostalgia and homesickness. In *Roughing it in the Bush* Susanna Moodie describes her reaction to seeing harebells growing from the grave of an Indigenous man during a trip to Stony Lake during the summer of 1835: "It was completely covered with stones, and from among the crevices had sprung a tuft of blue harebells, waving as wild and free as if they grew among the bonny red heather on the glorious hills of the North or shook their tiny bells to the breeze on the broom-encircled commons of England." She continues, "I gathered those flowers, and placed them in my bosom, and kept them for many a day; they had become holy, when connected with sacred home recollections, and the never-dying affections of the heart which the sight of them recalled."[31] Notably, Moodie bestows this wildflower with divine status because it reminds her of England and not because it was growing from an Indigenous man's resting place. Recognizing the strong emotional reso-

10.3
Slips, unknown artist, c. 1600, linen canvas, embroidered with silks in tent stitch.

nance that flowers had in the colonies, British authorities sent native flowers to homesick subjects abroad in the mid-nineteenth century.[32] The Fallowfield quilt can be contextualized within this broad cultural identification with flowers felt by many in the British colonies.

The nostalgic dimension of the Fallowfield quilt is reinforced by the two flowering plants that figure most prominently on the quilt top: roses and strawberries. Wild roses and strawberries were common motifs in historical slips, and both figure in the set of slips at the ROM. Roses appear in fifteen of the embroidered squares in the Fallowfield quilt as well as in its central medallion and are the most repeated motif on the quilt's top. According to textile historian Lanto Synge, roses have been "especial favorites" in every period of British embroidery.[33] Strawberries appear only on four of the quilt's eighty-one embroidered squares, but they festoon its central medallion and thereby make a big visual impact. One of the most prominent plant motifs in British embroidery, strawberries came to symbolize purity. Synge speculates that this came about in part because they were gathered from the woods, "unspoiled by less natural and coarser vegetables grown in the garden."[34] The strawberry's dual association with the British Isles and purity would have seemed suitable for Elizabeth Bell's wedding quilt.

There is one piece of visual evidence that seems to confirm Bell's contribution to the Fallowfield quilt, namely, a design worked on five of the quilt's blocks depicting a bloom of three elliptic maroon petals crowned by three pronounced sepals that strongly resembles the red trillium flower indigenous to the temperate woodlands of eastern North America. Five trillium species are native to the Ontario region, and the red petals in this embroidery design resemble those found on the red trillium.[35] Although Indigenous Peoples had long utilized this plant to hasten childbirth and ease delivery pains, as well as to treat insect and rattlesnake bites,[36] the trillium was unfamiliar to Europeans and has no precedent in Old World embroidery or textile design. If this design was inspired by the red trillium, then it must have been produced in Canada West by Elizabeth Bell rather than by her mother in Ireland. This instance of plant identification in the Fallowfield quilt leads one to conclude that Elizabeth Bell's interest in flowers formed part of her effort to develop a sense of belonging in a new land far from her childhood home.

Since few regional botanical inventories existed in the 1840s, Elizabeth Bell likely captured details of her trillium by direct observation. Elizabeth Bell may well have seen the flower and examined its particulars near her home in the 1840s, for a red trillium appears in the *List of Plants Collected by Mr. Billings in the Vicinity of the City of Ottawa during the summer of 1866*.[37] It is possible that she produced a drawing or sketch of a red trillium in nature and then translated this into an embroidery design by copying or tracing it onto her fabric at home.[38] By producing her trillium flower design from nature rather than from print or textile sources, Elizabeth Bell exemplified the enthusiasm for natural observation that, as art historian Kristina Huneault writes, distinguishes female Canadian naturalists and botanists from their European counterparts: "Canadian women adopted various means of emphasizing that their botanical drawings were representations not of types but of their own first-hand encounters with the natural world. While leading European botanical artists worked as much from other illustrations as from life, Canadian women stressed that they did not do so."[39] Thus, Maria Morris Miller inscribed *drawn from nature* onto each plate of *Wild Flowers of Nova Scotia* (1840), and Montreal-based Anne Ross McCord specified that her botanical illustrations were copied from nature. This emphasis on natural observation and on individual plants differed from the focus within botanical science on type specimens.

Elizabeth Bell's trillium flower situates her within the history of Canadian women who undertook the task of identifying plants that were unfamiliar to settlers. Significantly, one of the earliest references to the trillium flower in English Canadian literature is found in the diary of Elizabeth Simcoe, the wife of Upper Canada's first lieutenant governor. Recording her journey through the region between 1791 and 1796, Simcoe wrote, "I walked down the Hill in

10.4
Detail of Fallowfield quilt, attributed to Elizabeth Bell. 1840s, figure 10.1.

the Evening, and gathered Dragon's blood, Lychnis de Canada, Tryliums, toothache plant, Liquorice, wild lilies etc."[40] She details the wildflowers, natural curiosities, and topographies that differentiated the Canadian landscape from England's. By so doing, she may be writing in support of her husband's professional objective to survey and familiarize himself with the colonial territory in his charge.[41]

While plant identification is typically characterized as the pursuit of elite women like Simcoe, the Fallowfield quilt shows that this activity had broader appeal and application. The quilt presents the possibility that settler women in Ontario's farming and labouring class also worked to familiarize themselves with indigenous plants and encoded their botanical knowledge into their homecrafts. As a young woman occupied with the demands of rural settlement life, Bell may not have had the time or inclination to write about her encounters with the natural world, much less assemble an herbarium or flora, and yet the embroidered trillium flowers on her wedding quilt top recount an early summer walk and the delight stirred by a peculiar red bloom on the forest floor.

MARGARET ANN MCCRUM'S QUILT, INDIGENOUS MATERIAL CULTURE, AND GARDEN PLOTS

Like Elizabeth Bell, Margaret McCrum created a flowered quilt that signified her sense of self and belonging. She made hers sometime in the 1860s or 1870s in the Oxford Township of Ontario's Leeds and Grenville United Counties, which is approximately eighty kilometres south of where Elizabeth Bell lived as a married woman with her growing family. Margaret Ann McCrum did not date her quilt or indicate her age at its completion, but we can date her handiwork to sometime between the beginning of her adolescence in 1860 and her marriage to Samuel Thomas Gibson (1849–1884) in 1881 because she embroidered her maiden name onto the quilt in three places.[42]

Margaret McCrum's quilt is made in a four-block or quadrant-style, which means that its design features four oversized decorative units. The quadrant-style is believed to have developed in Pennsylvania in the 1840s, and it is most common in areas of North America settled by German and Scotch-Irish immigrants.[43] Margaret McCrum constructed her quilt by sewing together ten large blocks cut from lengths of plain-woven black wool cloth. The quilt measures approximately the size of a double bed and is embroidered with naturally dyed wool thread worked using a combination of stem and satin stitches. In each of the quilt's four quadrants there is an urn or pot of flowering plants rendered in bright, lively colours that pop against the quilt's black ground. Margaret McCrum has worked her signature, or some variation on it, beneath three of four of these flowerpots. Birds also figure in the quilt's design. Chickens, roosters, and ducks wander at the quilt's border, and swallows and robins feed and rest amongst the flora. The quilt's edge is decorated with an undulating vine blooming with different types of flowers, a border that is common in quadrant-style quilts. A complementary trail of flowering plants runs through the quilt's centre.

While Elizabeth Bell's quilt design references a historical British embroidery form and reinforces her ties to Britain, Margaret McCrum's quilt design expresses her identification with North America insofar as it borrows formal elements from both American and Indigenous material cultures. Like the quadrant style she has used to visually organize her quilt top, Margaret McCrum's embroidery designs derive in part from American decorative conventions. American coverlets and quilts began featuring urns, baskets, and pots of flowers as early as the 1830s. The trend began with Jacquard coverlet weavers, who introduced urns of garden flowers into the borders of their designs in response to the swelling interest that many American women were exhibiting in flowers and gardening. By the 1840s, these motifs had become much larger and they were moved from the borders of industrially manufactured coverlets to their centrefields. At midcentury, American women began incorporating urns or pots of flowering plants into the decorative schemes of their own applique quilts.[44] The popularity of these motifs developed alongside the popularity of the quadrant quilt style and as a result many early quadrant-style quilts use the same bilateral symmetry favoured by coverlet weavers and feature large flowerpot motifs. American quilt makers at this time were partial to green and red colour schemes on a white ground, and many of the quadrant style quilts with flowerpot designs were produced in this bold but limited palette. The composition and content of Margaret McCrum's quilt were apparently influenced by American applique quadrant-style quilts she may have seen reproduced in print.

The black ground and bright colour scheme of McCrum's quilt are distinctly different, however, from the American style, and they suggest another design tradition in her work, namely, the Woodland Indigenous beadwork tradition. This tradition is exemplified by a beaded black bag in the Thaw Collection at the Fenimore Art Museum in Cooperstown, New York.[45] Dating to the mid-nineteenth century and attributed to Mi'kmaw or Wəlastəkwewiyik traditions within the Algonquian language group, the colourful beadwork on the front of the bag depicts floral and abstract motifs made all the more striking by a black ground. Black trade cloth was favoured for beadwork in part because it resembled the black-dyed hides that had been used in the Woodlands artistic traditions for centuries.[46] Beaded bags and purses formed the bulk of the Indigenous souvenir trade in northeastern North America during the Victorian period, and Margaret McCrum may well have had one or more such items in mind when she decided to use a black ground for her embroidered garden.[47]

Significantly, more than twenty of the flowers that Margaret McCrum has embroidered onto her quilt top seem to derive from Woodlands Indigenous iconography. Distinct from the other floral designs on Margaret McCrum's quilt top, the motif in question is composed of circles or sequences of concentric circles that are either edged or filled with zigzagging patterns. In the Woodlands Indigenous artistic tradition, the sun was historically pictured as a circle or series of concentric circles edged with projecting lines or zigzagging patterns representing rays.[48] Many of the beaded items produced by Woodlands Indigenous artists for commercial and domestic use during the second

10.5
Bag, c. 1860–90, unidentified artist, Mi'kmaq or Wəlastəkwewiyik, New Brunswick, velvet, glass beads, silk bindings and lining, thread.

half of the nineteenth century feature floral designs that incorporate the sun motif.[49] Consider, for example, the single open-faced flower motif on the front of the aforementioned beaded bag. Its concentric rings of linework, including a red zigzagging line near its centre, may refer to the sun and its sacred power. The similarity between this motif and the uncharacteristic flowerheads that pepper the top of Margaret McCrum's quilt further suggests that she modelled aspects of her embroidery design on Woodlands Indigenous beadwork.

The appropriation of Indigenous visual culture by Margaret McCrum's embroidery style is reinforced by her subject, the garden, which served as a metaphor for the British settlement of Indigenous land in Canada during the first half of the nineteenth century. While nineteenth-century Americans arguably celebrated the natural world as a source of inspiration and morality, their Canadian counterparts seem to have been more inclined to fear the wilderness as hostile and threatening. Cultivated landscapes were the ideal in early English Canadian culture because they represented control and security. For example, when British author Anna Jameson describes an admiral's house in Upper Canada's London District in her autobiographical *Winter Studies and Summer Rambles in Canada* (1838), she alludes to the important role that gardens played in demarcating British settlements from the Canadian wilderness: "The woods are yet close up to the house; but there is a fine well-cultivated garden and the process of clearing and log-burning proceeds all around with great animation."[50] For Jameson, the threat of the forested surround is mitigated by the reassuring garden plot and its promise of further development. Given the correspondence between the garden and civilization in the British worldview, it is no wonder that gardens were features of even the earliest homesteads in Upper Canada.

McCrum was an unmarried woman when she made her quilt in the 1860s or 1870s, and her embroideries were therefore likely inspired by her mother's garden, rather than her own. What would her mother's garden have looked like? Writing for a 2014 issue of *Historic Gardens Review*, Brian Malcolm characterizes Ontario's garden history a story largely untold.[51] Nevertheless, we can come to some conclusions by considering what we know of the McCrum family farm and the history of gardens in the southeastern Ontario region. Elizabeth Fletcher McCrum and Edward McCrum registered as residents of the Township of Oxford in Canada West's Grenville County in 1851, when they were both twenty-nine years old and their eldest child, Margaret McCrum, five.[52] The enumerator notes that both McCrums were born in Ireland, and their children were born in Canada. He also logs Edward McCrum as a *farmer*, an indication that the family homesteads. The McCrums presumably settled on their farm in Oxford County as a young couple anticipating the arrival of their first child. Although horticulture and garden design were not concerns for the settlers who arrived in Canada before the mid-nineteenth century, informal flower gardens formed an important part of their homesteads and it is probable that the McCrums established their garden within months of their daughter's birth.[53] A garden on the farm would have been tended by Elizabeth McCrum because such plots were typically the purview of women.[54]

It is most likely that Elizabeth McCrum first planted an island bed of wildflowers. Island bed gardens were the earliest, crudest garden types in Upper Canada, so called because they were initially mounds of humus and rubbish near the house that were converted into informal plots for growing vegetables and flowers.[55] American seed merchants began advertising in Upper Canada soon after the United Empire Loyalists migrated north, and by the 1830s milliners and general stores were carrying flower seeds. Nevertheless, many of the first British women to settle in the region initially transplanted berries and flowers growing in the forests and wetlands around them.[56] Wild phlox, training rose, honeysuckle, and cowslips were amongst some of the wildflowers that could be found in these early Upper Canadian gardens.[57] These island bed gardens eventually developed into dooryard or parlour gardens, fenced plots just outside the house for decorative plants and trees. With the development of dooryard gardens, vegetable and herb gardens were moved to the back of the house.[58] Dooryard gardens were considered fashionable until the beginning of the 1840s and remained common in Ontario's small towns for years thereafter.[59] Island beds and parlour gardens remained customary in the region throughout the nineteenth century.

Margaret McCrum's embroidered quilt seems to picture an island bed that has been improved in accordance with the regional gardening trends that developed during the second half of the nineteenth century. The chickens and geese that wander through the scene suggest that the garden she has represented is not fenced in, making it an island bed rather than a dooryard or parlour garden. The flowerpots or urns that punctuate the otherwise disorderly plot may simply be inspired by the American coverlet and quilt designs mentioned earlier. Or else, they might reflect the growing interest in formal garden designs that developed in Ontario after the mid-century. By the 1850s, garden plans were being designed by Canadian nurserymen and architects, and flowerbed layouts were being published in garden manuals.[60] Manufacturers responded to these cultural developments, producing inexpensive cast iron chairs, benches, tables, and urns in popular Baroque and Rococo styles.[61] Canadian women duly began augmenting their lawns and gardens with walkways, furniture, and ornaments.[62] Given the likely timeframe for McCrum's quilt making, it is possible that her own quilt depicts an updated version of her mother's garden.

The advent of garden furniture and fixtures coincided with the development of regional horticultural societies and seed houses, as well as the introduction of new garden flowers.[63] While Elizabeth McCrum may have initially planted local wildflowers, her daughter's embroideries suggest that she eventually added a number of the garden-variety flowers that would have been common in Ontario's front yards during the second half of the nineteenth century, including peonies, roses, and amaranths. Flower varieties often circulated through informal networks of female friends, rather than distributed as commodities, and it is possible that Elizabeth McCrum's garden developed as her local network of friends grew.[64] The peonies in her daughter's embroidered scene are tightly closed pink buds, recognizable by their distinctive round shape. Peonies bloom in May and June, making this a springtime garden scene full of anticipation. In the horticulture section of the July 1849 issue of *The Canadian Agriculturalist*, the peony is lauded for its effortless beauty: "the species of the family are, in most cases, easily cultivated, hardy, showy and flower early."[65] The peony's dramatic annual display is precisely what made it so popular in island and dooryard gardens.

Margaret McCrum also incorporated representations of rosebuds into her embroidered quilt top. Roses were common in nineteenth-century Canadian gardens. Many rose varieties that flourished in early Canadian gardens, including the damask and sweetbrier roses, came to Ontario from Massachusetts with the United Empire Loyalists, and perhaps Elizabeth Fletcher McCrum grew her roses from a cutting gifted to her by a more established neighbour.[66] Roses were amongst the first flowers planted by settlers because their leaves, petals, flowers, and hips were chief ingredients in a number of herbal remedies.[67] During the earliest years of Upper Canada's settlement, growing and gathering medicinal plants as well as making poultices, salves, and ointments would have been tasked to women, and it is possible that Margaret McCrum associated roses with her mother's care.[68] Indeed, roses signified familial ties between mothers and daughters in nineteenth-century Ontario, where it was customary for married women to christen their gardens by planting cuttings from their mother's roses.[69]

In addition to the peony and rose, Margaret McCrum repeatedly represents a long, floppy, fuzzy flower that may very well be *Amaranthus caudatus*, a plant popular amongst early Canadian gardeners and commonly known as Love-Lies-Bleeding. This flower, which produces ropes of red flowers, was historically grown as a grain in South America, a fact that might explain why Margaret McCrum depicts birds feeding at this particular plant.[70] Europeans introduced the amaranths into their gardens as an ornamental in the sixteenth century, and Carl Linnaeus first encountered it in George Clifford's garden in Holland in 1730s.[71] As a relatively robust flower, amaranths were favoured in Upper Canada's dooryard gardens.[72] Its enduring appeal throughout the Confederation era is evident in the *John A. Bruce and Co. Seed Catalogue* issued from Hamilton, Ontario, for the spring of 1878, which describes the amaranthus family as "an extremely graceful and interesting character, producing a striking effect whether grown for the conservatory of out-door flower garden."[73] The garden flowers that Margaret McCrum represents on her quilt thus recall southeastern Ontario's historical gardens, which were understood in relation to British notions of discipline and taste.

CONCLUSION

Canadian quilt patterns and designs have typically been characterized in terms of symbolic meanings, as when quilt historian Mary Conroy reads floral motifs on quilts as representing ideals related to family and fortune, including marriage and abundance. "The designs for [marriage quilts]," she writes, "are almost always symbolic of love, longevity and fertility ... hearts, grapes, flowers and doves."[74] Similarly, Ruth McKendry characterizes floral quilt motifs as totems of "fruitfulness and the annual resurgence of growth."[75] This study, by contrast, challenges the notion that florals on quilts are purely symbolic or ornamental by showing that some flower designs were based on actual plants. Both Elizabeth Bell and Margaret McCrum, as has been shown, studied their surroundings and recorded specifics of their environment in their needle arts, using their own observations of nature.

The quilts made by Elizabeth Bell and Margaret McCrum reflect the scientific sensibility of their time. As historian Suzanne Zeller has observed, the purpose of science during the Victorian period "was to explore and to exploit new lands all over the world," and the first scientific inventory projects in British North America were duly formed to assess the land's commercial and agricultural prospects.[76] Science thus contributed to the project of nation building in Canada. Moreover, "[science] refocused Canadians' vision of the land they inhabited, broadening their conceptual horizons and encouraging them to carve out a place for the developing society in the larger world."[77] The importance of science to Canadian identity and its relationship to the land is exemplified by the national landscape painting tradition at the dawn of Confederation in the mid-nineteenth century, an era that saw John A. Fraser, Otto Jacobi, and Lucius O'Brien, amongst others, painting Canada's wild terrain in objective or scientific terms.[78] This approach was seen to depart from Romanticism, which favoured landscape ideals or types, and to distinguish Canadian culture from European culture.

Given the influence that Victorian science had on how Canadians viewed the natural world around them, it makes sense that the floral motifs on Elizabeth Bell and Margaret McCrum's quilts are informed by the Western scientific tradition and its ideals. Elizabeth Bell's quilt is scientific in its references and imagery. Her embroidered linen quilt squares descend from the early modern slips that were modelled on botanical illustrations in medieval herbals. Perhaps inspired by this history, Elizabeth Bell turned her attention to the natural world and took notice of the wildflowers around her family's homestead. The red trilliums that she worked onto five of her quilt blocks are tokens of personal discovery and everyday wonder, much like the pressed flowers and leaves gathered on nature walks. Moreover, the practices of botanical identification and plant collecting that are entangled with her trillium flower design are fundamentally settler-colonial exercises insofar as they were part of a larger effort to supplant Indigenous knowledge with European inventories and classification systems.[79] Elizabeth Bell's embroidered quilt can be understood, therefore, as a modest contribution to the effort to record and arrange the North American wilderness on Western terms.

Margaret McCrum's embodies this same ideal or objective insofar as it pictures a garden plot, the British model for controlled, useful land. McCrum marries form and content in her quilt top, referencing American coverlets and quilts that were inspired by the rise of horticulture and gardening in nineteenth-century America. The fertile land in nineteenth-century Ontario was likened to Eden, and gardening and farming were imbued with religious significance.[80] It would thus have been suitable for Margaret McCrum to associate her embroidered flowers with both Western science and virtue. Inadvertently, her quilt also stresses the settler-colonial dimension of Christianizing land cultivation by incorporating a motif from the floral beadwork then practised by northeastern Indigenous artists. Contemporary viewers might read this motif in Margaret McCrum's quilt as a regrettable symbol of the Indigenous Peoples that were displaced by the advancement of productive British landscapes in Canada.

In her study of early American "homespun," historian Laurel Thatcher Ulrich describes needlework as a field of personal expression that simultaneously operates as a site for cultural production.[81] Seen from this perspective, we can appreciate the artistic and creative aspects of embroidery by Victorian Canadian women and also think critically about the worldview that their embroidered projects represent. The flowers worked by Elizabeth Bell and Margaret McCrum are close studies of the natural world that attest to a desire to habituate to and connect with the Canadian land. At the same time, the design and science histories contained within their quilts can be read as contributing to the nation-building project that objectified wilderness and marginalized Indigenous Peoples. The complex meanings of the flowers represented on these quilts thus enrich our understanding of the past, and the motifs examined here show Victorian Canadian women in nineteenth-century Ontario wielding their art of floral decoration to contribute to emergent notions of Canadian national identity.

NOTES

1 This attribution has been made by Dr Lisa Binkley, who generously provided me with the chapter on this quilt from her forthcoming book, *Canadian Quilts and Their Makers*, 34–49. Elizabeth Bell's dates are confirmed by the inscription on her monument at Beechwood Cemetery in Carlton County, Ontario, which she shares with her husband, Hugh Davidson (1821–1891).

2 Margaret Ann McCrum's dates are confirmed by the inscription on her monument at Bishop Mills Cemetery in Leeds and Grenville United Counties, Ontario, which she shares with her husband, Samuel Thomas Gibson (1849–1884).

3 Canadian quilt historian Ruth McKendry purchased Elizabeth Bell's quilt from the quilt maker's granddaughters, Tena Davidson (1897–1989) and Ida Davidson (1901–1989), in 1979; and the Canadian Museum of History received Margaret McCrum's quilt from the quilt maker's great-great-niece, Grace Blair (1919–2019), in 2011.

4 Although Elizabeth Bell is an Irish immigrant and Margaret McCrum the immediate descendant of Irish immigrants, the Kingdoms of Great Britain and Ireland were unified as the United Kingdom of Great Britain and Ireland during their lifetimes (1801–1922). Additionally, both were from Presbyterian families that likely identified with British culture. (See Holmes, "Presbyterian Religion, Historiography, and Ulster Scots Identity.")

5 McKendry, "The Use of Embroidery," 10–11. See also McKendry, "The Quilt in Upper Canada," 10–14; and *Classic Quilts*, 39.

6 *A Picture Book of Flowers in English Embroidery*. London: V&A Museum, 1938. n.p.

7 Parker, *The Subversive Stitch*, 60.

8 Staples, "Embroidered Furnishings," 24.

9 Ibid., 23–4.

10 *A Picture Book of Flowers*, n.p.

11 Irwin and Brett, *The Origins of Chintz*, 1.

12 Synge, *Art of Embroidery*, 71.

13 *A Picture Book of Flowers*, n.p.

14 Shteir, "Iconographies of Flora," 19–21.
15 Shteir, *Cultivating Women, Cultivating Science*, 83–4.
16 Huneault, *I'm Not Myself at All*, 152.
17 Ibid., 180, 196.
18 Traill, *The Backwoods of Canada*, 88.
19 Huneault, *I'm Not Myself at All*, 156, 180.
20 The year of Elizabeth Bell Davidson's immigration to Canada is documented in the 1901 Canada Census.
21 "Mrs. Davidson Dead, Aged 96." *The Evening Journal* (Ottawa, ON), 21 June 1913, 1.
22 In English samplers and decorative schemes, rosebuds are typically represented with sepals that reach beyond the bud and are often topped by spikey or fuzzy details.
23 Synge, *Art of Embroidery*, 71.
24 *A Picture Book of Flowers*, n.p.
25 Levey, "Embroidery, Renaissance to Rococo," 16.
26 Arber, *Herbals*, 11–12.
27 Kendrick, *English Needlework*, 87.
28 Synge, *Art of Embroidery*, 73.
29 Ibid., 71.
30 Brett, *English Embroidery*, 7.
31 Moodie, *Roughing It in the Bush*, 173.
32 Dunlap, *Nature and the English Diaspora*, 54–5.
33 Synge, *Art of Embroidery*, 73.
34 Ibid., 74.
35 The three sepals that now read as dark brown would originally have been a bright shade of green, the recipe for which would have required goldenrod and indigo. Goldenrod grows abundantly in Ontario, and indigo, a blue dye historically made from the leaves of tropical plants and largely sourced in India, was readily available at general stories in Canada (McCalla, *Consumers in the Bush*, 173).
36 Langdon, *Pioneer Gardens at Black Creek Pioneer Village*, 38.
37 *List of Plants Collected by Mr. Billings in the Vicinity of the City of Ottawa during the summer of 1866* (Ottawa: Plant Research Institute Canada Department of Agriculture, 1968), 13.
38 Bell's process is difficult to discern, but evidence from colonial America points to practices of women translating their own drawings into embroideries. The sketchbook kept by Ann Flower in eighteenth-century Philadelphia, for example, is filled with drawings of flowers, birds, and animals that, Amanda Isaac posits, formed the basis for her exceptional needlework. The trillium flower on Elizabeth Bell's quilt speaks to the equivalent effort that Elizabeth Bell made to advance beyond the

conventional idioms of English needlework and adapt her handicraft to her Canadian context. See Isaac, "Ann Flower's Sketchbook."

39 Huneault, *I'm Not Myself at All*, 196.

40 Simcoe, *Mrs. Simcoe's Diary*, 230.

41 Gnarowski, "Foreword," 8, 13.

42 Gibson and McCrum *Indexes to Marriages: 1873–1932*.

43 Crews and Duc, *American Quilts in the Industrial Age, 1760–1870*, 428.

44 Applique quilts are produced by sewing cut fabric pieces onto larger plain grounds; Ibid., 438.

45 The Woodlands is the forested region that once extended from the Atlantic coast to the Great Plains, and from Lake Superior's northern shores down to the Gulf of Mexico. Before the arrival of Europeans, this territory was populated by Indigenous communities belonging to the Algonquian, Iroquoian, Siouan, and Muskogean language groups that, although culturally diverse, have common cosmological and artistic traditions. These language groups are broadly referred to as the Woodlands Indigenous Peoples (Dubin, *Floral Journey*, 66–7).

46 Speck, "The Historical Approach to Art in Archeology in the Northern Woodlands," 173.

47 Phillips, *Trading Identities*, 247.

48 Dubin, *Floral Journey*, 67; Phillips, *Patterns of Power*, 28.

49 Phillips, *Trading Identities*, 196.

50 Jameson, *Winter Studies and Summer Rambles in Canada*, vol. 2, 126.

51 Malcolm, "Speaking Out for Ontario's Historic Gardens," 11.

52 McCrum family, *1851 Census of Canada East, Canada West, New Brunswick, and Nova Scotia*.

53 Malcolm, "Speaking Out for Ontario's Historic Gardens," 1.

54 Langdon, *Pioneer Gardens at Black Creek Pioneer Village*, 14.

55 Skinner, "With a Lilac by the Door," 36.

56 Thomson, *Village Life in Upper Canada*, 72; Minhinnick, *At Home in Upper Canada*, 14.

57 Thomson, *Village Life in Upper Canada*, 72.

58 Langdon, *Pioneer Gardens at Black Creek Pioneer Village*, 15.

59 Minhinnick, *At Home in Upper Canada*, 8.

60 Ibid., 15.

61 Ibid., 16.

62 Skinner, "With a Lilac by the Door," 36; Minhinnick, *At Home in Upper Canada*, 15.

63 Ibid., 16.

64 Ibid., 2.

65 "Cultivation of Hardy Plants," *The Canadian Agriculturalist* 1, no. 7 (1849): 184.

66 *Plants of Pioneer and Early Days in Ontario* [1970] (Toronto: Paragon Press Ltd, 1986), 21–2; Oliver, *Outdoor Roses in Canada*, 7.
67 Langdon, *Pioneer Gardens at Black Creek Pioneer Village*, 32.
68 Ibid., 31.
69 Oliver, *Outdoor Roses in Canada*, 7.
70 Sauer, "The Grain Amaranths," 127.
71 Ibid., 126, 130.
72 Minhinnick, *At Home in Upper Canada*, 5.
73 *John A Bruce & Co.'s Illustrated and Descriptive Seed Catalogue for the Spring of 1878*, 59.
74 Conroy, *300 Years of Canada's Quilts*, 24
75 McKendry, *Quilts and Other Bed Coverings*, 111.
76 Zeller, *Inventing Canada*, 4–5.
77 Ibid., 6.
78 Harper, *Painting in Canada*, 180–1.
79 Huneault, *I'm Not Myself at All*, 163.
80 Smith, "Farms, Forests and Cities," 71–2.
81 Ulrich, *The Age of Homespun*, 40.

BIBLIOGRAPHY

Archives

Gibson, Samuel, and Margaret Ann McCrum (m. 1881). *Indexes to Marriages: 1873–1932*, Grenville County, Township of Oxford, microfilm reference MS 934, reel 5, page 69, no. 7, Archives of Ontario, Toronto, Ontario, Canada.

Davidson, Elizabeth Bell. *1901 Canada Census*. Ottawa, Ontario, microfilm reference T-6488, page 8, family no. 64, line 2. Library and Archives Canada, Ottawa, Ontario, Canada.

McCrum family. *1851 Census of Canada East, Canada West, New Brunswick, and Nova Scotia*, Grenville County, Canada West, microfilm reference C-11724, page 103, lines 2-6, Library and Archives Canada, Ottawa Ontario, Canada.

Books and Articles

Arber, Agnes. *Herbals, Their Origin and Evolution: A Chapter in the History of Botany*. Cambridge: Cambridge University Press, 1912.

Binkley, Lisa. *Canadian Quilts and Their Makers*. Vancouver: UBC Press, forthcoming.

Brett, Katherine B. *English Embroidery: Sixteenth to Eighteenth Centuries*. Toronto: Royal Ontario Museum, 1972.

Conroy, Mary. *300 Years of Canada's Quilts*. Toronto: Griffen House, 1976.

Crews, Patricia Cox and Carolyn Duc. *American Quilts in the Industrial Age, 1760–1870: The International Quilt Study Center and Museum Collections.* Lincoln and London: University of Nebraska Press, 2018.

Dubin, Lois S. *Floral Journey: Native North American Beadwork.* Seattle and London: University of Washington Press, 2014.

Dunlap, Thomas R. *Nature and the English Diaspora: Environment and History in the United States, Canada, Australia, and New Zealand.* Cambridge: Cambridge University Press, 1999.

Ellen, Roy and Reka Komaromi. "Social Exchange and Vegetative Propagation: An Untold Story of British Potted Plants." *Anthropology Today* 19, no. 1 (February 2013): 3–7.

Gnarowski, Michael. "Foreword." In *Mrs. Simcoe's Diary*, edited by Mary Quayle Innis, 7–17. 1965. Dundurn Press: Toronto, 2007.

Harper, J. Russell. *Painting in Canada: A History*. 1966. Toronto: University of Toronto Press, 1970.

Holmes, Andrew R. "Presbyterian Religion, Historiography, and Ulster Scots Identity, c. 1800 to 1914." *The Historical Journal* 52, no. 3 (September 2009): 615–40.

Huneault, Kristina. *I'm Not Myself at All: Women, Art, and Subjectivity in Canada.* Montreal and Kingston: McGill-Queen's University Press, 2018.

Irwin, John and Katharine B. Brett, *The Origins of Chintz.* London: Her Majesty's Stationary Office, 1970.

Isaac, Amanda. "Ann Flower's Sketchbook: Drawing, Needlework, and Women's Artistry in Colonial Philadelphia." *Winterthur Portfolio* 41, no. 2/3 (Summer/Autumn 2007): 141–60.

Jameson, Anna. *Winter Studies and Summer Rambles in Canada* (1838). Toronto: Coles Publishing Company, 1972.

Kendrick, Albert Frank. *English Needlework.* London: A & C Black Ltd., 1933.

Langdon, Eustella. *Pioneer Gardens at Black Creek Pioneer Village.* Toronto: Holt, Rinehart and Winston of Canada Ltd, 1972.

Levey, Santina. "Embroidery, Renaissance to Rococo." In *The Victoria & Albert Museum's Textile Collection: Embroidery in Britain from 1200 to 1750*, edited by Donald King and Santina Levey, 15–20. London: V&A Publications, 1993.

Malcolm, Brian. "Speaking Out for Ontario's Historic Gardens." *Historic Gardens Review*, no. 30 (April 2014): 11–14.

McCalla, Douglas. *Consumers in the Bush: Shopping in Rural Upper Canada.* Montreal and Kingston: McGill-Queen's University Press, 2015.

McKendry, Ruth. *Classic Quilts.* Toronto: Key Porter Books, 1997.

– "The Quilt in Upper Canada." In *The Heritage Quilt Collection*, edited by Dorothy Farr, 10–14. Kingston: Agnes Etherington Art Centre, 1992.

– *Quilts and Other Bed Coverings in the Canadian Tradition.* Toronto: Van Nostrand Reinhold, 1979.

– "The Use of Embroidery on Quilts in Canada," *Embroidery Canada* 8, no. 2 (February 1981): 10–11.

Minhinnick, Jeanne. *At Home in Upper Canada.* Toronto and Vancouver: Clarke, Irwin & Company, 1970.

Moodie, Susanna. *Roughing It in the Bush; or Life in Canada* (1852). London: Richard Bentley, 1857.

Oliver, R.W. *Outdoor Roses in Canada.* Ottawa: Department of Agriculture, 1950.

Parker, Rozsika. *The Subversive Stitch: Embroidery and the Making of the Feminine.* London: The Women's Press, 1984.

Phillips, Ruth. *Patterns of Power: The Jasper Grant Collection and Great Lakes Indian Art of the Early Nineteenth Century.* Kleinburg: The McMichael Canadian Collection, 1984.

– *Trading Identities: The Souvenir in Native North American Art from the Northeast, 1700–1900.* Montreal and Kingston: McGill-Queen's University Press, 1998.

Rowe, Ann Pollard. "Crewel Embroidered Bed Hangings in Old and New England." *Boston Museum Bulletin* 71, no. 365/366 (1973): 102–64.

Sauer, Jonathan D. "The Grain Amaranths and Their Relatives: A Revised Taxonomic and Geographic Survey." *Annals of the Missouri Botanical Garden* 54, no. 2 (1967): 103–37.

Schuette, Marie, and Marie Muller-Christensen, *A Pictorial History of Embroidery.* New York: Praeger, 1964.

Shteir, Ann B. *Cultivating Women, Cultivating Science.* Baltimore: The Johns Hopkins University Press, 1996.

– "Iconographies of Flora: The Goddess of Flowers in the Cultural History of Botany." In *Figuring It Out: Science, Gender, and Visual Culture*, edited by Ann B. Shteir and Bernard Lightman, 3–27. Hanover: Dartmouth College Press, 2006.

Simcoe, Elizabeth Posthuma. *Mrs. Simcoe's Diary.* Edited by Mary Quayle Innis. 1965. Dundurn Press: Toronto, 2007.

Skinner, Helen Ross. "With a Lilac by the Door: Some Research into Early Gardens in Ontario." *Bulletin of the Association for Preserving Technology* 15, no. 4 (1983): 35–7.

Smith, Alan. "Farms, Forests and Cities: The Image of The Land and the Rise of the Metropolis in Ontario, 1860–1914." In *Old Ontario: Essays in Honour of J.M.S. Careless*, edited by David Keane and Colin Read, 71–94. Toronto: Dundurn Press Limited, 1990.

Speck, Frank. G. "The Historical Approach to Art in Archeology in the Northern Woodlands." *American Antiquity* 8, no. 2 (October 1942): 173–5.

Staples, Kathleen. "Embroidered Furnishings: Questions of Production and Usage."

In *English Embroidery from the Metropolitan Museum of Art, 1580–1700*, edited by Andrew Morrall and Melinda Watt, 23–37. New York: Metropolitan Museum of Art, 2008.

Synge, Lanto. *Art of Embroidery: History of Style and Technique*. London: The Royal School of Needlework, 2001.

Thomson, Gary. *Village Life in Upper Canada*. Belleville: Mika Publishing Company, 1988.

Traill, Catharine Parr. *The Backwoods of Canada*. 1836. Ottawa: Carleton University Press, 1997.

– *The Female Emigrants Guide, and Hints on Canadian Housekeeping*. Toronto: Maclear and Company, 1854.

Turgeon, Laurier G. "Material Culture and Cross-Cultural Consumption: French Beads in North America, 1500–1700." *Studies in the Decorative Arts* 9, no. 1 (Fall–Winter 2001–02): 85–107.

Ulrich, Laurel Thatcher. *The Age of Homespun: Objects and Stories in the Creation of an American Myth*. New York: Vintage Books, 2001.

Zeller, Suzanne. *Inventing Canada: Early Victorian Science and the Idea of a Transcontinental Nation*. Montreal and Kingston: McGill-Queen's University Press, 2009.

1.2 *Above*
Dalhousie Castle and parkland, ca. 1824, after James Nasmyth.

1.4 *Left*
Lady Dalhousie with her third son, the Hon. James Ramsay, by William Douglas, 1816.

1.5
Watercolour of pink flowered plant by Lady Dalhousie, ca. 1820.

1.7 *Above*
Gentianella quinquefolia (L.) Small subsp. *quinquefolia* (present identification), collected by Lady Dalhousie in Sorel on 3 September 1827.

1.8 *Left*
Herbarium specimen of the fern *Asplenium dalhousiae*, collected by Lady Dalhousie in 1831 and named after her by William Jackson Hooker.

1.9
Portrait of Lady Dalhousie by Sir John Watson Gordon, 1837.

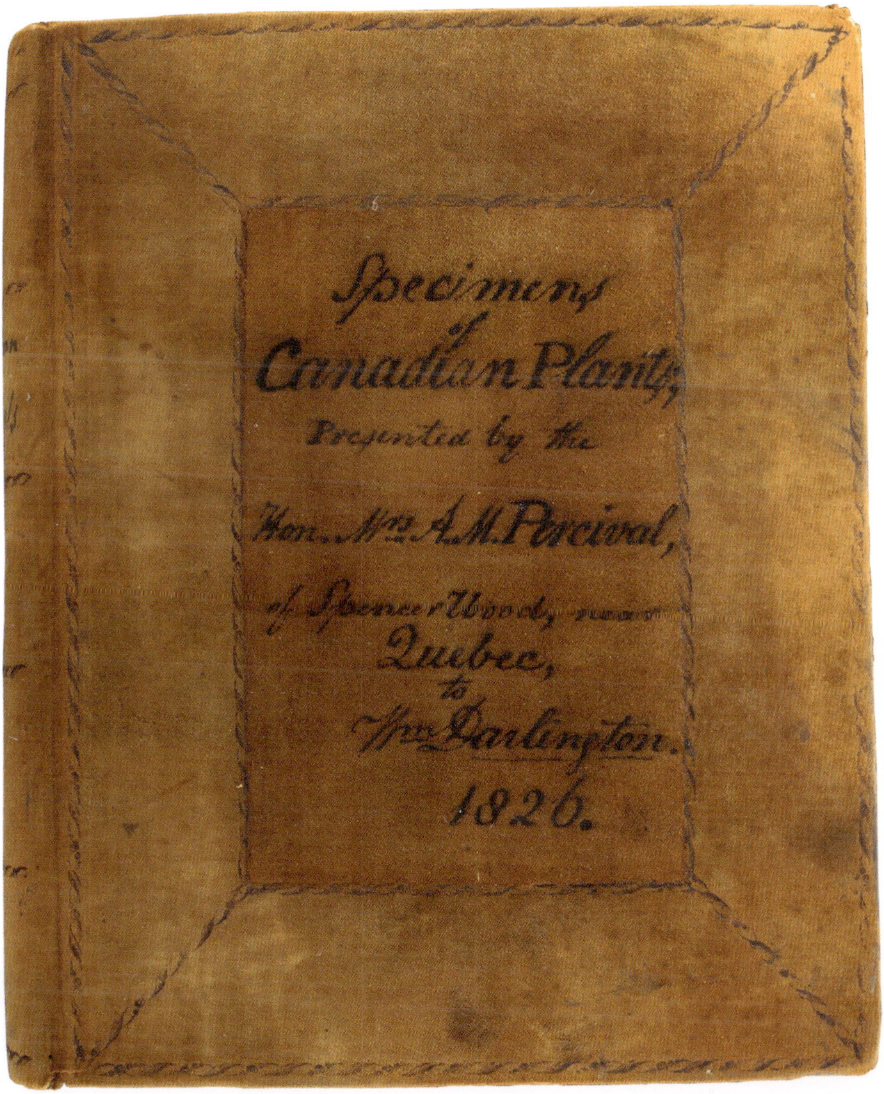

2.4
Cover of *Specimens of Canadian Plants presented by the Hon. Mrs A.M. Percival of Spencer Wood, near Quebec, to Wm. Darlington, 1826.*

2.5
Plant specimens of *Gentiana amarelloides* var. *alba* from *The Countess of Dalhousie Herbarium*, 1823, folio 75.

2.6
Herbarium specimen of *Orchis discolor??*, collected by Lady Dalhousie 21 July 1828.

2.7
Herbarium specimens of *Arethusa bulbosa*, collected in Sorel by Lady Dalhousie, from the album Anne Mary Perceval sent to botanist William Darlington in 1826.

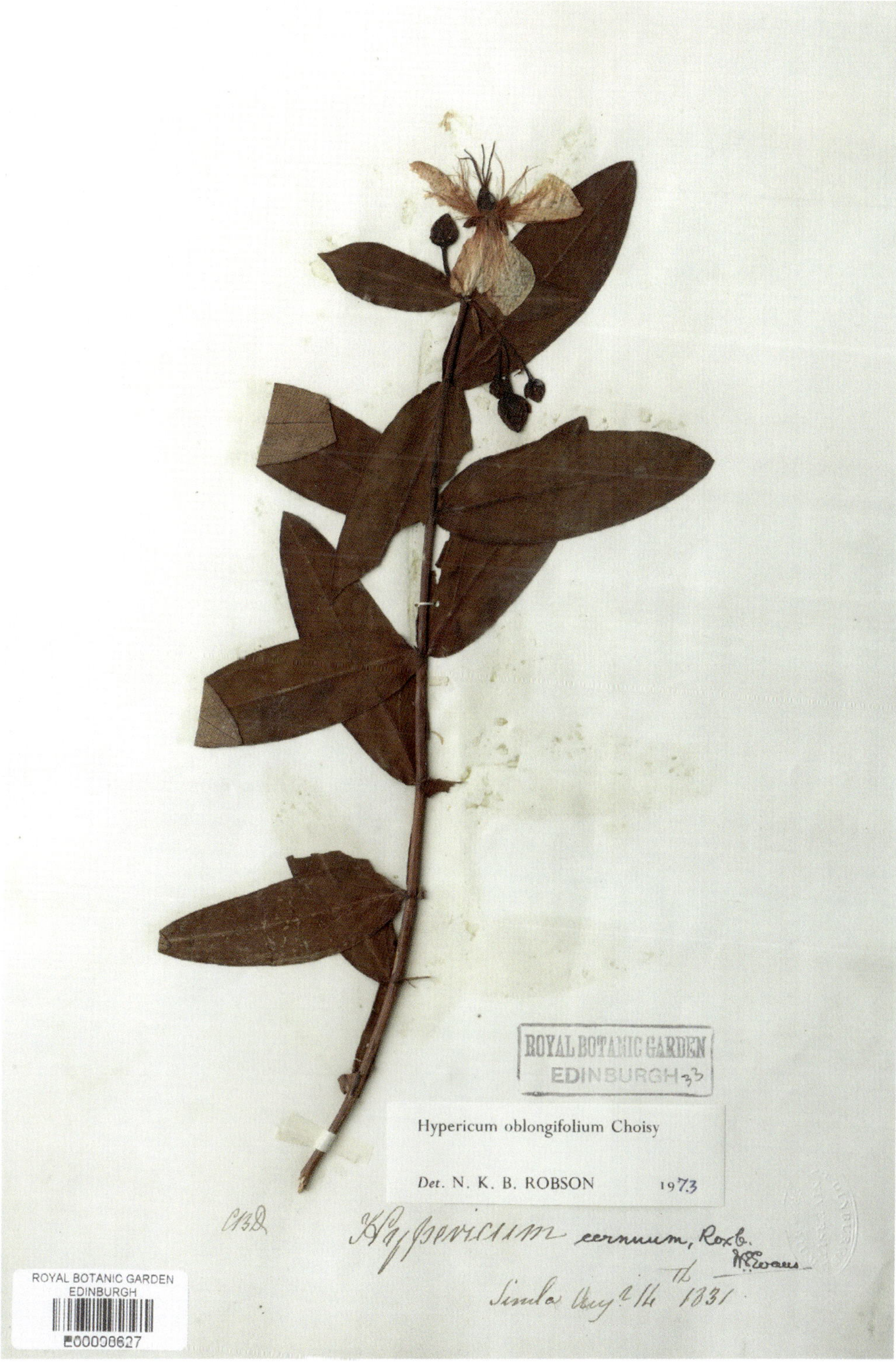

3.3
A specimen from Dalhousie's donated Indian Herbarium as now found in the Royal Botanic Garden Edinburgh: *Hypericum oblongifolium Choisy*, collected at Simla, 14 August 1831.

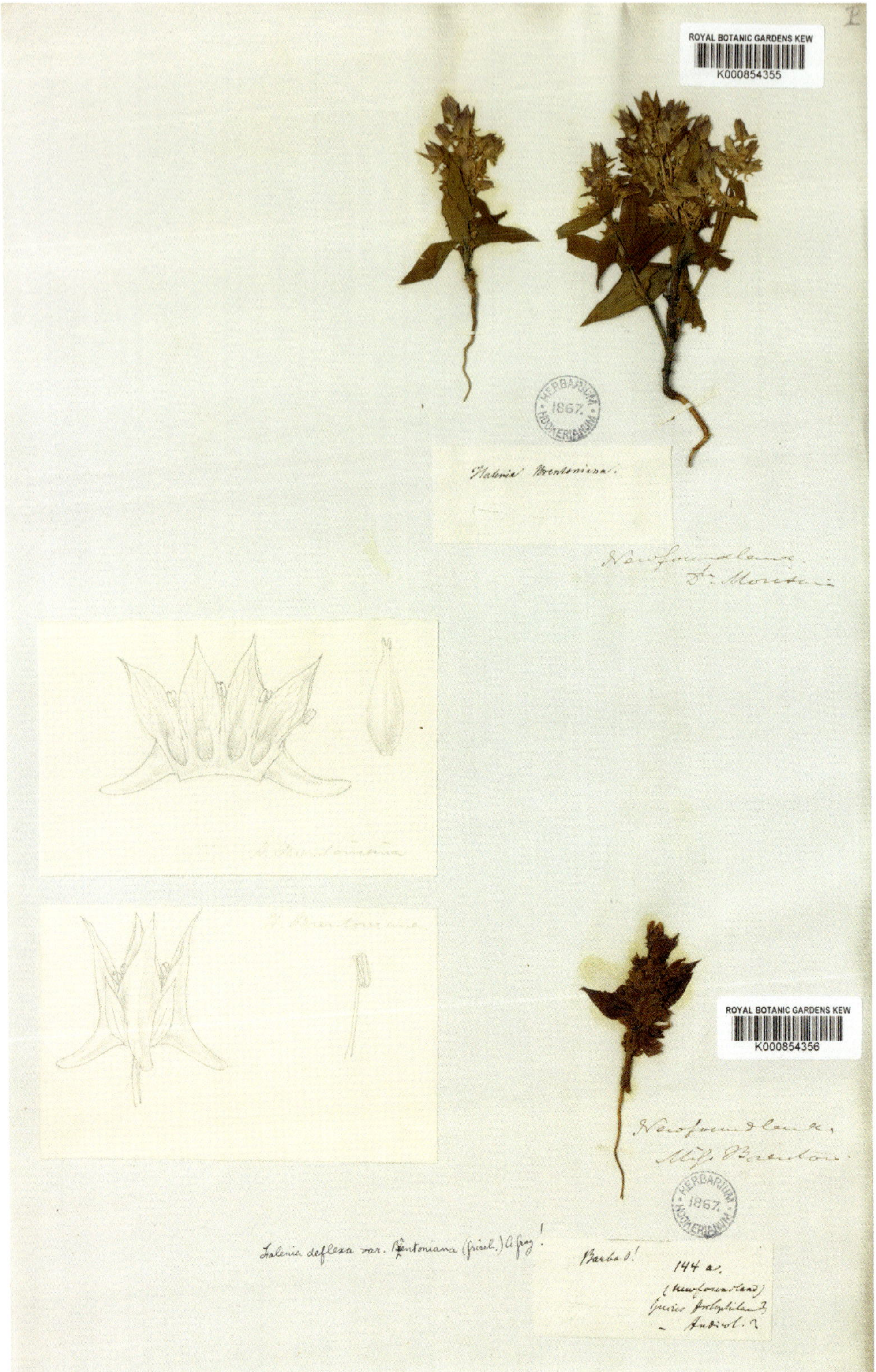

4.2
Mary Brenton's specimen of *Halenia brentoniana*, Herbarium, Royal Botanic Gardens, Kew, K000854356.

4.5 *Above*
"Goldenrod," *Log of the HMS Royal George* [1808]. This goldenrod is probably *Solidago rugosa* var. *rugosa* (Semple).

4.6 *Left*
Specimen of goldenrod collected by "Miss Brenton" in Newfoundland, now known as *Solidago juncea*, Herbarium, Royal Botanic Gardens, Kew, K000890973.

5.3
"Australian Flowers" by Samuel Begg, *Illustrated Australian News*, Melbourne, Victoria, Australia, 27 July 1881, 145.

5.5
Aneilema calandrinioides F.Muell., in Queensland, watercolour painting by Ellis Rowan, nla.obj-138801317.

5.6
Sarchochilus hartmannii F.Muell., delineated and lithographed by W.H. Fitch, *Curtis's Botanical Magazine* 114, series 3(44) (1888): tab. 7010.

6.2
A specimen of Eastern Wild Columbine (*Aquilegia canadensis* L.) collected by Alice Hollingworth in Muskoka District.

6.6 *Left*
Wild Clematis or Virgin's-bower (*Clematis virginiana* L.), an attractive native vine recommended by Alice Hollingworth for use in home landscaping.

7.1 *Below*
Colour portrait of Catharine Parr Strickland as a young woman in Suffolk (likely painted by Thomas Cheesman).

7.11
Agnes Fitzgibbon, Showy Orchis, *Canadian Wild Flowers*, Plate VII.

8.3 *Left*
Catharine Parr Traill, pressed specimen of Maiden Hair Fern, c. 1888.

8.4 *Below*
Cotton, *Oliver and Boyd's Object-Lesson Cards*, Vegetable Kingdom, No. 13, London: Simpkin and Co., 1880.

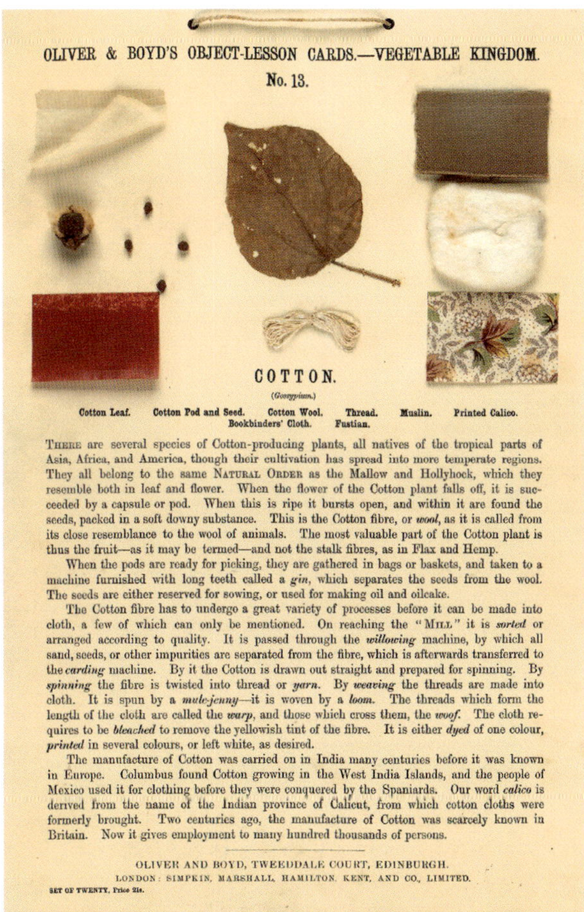

OLIVER & BOYD'S OBJECT-LESSON CARDS.—VEGETABLE KINGDOM.

No. 13.

COTTON.

(Gossypium.)

Cotton Leaf. Cotton Pod and Seed. Cotton Wool. Thread. Muslin. Printed Calico. Bookbinders' Cloth. Fustian.

There are several species of Cotton-producing plants, all natives of the tropical parts of Asia, Africa, and America, though their cultivation has spread into more temperate regions. They all belong to the same Natural Order as the Mallow and Hollyhock, which they resemble both in leaf and flower. When the flower of the Cotton plant falls off, it is succeeded by a capsule or pod. When this is ripe it bursts open, and within it are found the seeds, packed in a soft downy substance. This is the Cotton fibre, or ***wool***, as it is called from its close resemblance to the wool of animals. The most valuable part of the Cotton plant is thus the fruit—as it may be termed—and not the stalk fibres, as in Flax and Hemp.

When the pods are ready for picking, they are gathered in bags or baskets, and taken to a machine furnished with long teeth called a ***gin***, which separates the seeds from the wool. The seeds are either reserved for sowing, or used for making oil and oilcake.

The Cotton fibre has to undergo a great variety of processes before it can be made into cloth, a few of which can only be mentioned. On reaching the "Mill" it is ***sorted*** or arranged according to quality. It is passed through the ***willowing*** machine, by which all sand, seeds, or other impurities are separated from the fibre, which is afterwards transferred to the ***carding*** machine. By it the Cotton is drawn out straight and prepared for spinning. By ***spinning*** the fibre is twisted into thread or ***yarn***. By ***weaving*** the threads are made into cloth. It is spun by a ***mule-jenny***—it is woven by a ***loom***. The threads which form the length of the cloth are called the ***warp***, and those which cross them, the ***woof***. The cloth requires to be ***bleached*** to remove the yellowish tint of the fibre. It is either ***dyed*** of one colour, ***printed*** in several colours, or left white, as desired.

The manufacture of Cotton was carried on in India many centuries before it was known in Europe. Columbus found Cotton growing in the West India Islands, and the people of Mexico used it for clothing before they were conquered by the Spaniards. Our word ***calico*** is derived from the name of the Indian province of Calicut, from which cotton cloths were formerly brought. Two centuries ago, the manufacture of Cotton was scarcely known in Britain. Now it gives employment to many hundred thousands of persons.

OLIVER AND BOYD, TWEEDDALE COURT, EDINBURGH.
LONDON: SIMPKIN, MARSHALL, HAMILTON, KENT, AND CO., LIMITED.

SET OF TWENTY, Price 21s.

9.1
A page from one of Sophie Pemberton's two 1895 albums of wildflower drawings. She gave this albun, entitled *Wild Flowers BC*, as a gift to her brother Frederick and his wife Mary Ann. Scientific nomenclature was included on the backs of pages in both albums. The combination of images, Latin nomenclature, and snippets of poetry was familiar from floral dictionaries of the day. (See figure 9.4.)

9.2
Millicent Mary Chaplin, *Rudbeckia*, 1840–42. The 1830s and '40s saw the emergence of women's botanical albums in Canada, with early examples by Maria Morris Miller in Halifax and Millicent Mary Chaplin and Fanny Bayfield in the city of Quebec. Albums were especially popular among English-speaking women associated, whether through commercial or familial ties, with British military officers trained in natural history.

9.6
Frederick Traugott Pursh's original watercolour of *Digitalis purpurea*, pictured here, does not appear in his classic volume *Flora Americae Septentrionalis*, but an engraved version is preserved in the library of the American Philosophical Society. The engraved version contains additional anatomical information, suggesting that the scientific cutaways and magnified details typically absent from Canadian women's albums were sometimes added by engravers at a later stage.

9.7
Pemberton's large canvas, *Spring*, was exhibited at the Royal Academy in 1903. It continued the thematic pairing of women and flowers which had also marked *Daffodils*, her inaugural contribution to the Academy exhibition in 1897.

9.8 *Left and right*
Visual similarities and differences repeat like a refrain across the various folios of Pemberton's 1895 albums. Pictured here are her *Aster douglasii* (left) and *Sidalcea hendersonii* (right), both from Ada's album in the BC Archives.

9.9
St John artist Mary Rebecca Wilkinson included this *Lady's slipper, Cypripedium acaule,* in her 1868 album, *Wildflowers of New Brunswick*, a gift commissioned from the artist for Margaret Medley, a pioneering nurse and the wife of the local bishop. Despite differences in technique and style that develop over time, this album of wildflower watercolours shares a basic pictorial organization with those by Sophie Pemberton and Millicent Mary Chaplin.

9.10 *Left*
Sophie Pemberton, *Spirea discolor*, 1895, from the album for Ada in the BC Archives.

9.12 *Below*
Like her *Spirea discolor*, Pemberton's *Stachys ciliata*, 1895, shows insect holes and dying leaves. This emphasis on the particularities of individual specimens is common to many Canadian women album makers and contrasts with the idealized floral types so characteristic of scientific publications.

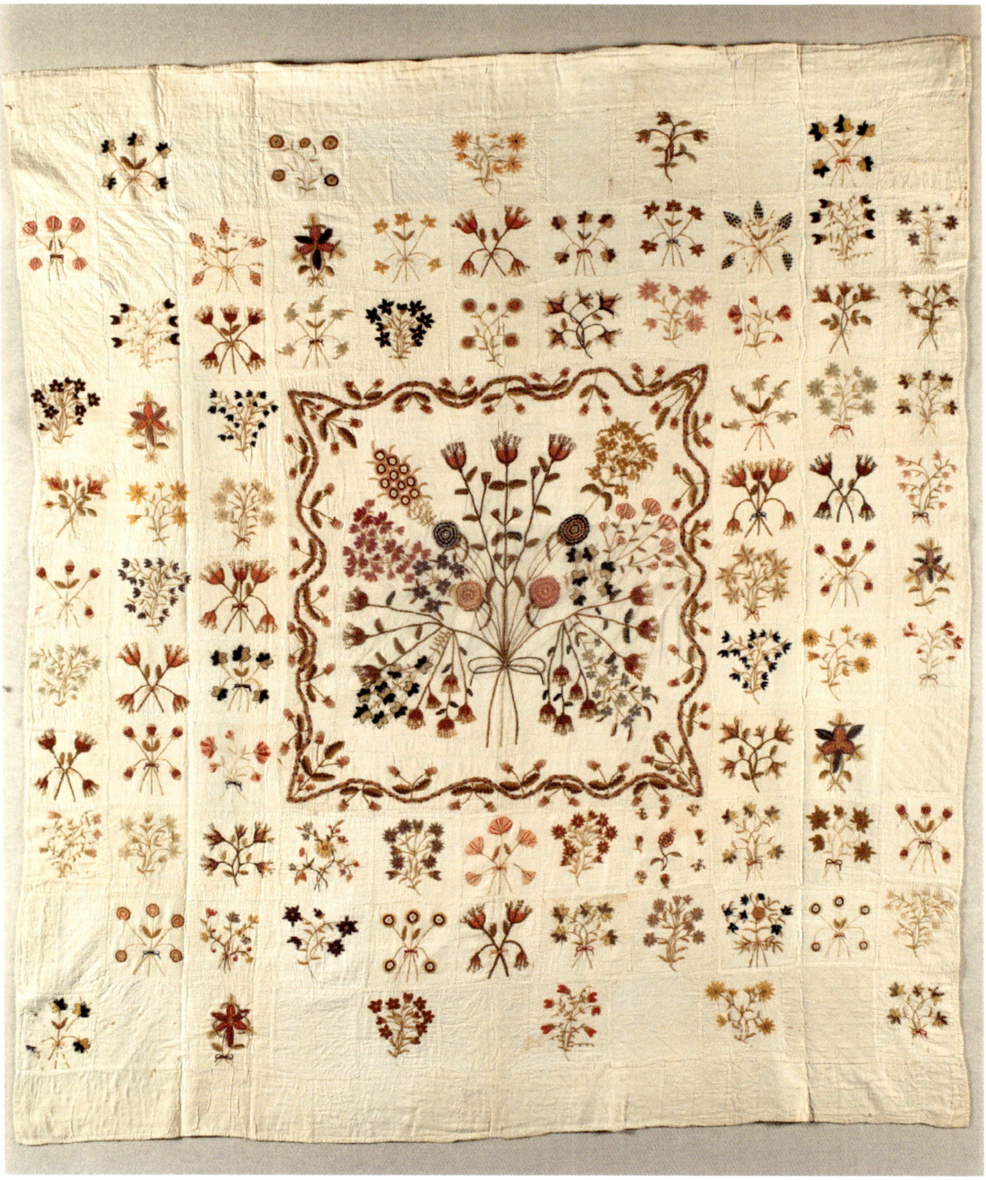

10.1
Fallowfield quilt, attributed to Elizabeth Bell, 1840s.

10.2
Coverlet, Margaret Ann McCrum, 1860–81.

10.3
Slips, unknown artist, c. 1600, linen canvas, embroidered with silks in tent stitch.

10.4
Detail of Fallowfield quilt, attributed to Elizabeth Bell. 1840s, figure 10.1.

10.5

Bag, c. 1860–90, unidentified artist, Mi'kmaq or Wəlastəkwewiyik, New Brunswick, velvet, glass beads, silk bindings and lining, thread.

11.2
Design attributed to William Mundie for a botanical garden for the University of Toronto. North is up. The stream depicted is Taddle Creek, complete with McCaul's Pond. The building in middle left is the original University of Toronto building, now University College. The centre of the circular compass rose ornamental garden would have been close to the southeast corner of Hart House. North of this are assorted plant beds and a circular conservatory with two rectangular wings. A rectangular glasshouse is also portrayed to the east of the compass rose garden design.

11.7
Kate Crooks's herbarium specimen of Rose Pink, *Sabatia angularis*, collected in Hamilton, 1865.

12.1
The Canadian Horticulturist 4, no. 2 (February 1881): i and 26–8. In the early 1880s colour plates were introduced to *The Canadian Horticulturist* magazine. In this print of Autumn Berries, the anonymous artist has included berries known as Climbing Bitter-sweet, Dogwood, European Barberry, Whorled Winterberry (Black Alder), Virginia Creeper, European Spindle Tree, Snowberry, and American Holly, each of which was explained to readers in an accompanying article.

12.4
Frontispiece from the fourth edition of *Canadian Wild Flowers* by Agnes Fitzgibbon (née Moodie and later Chamberlin) and Catharine Parr Traill, published by William Briggs in Toronto in 1895. The first edition was published by subscription in 1868. John Lovell published the second and third editions in Montreal.

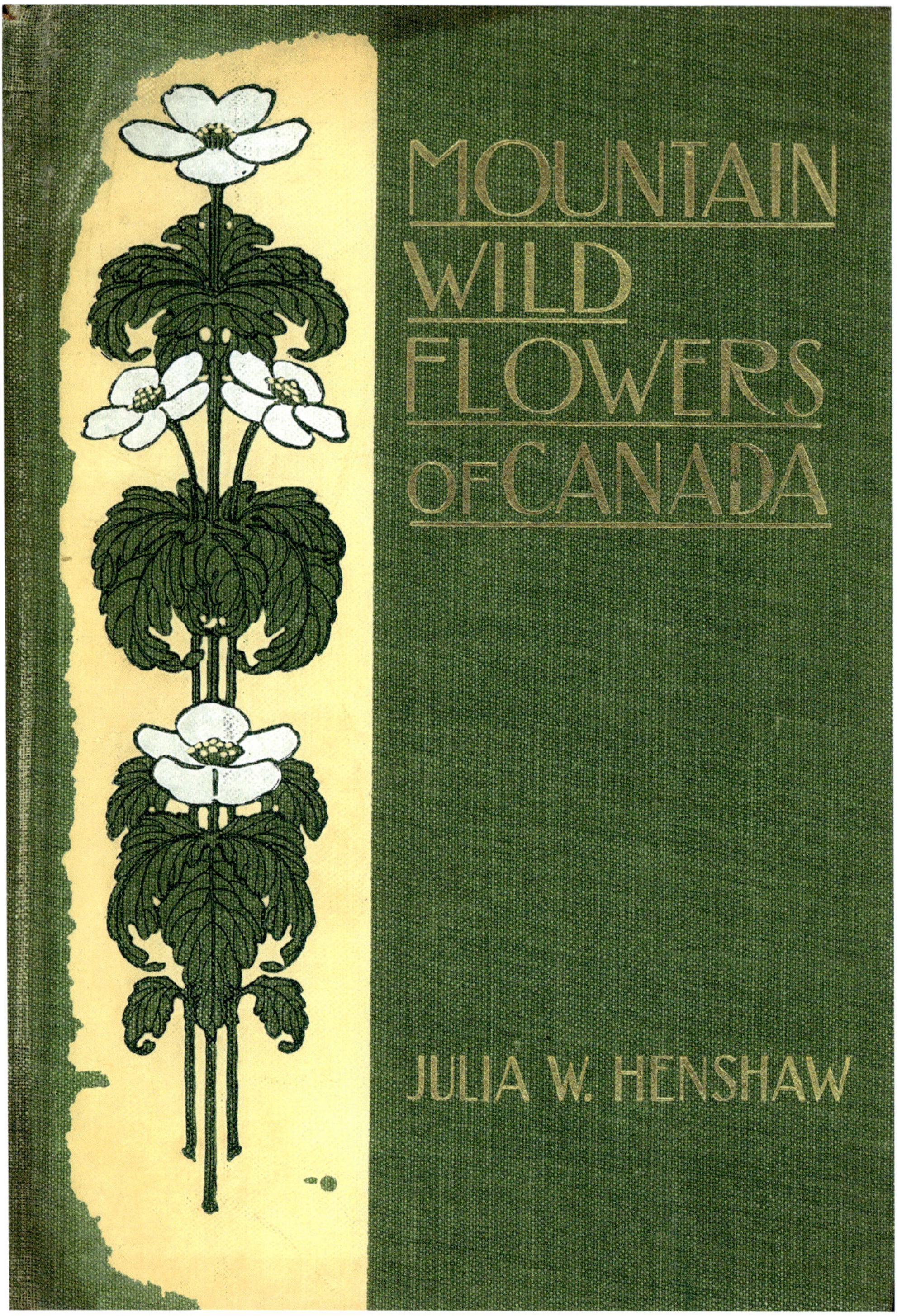

12.6
Maud Going, *With the Wild Flowers*, 1907. Maud Going first published *With the Wild Flowers* in 1894. This is the cover of the reprinted 1907 edition.

PART FIVE

Expanding Public Practices

11

Botanical Gardens in Nineteenth-Century Canada

Individuals and Institutions

DAVID GALBRAITH

In nineteenth-century Canada the developing science of botany was just one way that people encountered plants, learned about them, and codified, used, and transmitted that knowledge. Other ways of knowing were equally valid – notably, the practical, ceremonial, medicinal, and culinary traditions of Indigenous Peoples in North America, as well as knowledge about growing plants for food and other economic uses within settler culture. But botany itself was being rapidly elaborated as an accepted, growing science and by mid-century interest in plants was also expanding as a respectable topic of leisure interest and social exchange. Efforts to bring botanical gardens into being in Canada in the nineteenth century were expressions of these developments and became part of the landscape of a number of cities and educational institutions as locations for activities that promoted interest in plants, albeit with only short-term success. In Canada as in England and elsewhere, botanical gardens began to shift from being primarily university-based or government-directed research and training facilities into institutions with a more public face. They became parks for the enjoyment of exotic and beautiful specimens, exposed the public to botanical names and knowledge, fuelled interest in science and in plant diversity, and served as demonstration platforms for horticulture.[1] In relation to the focus of this book about women and botany in nineteenth-century Canada, botanical gardens were one way that women could have access to botany in its scientific aspects.

There have been few attempts to document the history of botanical gardens in Canada, and little has been done to place their early development into any kind of social or historic frame. This chapter focuses on botanical gardens as one facet of the history of plant knowledge in nineteenth-century Canada and also places the various projects to develop such facilities with reference to their eighteenth-century antecedents. It assembles materials toward this history by calling upon unstudied documents from the nineteenth century and, compiling information from new perspectives, it focuses on individuals and institutions. Records from published and archival sources offer insights into where and when Canadian botanical gardens developed, some reasons for their creation, and involvement by women in their development. Records also offer insights into signs of interest shown by proponents of these projects in the use of plants by Indigenous Peoples. One key resource is the *Annals of the Botanical Society of Canada*, published from 1860 to 1862, which brings into visibility the development of the short-lived Kingston Botanic Garden at the University of Queen's College, as well as the presence and participation of women in activities of the Botanical Society of Canada during the 1860s.

ANTECEDENTS

Since the time of the early botanical gardens in Italy and in later centuries in Germany, Switzerland, the Netherlands, France, and England, research on plants has intermingled with exchanges of living plant material, often in the form of seeds.[2] In eighteenth-century North America, one of the earliest gardens intended to contribute to botanical knowledge was created by Jean-François Gaultier (1708–1756), a physician, meteorologist, and botanist who lived in Quebec before the Seven Years War and assembled and sent plant specimens from New France to France.[3] Gaultier had use of the garden at the Intendant's Palace in the city of Quebec, the administrative centre of the colony.[4] Recent archaeobotanical studies of the site have identified more than sixty plant species present in the garden, a combination of European and North American plants with identifiable economic, medicinal, and culinary uses.[5] The early botanic enterprise in Quebec appears to have been lost following Gaultier's death in 1756 and then the defeat of France by the British during the Seven Years War a few years later.

Gaultier's counterpart in the English colonies in early eighteenth-century America was John Bartram (1699–1777), a largely self-trained botanist, horticulturist, and explorer who established a nursery garden on forty-one hectares on the then outskirts of Philadelphia. Bartram received political and scientific luminaries in the garden, corresponded widely, and in 1765 was named the Royal Botanist in the American Colonies by King George III.[6] Although various later authors have designated Bartram's garden the first botanical garden in North America, projecting their understanding of the term onto this early garden, it does not appear that Bartram himself referred to his enterprise this way and it did not function as we would describe a botanical garden today. Instead, it was an early garden prepared by a botanist for the purposes of the study and dissemination of plants – a botanist's garden with commercial purpose. Also in Philadelphia at midcentury, the managers of the Pennsylvania Hospital approved a proposal for a "Botanical Garden" as a source of medicinal plant material, essentially a physic garden. The hospital, established by Benjamin Franklin and Dr Thomas Bond, was chartered in 1751 and opened at a temporary site the following year. The garden was not built as planned, however, because of financial constraints.[7]

When the American Colonies established their own government in the 1780s, botanists lost some contact with English science, turning internally and also to colleagues in France.[8] In New York City in 1801, David Hosack (1769–1835), a physician and professor of botany at Columbia College, built the eight-hectare Elgin Botanic Garden five kilometres outside of the city limits. This was possibly the first garden in continental North America actually called a "botanic garden" and was well documented at the time.[9] Hosack himself published two editions of his *Hortus Elginensis*, a combination of a catalogue of botanical species and a visitors' guide, in 1802 and 1811. His goal, he explained in the preface, was to establish a botanic garden in the United States "as a repository of the native plants of this country, and as subservient to the purposes of medicine, agriculture, and the arts." As a professor at Columbia he had tried unsuccessfully to convince the state legislature to take up the "expensive and arduous" task of creating a public botanical garden and decided to build one himself. He hoped that "when the nature of the institution should be better, and more generally known, and its utility fully ascertained, it would receive the patronage and support of the public."[10] After nearly a decade of supporting the project himself, Hosack was forced to sell the land to the State of New York in 1810 at a considerable loss. In 1814 the legislature of New York turned the site over to Columbia College, initiating a long decline for the garden and its dissolution in 1823.[11] What began as a grand dream for North America's first botanical garden, envisioned by a dedicated and eminently qualified individual willing to stake personal resources in the project, failed to engage the interest of either public officials or an institution of higher education. This pattern would repeat itself.

THE KING'S BOTANICAL GARDEN FOR NORTH AMERICAN PLANTS

While the Elgin Botanic Garden was falling into obscurity in New York, the earliest known botanical garden named as such in British North America was in early development near Montreal. Christian Broun Ramsay, 9th Countess of Dalhousie, and her husband, George Ramsay, the 9th Earl of Dalhousie, governor of British North America 1822–28, undertook to create a research garden that they referred to as the King's Botanical Garden for North American Plants. As other chapters in this volume detail,[12] Lady Dalhousie was an interesting and influential figure in early Canadian botanical exploration. In addition to collecting and preserving plants as herbarium specimens, she collected seeds and live plants, and sent seeds for her own garden in Scotland at Dalhousie Castle outside Edinburgh. In the 1820s, the Dalhousies established a garden to study the Canadian flora at Île Ste-Hélène, an island in the St Lawrence River near Montreal and the site of a British fort. This is just five kilometres from where Le Jardin botanique de Montréal is today. According to Alexander McFarlane, their gardener at Île Ste-Hélène, this botanical garden was intended to exceed any "collection of American plants (in cultivation) on this side of the Atlantic." To this end, Lady Dalhousie instigated a regular exchange of plants and specimens with McFarlane.[13]

Lady Dalhousie was the botanist in the family, but her husband took a strong interest in her efforts and supported them. Letters in the National Archives of Canada and other sources reveal that the botanical garden being developed by the Dalhousies was no passing thought. From the middle of 1822 until 1828 Dalhousie brought what powers he could as governor of British North America to the project, as well as personal funds. On 26 July 1822, Dalhousie wrote to Sir Peregrine Maitland, lieutenant governor of Upper Canada, asking for assistance with the botanical garden.[14] He stated that he was looking to correspond with a "Mr. Thos. Blair, Percy, Markham, York, with whom I am desirous to establish a correspondence, with a view to the formation of a Botanical Garden of Public nature on the Isle St. Helene [*sic*] at Montreal."[15] Dalhousie's extant letter to Thomas Blair indicates that he viewed Blair as a kindred spirit. "I will join with you most willingly in the pursuit you purpose in Botanical research, to the utmost of my power, not only in the desire to send home as much as I can to promote the study, but in the same conviction as yourself, that there is an endless variety of plants yet unknown in this region, and many both of medicinal qualities, and of dyes which may be acquired from the Indian natives, of the greatest important to science in general."[16]

The scale of what was intended was hinted at on 21 August 1822 when Dalhousie wrote to McFarlane, stating that his "object is to make a beginning of what may here-after be thought worthy of being established as 'The Kings Botanical Garden for American Plants' and that ought to stimulate your exertions." Further, he wrote, "I wish all within the walls to be laid off in compartments for Botanical collections and those principally of Canadian plants[,] shrubs and even young trees from seed."[17]

Dalhousie also gave some indication of his interest in the botanical garden as a means to study the use of plants as medicines by Indigenous Peoples. On 24 September 1826 he recorded in his diary having witnessed the skills of Indigenous healers. A twenty-two-year-old man who attempted to carry a large load had "sprained his loins" and been confined to bed for three months as a result: "[t]he surgeon did him no good but an old Indian woman[18] saw him, brought some herbs with her some days afterwards, and completely restored him – this is one of many instances I have met with, which point out the wisdom in Government establishing a Botanical garden, to search out from native Indians the medicinal herbs, and dye plants of this new world."[19]

Had the government of the day supported the King's Botanical Garden, the use of plants by Indigenous Peoples in Canada might be better documented and more widely known than it is today. At the same time, Dalhousie's actions were as a colonizer. Had the botanical garden survived and been developed, Indigenous traditional knowledge in the area and further afield would have been recorded, appropriated, and put to use by the British colonial enterprise.

Earlier in May 1826, Dalhousie had written to Henry Bathurst, 3rd Earl Bathurst, secretary of state for War and the Colonies, formally requesting that the government underwrite the new botanical garden to the amount of £200 per year. The letter reveals that Dalhousie had expended £151 on the garden in the previous year, and that the Board of Ordnance had objected to the expense:

> I availed myself at first of the walled garden & hothouses the whole purchased on the island of St. Helen's, opposite Montreal by the late Duke of Richmond: it was not within the ground required for Military Works, and the expense was limited to the Salary of a good Gardener, and the small pay of two or three men from the Troops. I need hardly say that it is an object highly desired to gain a knowledge of American plants; they are little known yet their qualities medicinal, and applicable to the sciences generally are varied and exceedingly interesting.

Dalhousie concludes:

> I did intend to have asked that the same encouragement might be given here, as is given to a similar Establishment at the Cape of Good Hope. I admit that it does not belong to Military Expenditures, and I shall strike it out after this summer. I have long ago removed all of the hothouses in order to reduce the expenses and to limit our collection to American plants alone. I would therefore now submit to Your Lordship that this Institution may still be carried on with every possible economy where it is, and that the expense, not exceeding £200 per annum, may be charged upon the Territorial Revenue of the Crown in this Province. It will be no burthen there and it is entirely at the disposal of the Crown. I entreat Your Lordship's answer to this, as I must otherwise abandon it.[20]

Lady Dalhousie tried to enlist the assistance of British troops at the fort, but the commander of the island's fort challenged her instructions to McFarlane for plants to be sent to her in Sorel. Were issues of military hierarchy at work here, or was the commander not prepared for his authority to be challenged by a woman?[21] Whatever the reason, the venture foundered as a result. Just as Hosack had failed in New York to create sufficient interest in his botanical garden for support to be forthcoming through public or government sources, so the Ramsays were unable to move the colonial British government to maintain a research garden in Quebec. They attempted to move the garden from Île Ste-Hélène to their summer residence at Sorel, on the south shore of the St Lawrence River downstream from Montreal. However, despite these efforts, by the next year the first botanical garden in Canada was abandoned. The Ramsays left Quebec for Great Britain in September 1828 and their garden at Sorel was largely forgotten.

"BOTANICAL GARDENS"? EARLY INITIATIVES 1830S–50S

A recurring theme in the history of botanical gardens in Canada in the nineteenth century is their relatively short life. Another is the investment of time and energy in proposals for gardens that were never developed. The 1830s were notable for at least two initiatives that would have included a botanical garden or something similar. In 1835 Toronto entrepreneur, politician, and naturalist Charles Fothergill proposed the "Lyceum of Natural History and the Fine Arts," a major natural history museum that would have included a botanical garden along with a zoological park, museum, and art gallery.[22] In Nova Scotia Andrew Downs, generally remembered as a zoologist, proposed a "park for the study and enjoyment of zoology, ornithology, and flora, as well as a library" as part of a plan in 1838 for a provincial natural history museum.[23] Neither proposal came to fruition. In 1841, the Nova Scotia Horticultural Society created the People's Garden in Halifax, proposed by Joseph Howe and intended as a beautiful public landscape park with an educational function. The project itself was overly expensive, and the society was forced to sell the garden to the city in 1874. The city of Halifax then brought the former People's Garden site together with adjacent lands to create the Halifax Public Gardens.[24] Both gardens were created with the intention of being an educational facility to promote horticulture for the public within Nova Scotia. They did not seem to have had research as a key part of their purpose and were not formally referred to as a "botanical garden" at that time.

By contrast, three botanical gardens under development in Toronto during the 1850s actually used the term "botanical garden" in their names. Prominent Toronto lawyer and politician George Allan created the first of these, a private botanical garden on his family's estate at Moss Park in the early 1850s. He offered it to the Toronto Horticultural Society (THS) in 1856 and later deeded it to the society. Although the Prince of Wales (later King Edward VII) officially opened the "Botanical Garden" in September 1860, by 1863 it was under management of the Toronto Horticultural Society and was referred to as the "Horticultural Gardens." This change of name suggests either a shift in the garden's intended function or a perception that the term "Horticultural" would be more widely understood by potential visitors. It was formally renamed Allan Gardens in 1901 following the death of its founder.[25]

The second, a 0.8 hectare botanical garden on the grounds of the Provincial Normal School, later renamed the Toronto Normal School, was planned by educator Egerton Ryerson to have a clear pedagogical purpose. The school was intended as a training centre for teachers from across Ontario, and the grounds also included a 1.2 hectare plot for experiments with agricultural crops. The botanical garden was part of the 1853 design for the school's landscape by William Mundie, one of the first practising landscape architects in Canada West. By 1866, the site included more than 200 specimens of plants native to Canada, including about thirty ferns, Cypripediums (Lady Slipper Orchids), and other orchid species (figure 11.1). The grounds were also noted for more than one hundred ornamental shrubs and trees.[26] It is not clear when the Toronto Normal School's botanical garden ceased to function, but it would be interesting to know the degree to which the curriculum for women teachers

11.1
Toronto Normal School with botanical garden in foreground. Photographer unknown. 1934.

included botany and horticultural topics. Other parts of the Normal School later developed into institutions in their own right. The agricultural garden at the school is considered to be the beginnings of the Ontario Agricultural College. Later in the nineteenth century, a museum collection established at the school was the immediate predecessor to the Royal Ontario Museum. The botanical garden itself, however, left no institutional offspring.

The third "botanical garden" was proposed at the University of Toronto shortly after its establishment as a secular university in 1850. A botanical garden for teaching purposes was to be in University Park along Taddle Creek, a largely undeveloped landscape that was the home of the original campus. Just east of today's Hart House, the site was marked on several maps of the period as "proposed botanical garden." It is believed that William Mundie designed this landscape and garden, and there is evidence that something was built on the site.[27] By 1860 the university lost interest in creating a botanical garden, however, and the concept was abandoned. The site was incorporated into Queen's Park and plans for the provincial legislature, and whatever was built of this garden was lost. Although botany as a science could have benefited from the botanical garden, the university took no steps toward realizing a garden for

11.2
Design attributed to William Mundie for a botanical garden for the University of Toronto. North is up. The stream depicted is Taddle Creek, complete with McCaul's Pond. The building in middle left is the original University of Toronto building, now University College. The centre of the circular compass rose ornamental garden would have been close to the southeast corner of Hart House. North of this are assorted plant beds and a circular conservatory with two rectangular wings. A rectangular glasshouse is also portrayed to the east of the compass rose garden design.

purposes of teaching, research, or both. Had it done so, a botanical garden might, for example, have provided plants as *materia medica* for students at the Trinity College Medical School, but this school was dissolved in 1856. John Court has examined a number of factors that diverted the university's attention from a projected botanical garden; among these were weak arguments for its necessity for teaching purposes, increases in the value of available land for other uses, and diverse options or models to provide the services of a research garden without the facility taking the form of a botanical garden.[28]

THE BOTANICAL SOCIETY OF CANADA AND ITS "APPLIANCES"

The development of botanical gardens in Canada appeared to advance considerably in December 1860 with the formation of the Botanical Society of Canada in Kingston, Canada West. This society was formed by George Lawson (1827–1895), professor of chemistry and botany at the University of Queen's College between 1858 and 1863, and other faculty members at Queen's. During its first meeting at the end of 1860, Dr John Litchfield, superintendant of the Rockwood Lunatic Asylum in Kingston and on the medical faculty at Queen's,[29] read a paper setting out the advantages of having a botanical garden as part of their work. While primarily directed at the benefits a botanical garden could bring to the Botanical Society, Litchfield, who may have been the main proponent of the botanical garden project, also made broader remarks about the potential for the society to have a strong role in public education. To these ends, members set about in 1861 creating its various facilities, including a library and herbarium in addition to the botanical garden.

The site of the botanical garden was the south-facing slope of the lawn of Summerhill, the first building on the Queen's campus. Although few descriptions are available of the garden and its activities, Lawson reported on some features in a letter to the Botanical Society of Edinburgh in 1863:

> Our Botanic Garden is making progress. We have about seven acres of land, which is being gradually opened up into flower borders, and many of our students and graduates are active in bringing roots from the woods, while the citizens of Kingston send contributions from their gardens and members at a distance contribute seeds and such are plants as come in their way. We are under great obligation to Professor Asa Gray, who, in the most liberal manner, sent a large collection of roots from the Botanic Garden at Cambridge, including many rare species that could not have otherwise been obtained.[30]

Two extant photographs that primarily portray Summerhill also include the gardens as foreground landscaping.

At that time, Queen's College was in large part a medical school. A botanical garden could have been a valuable tool for teaching medical students about the use of plants for medicinal purposes – *materia medica*. In theory, knowledge about plants as medicines might be useful for physicians in rural areas of the Province of Canada in the 1860s. However, the medical profession in

11.3
Summerhill, University of Queen's College, with Kingston Botanic Garden in foreground 1860s.

Canada in the mid-nineteenth century had little regard for plants as medicines,[31] and the botanical garden envisaged at Queen's College was first and foremost the creation of the Botanical Society and not of the university – although the two were wholly intertwined. The stated goals of the society were to advance the study and appreciation of botany as a science, and to engage interested members of the public.[32]

The opening address to the first meeting of the society, delivered by Queen's principal William Leitch on 7 December 1860, gives some idea of the energy and purpose being brought to this new organization:

> It may be said that now is scarcely the time to commence a Botanical Society, that the country is not yet far enough advanced, that botany is not sufficiently studied, to warrant the establishment of a Botanical Society. It is true that botany has been neglected in this country. While this is a reproach to Canada, it affords no reason why a society should not be established. On the contrary, it is a strong reason why an attempt should be made to form one. There is a patriotic feeling rising up in Canada, which is especially strong in the youth of the province, and every well-wisher of Canada must be delighted to see it. Here then is an opportunity, by the establishment of this Society, to wipe off a reproach

> that has long hung over the country, by prosecuting a path of research that has been neglected.[33]

Despite the strong start, the Kingston Botanic Garden did not survive long. Lawson left Queen's in October 1863 following a dispute with the college's administration and took up a faculty position at Dalhousie University in Halifax.[34] He continued his interest in economic botany through experiments at his own farm but did not attempt to create another botanical garden *per se*. The Botanical Society of Canada appears to have simply stopped functioning after Lawson's departure. The botanical garden had some existence for a few years longer[35] but was abandoned by 1870. The land remained undeveloped, but the area was occasionally used for experimental plantings by faculty. In the early twentieth century, the land was planted with a variety of trees, and in 1999 Queen's University designated the site as its Arboretum.[36] In 2009 the Arboretum was renamed to honour the Rev. William Snodgrass, principal of Queen's following the death of Principal Leitch in 1864.[37] One botanical specimen is known to survive from the Kingston Botanic Garden, an herbarium sheet of *Buglossoides arvensis* (Corn Gromwell, now identified as *Lithospermum arvense*), collected by George Lawson on 7 May 1862.[38] The species is an adventitious European plant thought to have arrived at the garden unintentionally as a seed among other specimens.

Intriguingly, the existence of the Kingston Botanic Garden coincided with the final two years of formal medical education of the first Indigenous person trained and licensed to practise medicine in Canada West. Peter Edmund Jones (1843–1909), known in Anishinaabemowin as Kahkewāquonāby,[39] completed his medical training in 1865 at Queen's. Jones was born in Brantford, Canada West, the son of Mississaugas of the Credit River Chief Kahkewāquonāby, or the Reverend Peter Jones (also known as Desagondensta in Mohawk, 1802–1856), and his wife Eliza Field (1804–1890). During his medical training at Queen's, Jones prepared a senior paper, now lost, entitled "The Indian Medicine Man,"[40] at a time when most such student papers were recitations of standard treatments or diagnostic practices. The second Indigenous person to complete Western medical training in Canada West was Oronhyatekha,[41] also known as Peter Martin (1841–1907), a Mohawk scholar, athlete, and entrepreneur originally from the Six Nations of the Grand River near Brantford. Oronhyatekha graduated in 1865 with his BM and his MD in 1866, from Toronto School of Medicine[42] just weeks after Jones. As Jones had begun his medical training at Toronto School of Medicine and moved to Queen's to complete it, he and Oronhyatekha would certainly have known each other. By the time

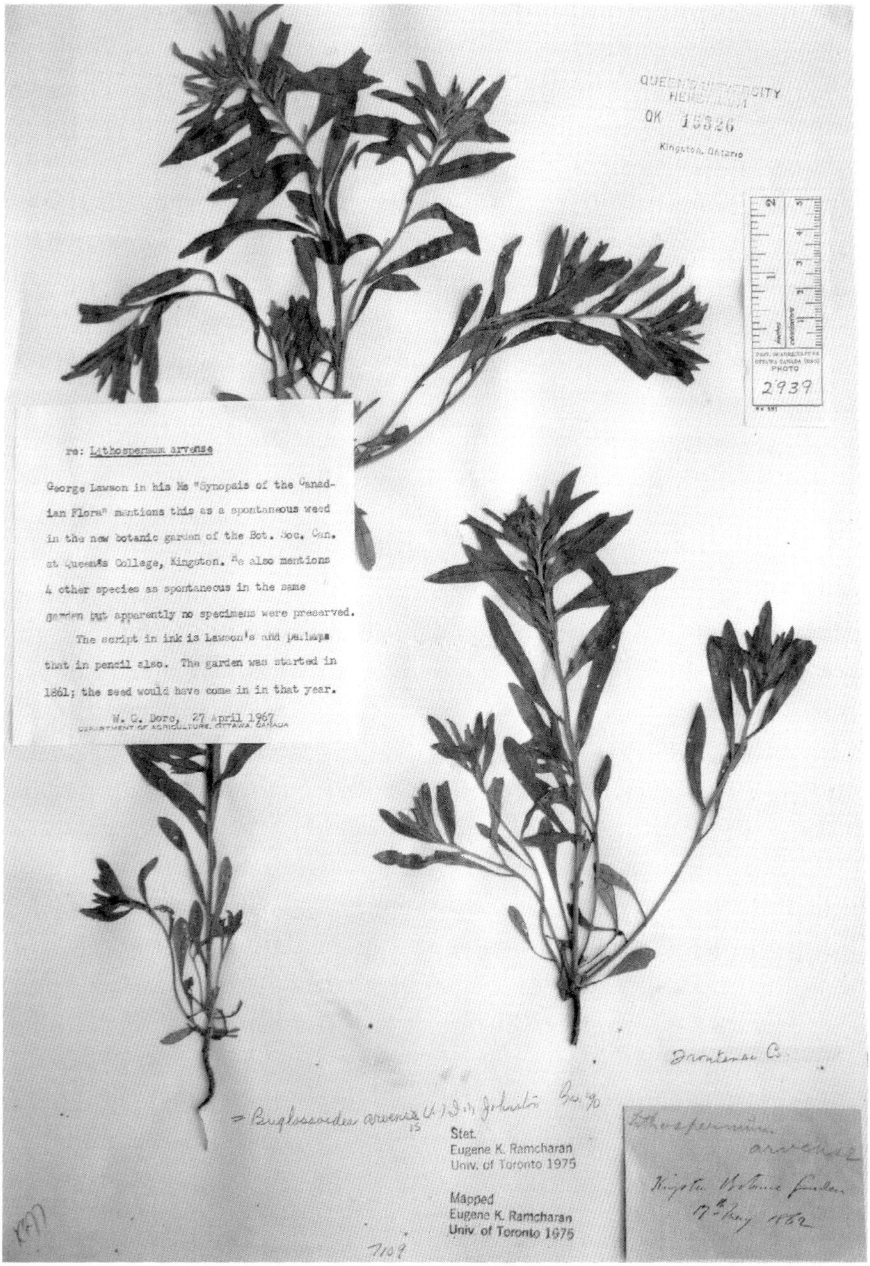

11.4
Herbarium specimen of *Buglossoides arvensis* collected by George Lawson at Kingston Botanic Garden.

Jones was at Queen's, however, George Lawson had left Kingston for Halifax and the botanical garden was on its way to abandonment.

It is tempting to wonder what could have happened if the foresight of the Dalhousies in the 1820s had resulted in a botanical garden dedicated to the documentation of Indigenous plant use and, specifically, medicines. The opportunity may nearly have arisen in the 1860s at Queen's College, but there is no evidence that the Botanical Society or its garden gathered or preserved Indigenous botanical knowledge in any way.

WOMEN AND THE *ANNALS OF THE BOTANICAL SOCIETY OF CANADA*

The purpose of the Kingston Botanic Garden was to promote scientific botany. It was a means to an end, "an appliance"[43] rather than a goal in itself. The Botanical Society of Canada developed other appliances as well. Foremost among these were a publication, the *Annals of the Botanical Society of Canada*, and regular meetings at which members and guests met and read contributions. The *Annals* published the society's bylaws, accounts of meetings, and correspondence, as well as some papers read before meetings. Its records provide a rich, albeit incomplete, glimpse into the development of the society from the first meeting in December 1860 until the twelfth recorded meeting in late 1862.[44]

The *Annals of the Botanical Society of Canada* is a resource for studying institutional features of science in Canada and brings individuals into visibility who were part of this story. In particular, it records the involvement of women who were encouraged from the beginning to participate in the society and become members. The original articles setting up classes of membership include the senior membership level of "Fellow" (evidently for men) and also a "Lady Member" category equivalent to fellow.[45] Lady members were expected to apply for membership on the same basis as fellows and, once elected, enjoyed the same privileges. Whether these categories of membership were as inclusive as they sound has been questioned by Kennedy.[46]

The *Annals* record that thirty women participated in some way in the society from 1860 to 1862. Most are listed as resident in Kingston and several among them appear from their names to be wives of college faculty or of male members of the society. A few other women are known now for having botanical interests in their own right. Catharine Parr Traill, for example, was added as a corresponding member in November 1861, and a letter from her was read at that meeting.[47]

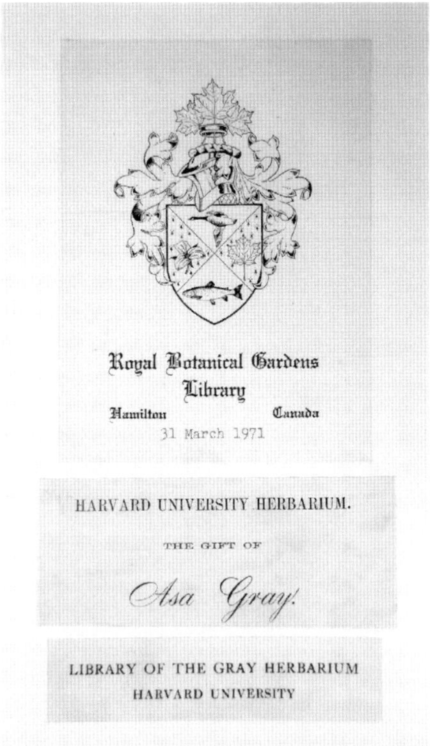

11.5 *Left*
Bookplates on the copy of the *Annals of the Botanical Society of Canada* owned by Royal Botanical Gardens (Canada). The plates indicate that the copy came from the Herbarium at Harvard University and may have been the personal copy of Asa Gray.

11.6 *Below*
Table of contents of volume 1, part III, of the *Annals of the Botanical Society of Canada*.

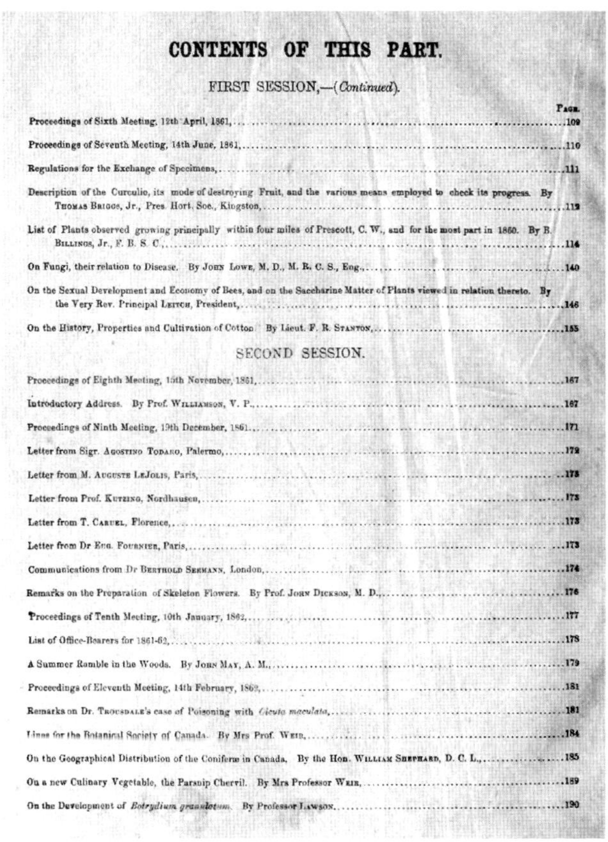

CONTENTS OF THIS PART.

FIRST SESSION,—(*Continued*).

SECOND SESSION.

Two women whose names appear repeatedly in the *Annals* are Lucy Stapley Lawson (d. 1871) and Catherine McGill Crooks (1833–1871), each warranting attention for botanical studies that go beyond the purposes of this chapter. Lucy Lawson, wife of Professor George Lawson, joined the society at the second meeting (11 January 1861) and read a paper at the third (15 February 1861). Her paper "On the Silk-Worm and other fibre-yielding insects, and the growth of their food plants in Canada"[48] was the first contribution to economic botany published in the *Annals*. A second paper by Lucy, "Further Observations on Silk Culture," was presented to the society's sixth meeting (12 April 1861) and published not in the *Annals* but in a local newspaper.[49] Lucy Lawson is also known to have contributed a paper in German, entitled "Der Canadische [*sic*] seidenspinner: Bombyx Cecropia," that was later published in *Zeitschrift für Akklimatisation*. She was not the only woman member of the Botanical Society to address the topic of silkworm culture. A Miss Gildersleeve conducted experiments in this area as well, and her paper, "Remarks on the Silk Obtained from Lettuce-fed Silkworms," read at the same sixth meeting, generated disagreement between the two women about their findings, with Miss Gildersleeve contending that silkworms raised on lettuce produce fine silk and Lucy Lawson strongly disagreeing.[50] As was the case for most women who appear in the *Annals*, Miss Gildersleeve appears without a first name, making individualization difficult. Restoring the individuality of these women through recovery of their full names may help to better understand their role in science and their social context and restore a degree of dignity not afforded by Victorian attitudes and practices. In some cases such recovery is possible through genealogical resources.[51]

Catherine ("Kate") McGill Crooks is first recorded in the *Annals* of the society as joining at the third meeting (15 February 1861). She is named in a paper printed in the *Annals* that year as collector of twenty-three individual plant specimens in the Galt area; the paper, "List of plants found growing in the neighborhood of Hamilton, during the years 1859 and 1860," was written by her brother-in-law Alexander Logie.[52] She read her own paper, "Remarks on the species of Oak, their history, habits and uses," during the seventh meeting of the society (14 June 1861).[53] Kate Crooks appears again in the *Annals* in notes from the eighth meeting (15 November 1861) as the last of eight authors on a paper read by John Macoun (later the first government-employed botanist in Canada) entitled "Returns of the periodical phenomena of vegetation during the season 1861."[54] Her brother-in-law Alexander was also a co-author on this paper, suggesting they had an ongoing collaboration on botanical studies. Further evidence of Kate Crooks's botanical work was an herbarium of "490 native plants, found mostly in the vicinity of Hamilton" that she compiled in 1862 for the International Exhibition in London, UK, and which earned her an honourable mention in awards.[55] The relationship between this collection and the plant list of similar size recounted in the 1861 paper by Logie is not clear; what became of her Hamilton herbarium after the exhibition is also not known. Following the collapse of the Botanical Society of Canada and the cessation of publication of its *Annals*, Kate Crooks seems to have disappeared from the scientific record. However, at least one herbarium specimen Crooks collected in 1865 has survived in the Herbarium at McGill University,[56] indicating that her botanical interests continued. In 1866, a year following her marriage and just weeks before the birth of her first child, Crooks (by then recorded as Mrs Smart) presented a collection of native plant specimens at the Provincial Exhibition in Toronto.[57] There is also evidence that Crooks had wider interests in natural history; she was named in 1862 as one of four women among thirty-six entomologists known in Canada.[58]

What is known of participation by women in the Botanical Society of Canada highlights a tension in the expectations of what the society was going to do. George Lawson had hopes that the society would be directed at advancing botany as a science rather than becoming a social or popular forum for the interested amateur, and he stated as much at the society's inaugural meeting: "Scientific societies on a broader basis have too often degenerated into popular institutions, calculated rather for the amusement of the many than for the encouragement and aid of the few who are engaged in the prosecution of original discovery." Nevertheless, he went on to soften the message by stating that the society should help to popularize botany, to "bring our members and the public into scientific modes of thought and expression, [rather] than to allow our Society to yield up its scientific character to suit the popular taste."[59]

In May 1891, nearly thirty years after he left Kingston for Halifax, George Lawson read an address to the Royal Society of Canada entitled "On the Present State of Botany in Canada, with suggestions as to promising lines of investigation, and a proposal for united effort in systematic observation throughout the several Provinces and Territories." Having helped to create one of Canada's first botanical gardens and the first society specifically to promote the advancement of the science of botany, Lawson's remarks do not reference a lack of botanical gardens in Canada as an issue. Instead he points out that there is no journal for communicating the results of botanical investigations in Canada and that "some method may be devised whereby immediate publication of every season's botanical field observations throughout Canada may be secured."[60] Lawson did not believe that a specific journal was

11.7
Kate Crooks's herbarium specimen of Rose Pink, *Sabatia angularis*, collected in Hamilton, 1865.

needed for botanists in Canada because other scientific outlets like the *Transactions of the Royal Society of Canada* were publishing on a range of subjects. He recommended, however, that an annual meeting of Canadian botanists should be organized.

LATER NINETEENTH-CENTURY DEVELOPMENTS AND BEYOND

Proposals for botanical gardens in other Canadian cities continued to be developed in the 1860s and beyond. In the city of Quebec in 1861 Louis-Ovide Brunet (1826–1876) was tasked with the creation of a botanical garden at Petit Séminaire de Quebec at Laval. The proposal for this garden involved both the municipality and the provincial government. Brunet was a serious botanist, and his commitment to the project went as far as spending a year in Europe visiting other botanical gardens to understand their management. However, the concept was abandoned around 1870 after nearly a decade of planning and before any construction or planting, and Brunet died "a sick man, disappointed by repeated failure to achieve success."[61] Léon Provancher (1820–1892), a botanist and entomologist considered the "Father of Natural History in Canada," also advocated for the development of a botanical garden in Quebec.[62]

A proposal for a botanical garden in Montreal was underway at about the same time. The *Annals of the Botanical Society of Canada* records correspondence from "J. Phayer, Jr., Esq., Secretary of the Montreal Agricultural and Horticultural Society," requesting any information from the Botanical Society that would be of assistance to the Montreal Agricultural and Horticultural Society in the preparation of their proposal;[63] this initiative does not appear to have gone forward, however. Another proposal for a botanical garden in Montreal was in development in the 1880s, with McGill University botanist David Pearce Penhallow (1854–1910) as the primary proponent.[64] Although this was not successful, McGill University took up the cause for a time and leased four and a half hectares of land along the Côte-des-Neiges for its own botanical garden in 1890. Some planting was undertaken, but this garden was abandoned by 1901. It would be two more decades before another proposal for a botanical garden in Montreal began to gain traction and result in the formation of Le Jardin botanique de Montréal.

An initiative for a government-supported botanical garden in Canada came into existence in 1886 in Ottawa with the passing of the Experimental Farms Act, under the Federal Department of Agriculture.[65] Sixty years after the Ninth Earl and the Countess of Dalhousie had attempted to bring the flora of

Canada to prominence as an important field for research that would yield practical benefits, the Dominion government established the Central Experimental Farm (CEF), with provision for space for an arboretum[66] and a botanical garden. Development of the Dominion Arboretum on the site began in 1889. Eventually a set of ornamental gardens and a wildlife garden would be located on the site as well as a garden in memory of Dominion Horticulturist W.T. Macoun. This institution now is a significant complex of gardens and collections covering forty hectares and is sometimes referred to as the Dominion Arboretum and Botanical Garden. The work of the CEF and the other national agricultural research stations was strongly aligned with the economics of improvements for agriculture, such as the large-scale production of food crops, whether through breeding, soil amendment, pest or disease management, or other fields. However, it would be short-sighted to rigidly separate the use of permanent gardens as research or teaching facilities for horticultural from those for agricultural subjects in the nineteenth century. The Dominion Arboretum and Botanical Garden stands today as the longest-surviving research garden in Canada.[67]

Thus, in the nineteenth century botanical gardens were built in Montreal (1822 and 1895), Toronto (1853), Kingston (1862), and Ottawa (1889). If the People's Garden (1837) is included as a botanical garden because of its intended educational purpose, Halifax belongs on the list as well. Additional proposals for projects to build botanical gardens were also discussed in Toronto (1837), Halifax (1837), Laval (1861), and Montreal (1860s and 1885). All the gardens that were realized included some combination of botanical research, education, and social amenity. Without a doubt, the first botanical garden in Canada was not that at University of Queen's College in Kingston, as has sometimes been claimed.[68] At least three gardens were built that predated the Botanical Society of Canada's efforts in the 1860s, and proposals for other gardens were in development elsewhere. At least twelve botanical gardens were either proposed or actually built in Canada during the nineteenth century, a fact that does not appear to have been indicated previously (see table 11.1). The list of these projects may serve as a resource for further historical work on botanical gardens in Canada.

Available evidence suggests that women were strongly involved in at least two of the botanical gardens built in nineteenth-century Canada: the King's Botanical Garden for North American Plants in Montreal in 1822 and the Kingston Botanic Garden in 1861. There is little evidence that those developing these early botanical gardens were interested in the use of plants by Indigenous Peoples in Canada, beyond George Ramsay's references. Unlike the situation

Table 11.1
Botanical gardens in Canada proposed or built before 1900

Garden name	Proponent	Location	Status	Extant
Garden of the Intendant's Palace	Gaultier	City of Quebec	Built, abandoned	17??–
King's Botanical Garden of North American Plants	Dalhousie	Montreal, Sorel	Built, abandoned	1822–28
Lyceum of Natural History and the Fine Arts	Fothergill	Toronto	Proposed	1835
Park for study and enjoyment of flora	Downs	Halifax	Proposed	1838
The People's Garden	Howe	Halifax	Built, modified	1841–
Toronto Normal School Botanical Garden	Ryerson	Toronto	Built, demolished	1856–?
Botanical Garden	Allan	Toronto	As Allan Gardens	1857–
Kingston Botanic Garden	Litchfield	Kingston	Built, demolished	1861–?
Laval University Botanic Garden	Brunet	City of Quebec	Proposed	1861–70
Botanical Garden at Montreal	Phayer	Montreal	Proposed	1861
Montreal Botanical Garden	Penhallow	Montreal	Proposed	1885
Central Experimental Farm Botanic Garden	Fletcher	Ottawa	Built, extant	1889–
McGill University Botanical Garden	Penhallow?	Montreal	Built, abandoned	1895–1901

Information compiled by David Galbraith.

in some British colonies where botanical gardens cooperated to develop economic crops from local species[69] or further establish a colonial presence as was the case in India,[70] botanical gardens do not appear to have been part of the formal mechanisms of colonization in British North America.

In contrast to isolated attempts to develop botanical gardens in the first two-thirds of the nineteenth century in Canada, the creation of five agricultural research stations – in most senses research gardens – under the aegis of Agriculture Canada marked a shift toward establishment of a national presence in research related to plant cultivation. Many of the components of the Central Experimental Farm were organized early on by Professor William Saunders, and by the end of the nineteenth century the Dominion Arboretum and Botanical Garden were well established. Saunders considered the lack of a botanical garden in Canada a shameful situation, one erased by the creation of the botanical garden at the Central Experimental Farm.[71] However, the impetus for dedicated research gardens in the service of agriculture did not translate into support for gardens to be used for other kinds of botanical research.

The development of public gardens of various kinds across Canada did not end with the nineteenth century. Many proliferated across twentieth-century Canada, and some related to botanical gardens. They included large horticultural display gardens created initially as private gardens (e.g., the Butchart Gardens in Victora, BC, in 1904, and Les Jardins de Métis/Reford Gardens in Mont Joli, QC, in 1926), university botanical gardens (e.g., University of British Columbia Botanical Garden in Vancouver, BC, in 1916; University of Alberta Botanic Garden, near Devon, AB, in 1959; and Memorial University of Newfoundland Botanic Garden in St John's, NL in 1971), and publicly supported botanical gardens arising out of the efforts of municipalities (e.g., Le Jardin botanique de Montréal, Montréal, QC, proposed in 1919; Royal Botanical Gardens, Hamilton and Burlington, ON, proposed in the late 1920s) and provincial leadership (Niagara Parks Botanical Garden and School of Horticulture, Niagara Falls, ON, in 1936). Closely allied to the rise of these public gardens has been the establishment of arboreta, often attached to a university (e.g., University of Guelph Arboretum in Guelph, ON, in 1970; Sherwood Fox Arboretum at Western University in London, ON, in 1981; and the Snodgrass Arboretum and Queen's University at Kingston, ON, in 1999).

Running through the stories of these projects in the nineteenth century is a repeated theme: an enthusiastic individual or an organized group proposes, and in some cases actually creates, a botanical garden with the hope that its utility would then be obvious to higher authorities such as municipal, provincial, or federal governments, or a university or college. Sometimes years and personal fortunes would be expended on the project, only to have it collapse and be abandoned when the wished-for patronage did not materialize. Some of this cycle rests, arguably, on a lack of understanding among the envisioned patrons of just what a botanical garden was supposed to *do* or *be*.[72] Was it to be a public park or a manicured landscape displaying exotic plants for public enjoyment? A serious research or educational centre? A museum of living plants, cataloguing and holding a diversity of plant life for research, even to economic ends? Which of these, if any, would justify the ongoing cost? The balance among these functions remains a pertinent issue for botanical gardens and allied institutions today.

ACKNOWLEDGMENTS

The author gratefully acknowledges the long history of First Nations, Métis, and Inuit peoples of Turtle Island and wishes to pay respects to his Indigenous colleagues and friends. An awareness of some of their history has shaped his understanding of how plants were known by various peoples and the degree to which colonial change in the nineteenth century led to many serious social issues that we face together today. Dale Kristensen, curator of the Queen's University Herbarium at Kingston, ON, provided access to the 1862 specimen of *Buglossoides arvensis* collected by George Lawson. Frieda Beauregard, curator of the herbarium at McGill University, provided the scanned image of Kate Crooks's herbarium sheet. Anna Oxford Soper, Queen's University Library, provided information about Kate Crooks's participation in the 1862 International Exhibition. The author is grateful to James Pringle, Erin Aults, Deborah Reid, and especially Ann Shteir for their extensive assistance, patience, and encouragement.

NOTES

1 Perhaps nothing signalled the solidification of purpose of botanical gardens more clearly than the reorganization of Royal Botanic Gardens, Kew on 1 April 1841. Before this, the fifteen-acre botanic garden had largely functioned as a research centre backed by royal patronage. That date signalled both the appointment of Sir William Hooker as director and transfer of responsibilities and funding from a hodgepodge of agencies to the chief commissioners of Woods and Works (Desmond, *Kew*, 167, and Thistleton-Dyer, "Historical Account of Kew to 1841," 326). Later in the nineteenth century emphasis (and funding) at Kew would shift back and forth between roles as scientific centre and public park (Desmond, *Kew*, 223–38).

2 Aplin et al., "Indices seminum," 94.
3 Boivan, "Gaultier, Jean-François."
4 On Gaultier's garden see Marchand, *Voyages dans l'amerique du Nord*, 77–8.
5 Bain et al., "Bugs, seeds and weeds."
6 On John Bartram, see Wulf, *The Brother Gardeners*, 149–69.
7 As part of the American Bicentennial celebrations in 1976, a Physic Garden was built at Pennsylvania Hospital in commemoration of the proposed Botanical Garden and has been maintained since (Trustees of the University of Pennsylvania, "History of Pennsylvania Hospital").
8 Ewan, "Early history," 38.
9 Johnson, *American Eden*, 5.
10 Hosack, *Hortus Elginensis*, 1.
11 Following the abandonment of Elgin Botanic Garden, the land was rented out by Columbia College to a series of small tenants. By the late nineteenth century, it was developed as a residential area. The site today is occupied by Rockefeller Center (Johnson, *American Eden*, 326–7).
12 See Reid, Cayouette and Khoo, and Vandenberg.
13 Reid, "Unsung Heroines of Horticulture," 108.
14 Library and Archives Canada: MG24 A12 Vol. 9(4).
15 Blair was active in the 1820s and 30s. He collected plants in Canada and New Hampshire, and contributed to the *Gardeners Magazine* (Desmond, *Dictionary of British and Irish Botanists*, 323–4).
16 Library and Archives Canada: MG24 A12 Vol. 9(4).
17 Great Britain. Colonial Office, Canada, Formerly British North American, Original Correspondence CO 42: C-11934, Images 1003 and 1004, Library and Archives Canada, online document, http://heritage.canadiana.ca/view/oocihm.lac_reel_c11934/1003?r=0&s=1.
18 Since the Dalhousies were based in the immediate area around Montreal, it may have been a Kanien'kehá:ka (Mohawk) woman who helped the man with the sprained loins.
19 The National Archives of Scotland, GD45: Papers of the Maule Family, Earls of Dalhousie GD45/3/548: From the diary of George Ramsay, 9th Earl of Dalhousie. Transcribed by Deborah Reid and provided to the author.
20 Great Britain. Colonial Office, Canada, Formerly British North American, Original Correspondence CO 42: C-11934, Images 1003 and 1004, Library and Archives Canada, online document, accessed 24 November 2019, http://heritage.canadiana.ca/view/oocihm.lac_reel_c11934/1003?r=0&s=1.
21 See Deborah Reid's discussion of this episode in her chapter in this volume.
22 Romney, "Fothergill, Charles."

23 Buggey, "Downs, Andrew."

24 Friends of the Public Gardens, *The Halifax Public Garden*. 10.

25 Crawford, "A Heritage Conservation Management Strategy," 4–26.

26 Anonymous. "Horticulture: The Normal School Grounds, Toronto," *The Canada Farmer* III, 14 (16 July 1866): 220. The 1866 account of the grounds of the Toronto Normal School published in *The Canada Farmer*, vol. III, may be the first contemporary published account of a botanical garden in Canada. The names of more than thirty species in the garden and on the grounds are mentioned, with indications of many more present on the site.

27 Bain, "William Mundie, Landscape Designer," 300.

28 Court, "An Erosion of Imagination," 180.

29 Litchfield, "Notes and Suggestions," 9. For more on Litchfield, see Gibson, "The Astonishing Career of John Palmer Litchfield," and Rasporich and Clarke, "Litchfield, John Palmer."

30 Lawson, "Extract from a Letter," 587.

31 Court, "An Erosion of Imagination," 174.

32 For detailed discussion of the interplay in the Botanical Society of Canada between advancing botanical science and serving as a social club for the upper echelons of Kingston society, see Connor, "To Promote the Cause of Science."

33 Leitch, "Opening Address," 3

34 Zeller, "Lawson, George."

35 Dore, "Canada's First Botanic Garden," 11.

36 University Planning, "Queen's University's Snodgrass Arboretum History," Queen's University at Kingston, accessed 26 November 2019, https://www.queensu.ca/planningandbudget/university-planning/arboretum.

37 Moir and Spragge. "Snodgrass, William."

38 Queen's University Herbarium (Herbarium Code OK), specimen number 15326.

39 Sherman, *Bridging Two Peoples*, xiii.

40 Ibid., 23.

41 Jamieson and Hamilton, *Dr. Oronhyatekha*, 119.

42 Ibid., 117.

43 Lawson, "Remarks on the Present State of Botany in Canada," 4.

44 The *Annals of the Botanical Society of Canada* were printed in three parts between December 1860 and February 1862 (vol I part I: 7 December 1860–8 March 1861; vol I part II: 8 March 1861–25 March 1861; vol I part III: 12 April 1861–14 February 1862) spanning 196 numbered pages. The production of the *Annals* (originally 300 copies were printed) was the most expensive undertaking of the Botanical Society of Canada.

45 Laws of the Botanical Society of Canada, read at the First Meeting of the Society, 7 December 1860. *Annals of the Botanical Society of Canada*, 15.

46 Kennedy, "Science Culture in the Nineteenth Century," is a detailed enquiry into participation by women in the Botanical Society of Canada.

47 *Annals*, 170. On Catharine Parr Traill, see the chapter by Michael Peterman in this volume.

48 Lawson, "On the Silkworm."

49 Kennedy, "Science Culture in the Nineteenth Century," 69.

50 Ibid., 59–61.

51 References in genealogical sources Gildersleeve and Horsey suggest that "Miss Gildersleeve" could have been one of two sisters, Lucretia Ann(e) Marie (1826–1909) or Emily Gertrude (1843–1898), but Lucretia, then approximately thirty-five years old, is more likely. While it is not yet possible to resolve her identity as an individual, the family context of Miss Gildersleeve is known.

52 Logie, "List of Plants," 90. Alexander Logie married Mary Ritchie Crooks, Kate's older sister, in 1852 (Huggard, *Ramsay Crooks of the American Fur Company*, 82).

53 *Annals*, 111.

54 Authors listed as "John Macoun, Belleville; B. Billings, Prescott; A.T. Drummond, B.A., Kingston; T. Dupuis, M.D., Odessa; Judge Logie, and Miss Crooks, Hamilton." *Annals*, 170.

55 Hodgins, "The Canadian Department at the International Exhibition," 76.

56 Soper, "In Ontario, a Quest to Rediscover."

57 Anonymous, "The natural history department at the recent exhibition," *The Canadian Farmer* III, 19 (1 October 1866): 293.

58 Bethune, "List of Entomologists in Canada." 200.

59 Lawson, "Remarks on the Present State of Botany in Canada," 8.

60 Lawson, "On the Present State of Botany in Canada," 61.

61 Rosseau, "Brunet, Louis-Ovide."

62 Perron, "Provancher, Léon."

63 *Annals*, 170.

64 See Zeller, "Lawson, George," and Penhallow, "On the Establishment of a Botanic Garden and Arboretum in Montreal."

65 Smith and Bramley, *Ottawa's Farm*, 15.

66 Differentiating between a botanical garden and an arboretum is sometimes difficult. Botanic Gardens Conservation International considers an arboretum simply to be a botanical garden where the plants in the collections consist of trees. Every other functional aspect of a botanical garden, such as development, labelling, and management of plant collections, public display, research, and conservation applies to an arboretum too.

67 Anstey, "One Hundred Harvests," 11. The Government of Canada established five experimental farms following passage of the Experimental Farm Station Act of 1886, in Ottawa, ON; Nappan, NS; Brandon, MB; Indian Head, NWT; and Agassiz, BC.

68 See Dore, "Canada's First Botanic Garden," and Tyler, "Botanical Garden."

69 Rackow and Lee, "Western Botanical Gardens," 291.

70 Baber, "The Plants of Empire," 1.

71 Hinchcliff, *Blooms*, 37. Whether a National Botanical Garden was desirable or needed in Canada has long been in contention. The Dominion Arboretum and Botanical Garden at the Central Experimental Farm has existed since the 1890s but has not been considered a truly national institution. See Chan, "A National Botanical Garden for Canada," on repeated failures of proposals for a national botanical garden in the twentieth century. Since the 1980s a group of citizens has been trying to establish a new botanical garden in Ottawa, referred to in the twenty-first century as the Canadensis Garden.

72 The question of just what a botanical garden does is not new, as Court observed regarding the University of Toronto's proposal in the 1850s ("An erosion of imagination," 180).

BIBLIOGRAPHY

Archives

National Archives of Scotland, Edinburgh
Archives, Queen's University, Kingston
Library and Archives Canada, Ottawa
Archives of Ontario, Toronto

Books and Articles

Aplin, David, Simon Linington, and Jan Rammeloo. "Indices seminum: Are They Really Worth The Effort?" *Sibbalda: The Journal of Botanic Garden Horticulture* 5 (2007): 93–107.

Anstey, T.H. "One Hundred Harvests: Research Branch Agriculture Canada 1886–1986." Research Branch Agriculture Canada Historical Series No. 27. Ottawa: Canadian Government Publishing Centre, 1986.

Baber, Zaheer. "The Plants of Empire: Botanic Gardens, Colonial Power and Botanical Knowledge." *Journal of Contemporary Asia* 46, 4 (2016): 659–79.

Bain, Allison, Julie-Anne Bouchard-Perron, Réginald Auger, and Daniel Simoneau. "Bugs, Seeds and Weeds at the Intendant's Palace: A Study of an Evolving Landscape." *Post-Medieval Archaeology* 43, 1 (2009): 183–97.

Bain, David. "William Mundie, Landscape Designer." *Journal of Garden History* 5, no. 3 (1985): 298–308.

Bethune, Charles J.S. "List of Entomologists in Canada." *The Canadian Naturalist and Geologist and the Proceedings of the Natural History Society of Montreal* VII, Article XIX (1862): 199–201.

Blackie, George S. "On the *Cornus florida* of the U.S." *Annals of the Botanical Society of Canada* (1861–62): 22–5.

Boivin, Bernard. "Gaultier, Jean-François." In *Dictionary of Canadian Biography*, vol. 3, University of Toronto/Université Laval, 2003–. Accessed 3 August 2017, http://www.biographi.ca/en/bio/gaultier_jean_francois_3E.html. 1974.

Botanical Society of Canada. *Annals of the Botanical Society of Canada*. Biodiversity Heritage Library. Accessed 24 November 2019, https://www.biodiversitylibrary.org/item/182352.

Buggey, Susan. "Downs, Andrew." In *Dictionary of Canadian Biography*, vol. 12, University of Toronto/Université Laval, 2003–. Accessed 2 August 2017, http://www.biographi.ca/en/bio/downs_andrew_12E.html 1990.

Chan, Allan P. "A National Botanical Garden for Canada – A History of Failures." In *Proceedings of the Symposium on a National Botanical Garden System for Canada. Royal Botanical Gardens Technical Bulletin #6*, edited by P.F. Rice, 22–7. Burlington, ON: Royal Botanical Gardens, 1972.

Connor, James T.H. "To Promote the Cause of Science: George Lawson and the Botanical Society of Canada, 1860–1863." *Scientia Canadensis* 10, 1 (1986): 3–33.

Court, John P.M. "An Erosion of Imagination: Unfulfilled Plans for a University Botanical Gardens and Taddle Creek, 1850 to 1884." *Ontario History: The Journal of the Ontario Historical Society* XCV, 2 (Autumn 2003): 166–92.

Coutu, Joan. "A Drive through Canadian History: People, Cars, and Public Art at Niagara Falls in the 1930s." In *Public Art in Canada: Critical Perspectives*, edited by Annie Gérin and James S. McLean, 45–64. Toronto, Ontario: University of Toronto Press. 2009.

Crawford, Pleasance. "A Heritage Conservation Management Strategy for Allan Gardens, City of Toronto." Unpublished report for Heritage/Preservation Services, Culture Division, Economic Development, Culture and Tourism Department, City of Toronto, 2004.

Dagenais, Michèle. "Le Jardin botanique de Montréal: une responsabilité municipale?" *Revue d'histoire de l'Amérique française* 52, 1 (Spring 1998): 3–22.

Desmond, Ray. *Dictionary of British and Irish botanists and horticulturists*. London: Taylor & Francis Ltd, 1977.

– *Kew: The History of the Royal Botanic Gardens*. London: The Harvill Press, 1995.

Dore, William G. "Canada's First Botanic Garden." *Greenhouse Garden Grass, Journal of the Plant Research Institute, Canadian Department of Agriculture* 6, 2 (1967): 6–14.

Ewan, Joseph. "Early History." In *A Short History of Botany in the United States*, edited by Joseph Ewan, 27–48. New York: Hafner Publishing Company 1969.

Friends of the Public Gardens. *The Halifax Public Gardens*. Halifax, Nova Scotia: Friends of the Public Gardens 1989.

Gibson, T. "The Astonishing Career of John Palmer Litchfield." *Canadian Medical Association Journal* 70 (1954): 326–30.

Gildersleeve, William Harvey. *Gildersleeves of Gildersleeve, Conn. And the Descendants of Philip Gildersleeve*. Meriden, Conn. Press of the Journal Publishing Co., 1941. Accessed 17 June 2019, https://archive.org/details/gildersleevesofgoogild/page/n10.

Hinchcliff, Richard. *Blooms: An Illustrated History of the Ornamental Gardens at Ottawa's Central Experimental Farm*. Alexandria, ON: Sanderling Press, 2016.

Hodgins, J. George, ed. "The Canadian Department at the International Exhibition (excerpt from *Canadian News*)." *The Journal of Education for Upper Canada* 15, IV Papers on Natural History (1862): 76.

Horsey, Edwin E. "The Gildersleeves of Kingston – Their Activities 1816–1930." Unpublished manuscript. Accessed 17 June 2019, http://images.maritimehistoryofthegreatlakes.ca/images/MHGL0001131851T.PDF.

Hosack, David. *Hortus Elgenensis: or a catalogue of plants, indigenous and exotic, cultivate in the Elgin Botanic Garden, in the vicinity of the City of New-York, Established in 1801 by David Hosack, M.D F.L.S.* Second edition, enlarged. T. & J. Swords, 1811.

Huggard, Robert E. *Ramsay Crooks of the American Fur Company. A biography, featuring: Ancestors, Children, Siblings; Including: The Ancestors of Emilie Pratte*. Accessed 4 February 2019, https://sites.google.com/site/ramsaycrooksbiography/.

Jamieson, Keith, and Michelle A. Hamilton. *Dr. Oronhyatekha: Security, Justice, and Equality*. Toronto: Dundurn Press, 2016.

Johnson. Victoria, *American Eden: David Hosack, Botany, and Medicine in the Garden of the Early Republic*. New York: Liveright, 2018.

Kennedy, Kerrie. "Science Culture in the Nineteenth Century: Women and the Botanical Society of Canada." *Resources for Feminist Research* 33, 3/4 (2010): 47–70.

Laking, Leslie. *Love, sweat and soil: A history of Royal Botanical Gardens from 1930 to 1981*. Burlington, ON: Royal Botanical Gardens Auxiliary, 2006.

Lawson, George. "Extract from a Letter from Professor Lawson, Queen's College, Kingston, Canada read 9th July 1863." *Transactions of the Botanical Society of Edinburgh* 7, 1860–1863 (1863): 587.

– "On the Present State of Botany in Canada, with Suggestions as to Promising Lines of Investigation, and a Proposal for United Effort in Systematic Observation throughout the Several Provinces and Territories." *Transations of the Royal Society of Canada*, section IV (1891): 17–20.

– "Remarks on the Present State of Botany in Canada, and the Objects to Be Attained by the Establishment of a Botanical Society." *Annals of the Botanical Society of Canada* 1, 1 (1860): 4–9.

Lawson, Lucy S. "Der Canadische seidenspinner: Bombyx Cecropia." *Zeitschrift für Akklimatisation*. 2 (1864): 37–9. (Cited in M.R.S. Creese and T.M. Creese. *Ladies in the Laboratory III: South African, Australian, New Zealand, and Canadian Women in Science: Nineteenth and Early Twentieth Centuries*. Lanham, MD: Scarecrow Press, 2010.)

– "On the Silkworm and Other Fibre-Yielding Insects, and the Growth of Their Food Plants in Canada." *Annals of the Botanical Society of Canada* 1861–1862 (1891): 43–8.

Leitch, William. "Opening Address, First Meeting of the Botanical Society of Canada." *Annals of the Botanical Society of Canada* 1, 1 (1861): 3.

Litchfield, John P. "Notes and Suggestions Relative to the Establishment of a Botanical Garden." *Annals of the Botanical Society of Canada* 1, 1 (1861): 9–11.

Logie, Alexander. "List of Plants Found Growing in the Neighborhood of Hamilton, during the Years 1859 and 1860." *Annals of the Botanical Society of Canada* 1, 2 (1861): 90–108.

Marchand, L.W. *Voyage dans L'amérique du Nord par Pierre Kalm, Naturaliste Suédois*. vol. 2. Analyse et tradiut par L. W. Marchand. Mémoires de la Société Historique de Montréal, 1880. Accessed 13 August 2019, https://www.biodiversitylibrary.org/item/121803.

Moir, John S., and Shirley C. Spragge. "Snodgrass, William." In *Dictionary of Canadian Biography*, vol. 13, 1994. University of Toronto/Université Laval, 2003–. Accessed 26 November 2019, http://www.biographi.ca/en/bio/snodgrass_william_13E.html.

Penhallow, David. "On the Establishment of a Botanic Garden and Arboretum in Montreal Under the Auspices of the Montreal Horticultural Society and Fruit Grower's Association of the Province of Quebec." Reprint from 10th Annual Report, 1885.

Perron, Jean-Marie. "Provancher, Léon." In *Dictionary of Canadian Biography*, vol. 12, 1990. University of Toronto/Université Laval, 2003–. Accessed 3 August 2017, http://www.biographi.ca/en/bio/provancher_leon_12E.html.

Rackow, Donald A., and Sharon A. Lee. "Western Botanical Gardens: History and Evolution." *Horticultural Reviews* 43 (2015): 269–310.

Rasporich, A.W., and I.H. Clarke. "Litchfield, John Palmer." In *Dictionary of Canadian Biography*, vol. 9, 1976. University of Toronto/Université Laval, 2003–. Accessed 13 August 2019, http://www.biographi.ca/en/bio/litchfield_john_palmer_9E.html.

Reid, Deborah A. "Unsung Heroines of Horticulture: Scottish Gardening Women, 1800 to 1930." Doctoral dissertation, University of Edinburgh, Edinburgh, UK, 2015.

Romney, Paul. "Fothergill, Charles." In *Dictionary of Canadian Biography*, vol. 7, 1988. University of Toronto/Université Laval, 2003–. Accessed 3 August 2017, http://www.biographi.ca/en/bio/fothergill_charles_7E.html.

Rousseau, Jacques. "Brunet, Louis-Ovide." In *Dictionary of Canadian Biography*, vol. 10, 1972. University of Toronto/Université Laval, 2003–. Accessed 3 August 2017, http://www.biographi.ca/en/bio/brunet_louis_ovide_10E.html.

Sherman, Allan L. *Bridging Two Peoples: Chief Peter E. Jones, 1843–1909*. Waterloo, ON: Wilfrid Laurier University Press, 2013.

Soper, Anna. "In Ontario, a Quest to Rediscover the Work of a Groundbreaking 19th-Century Botanist." *Atlas Obscura*, 23 May 2919. Accessed 26 November 2019, https://www.atlasobscura.com/articles/botanist-ontario-kate-crooks.

Smith, Helen, and Mary Bramley (photographer). *Ottawa's Farm: A History of the Central Experimental Farm*. Burnstown, ON: General Store Publishing House, 1996.

Taylor, Roy. "Botanical Garden." In *The Canadian Encyclopedia: Historica Canada*, 2006. Accessed 2 August 2017, http://www.thecanadianencyclopedia.ca/en/article/botanical-garden/.

Terpstra, Nicholas. "Local Politics and Local Planning: A Case Study of Hamilton, Ontario, 1915–1930." *Urban History Review/Revue d'histoire urbaine* 14, 2 (1985): 114–28.

Thistleton-Dyer, W.T. "Historical Account of Kew to 1841." *Bulletin of Miscellaneous Information (Royal Botanic Gardens, Kew)* 60 (1891): 279–327.

Trustees of the University of Pennsylvania. "History of Pennsylvania Hospital: Historical Timeline, 1751–1800." Philadelpha: Penn Medicine, 2017. Accessed 17 August 2019. http://www.uphs.upenn.edu/paharc/timeline/1751/.

Wulf, Andrea. *The Brother Gardeners: Botany, Empire, and the Birth of an Obsession*. New York: Vintage Books, 2008.

Wyman, Donald. "The Arboretums and Botanical Gardens of North America." *Chronica Botanica* 10, 5/6 (1947): 398–497.

Zeller, Suzanne. "Lawson, George." In *Dictionary of Canadian Biography*, vol. 12, 1990. University of Toronto/Université Laval, 2003–. Accessed 1 August 2017, http://www.biographi.ca/en/bio/lawson_george_12E.html.

12

Women, Citizen Science, and Botanical Knowledge in Ontario, 1870–1920

DAWN R. BAZELY AND KATHRYN MCPHERSON

INTRODUCTION

There is a common perception today in many parts of the international STEM (science, technology, engineering, and mathematics)[1] community that large segments of many publics lack science literacy skills,[2] which in turn has fuelled popular distrust and even rejection of peer-reviewed scientific research. For example, surveys show large numbers of people do not believe in evolutionary theory or vaccination science or that climate change (global warming) is caused by anthropogenic sources of greenhouse gas emissions, and that few Americans know or could name a living scientist.[3] In response to attacks on the credibility of scientists, university academics are facing increased expectations to be better communicators of their research and also be more relatable and accessible. STEM academics are being challenged to get out of our ivory towers, engage in public science,[4] and encourage the work of citizen scientists – those members of the general public who, often in collaboration with professional scientists, collect and analyze data pertaining to the natural world.[5] Social media is touted as a means of accelerating connections between professional scientists, citizen scientists, and broader society, and for allowing information and knowledge to flow both ways more effectively.[6] Social media is often described as an innovative, disruptive technology that, such as in the case of Twitter or species-identification "apps," can unexpectedly transform how people connect with each other and access information.[7]

Increasing equity, diversity, and inclusivity in STEM is another key component of contemporary campaigns to democratize science and increase science literacy. This includes recognizing the contributions of women, both as professional scientists and as amateurs, who have often gone unrecognized in STEM and in society more generally. Perhaps the most notorious examples of women scientists being deliberately erased are Rosalind Franklin – whose contributions to the discovery of DNA were not acknowledged – and astrophysicist Jocelyn Bell Burnell – whose supervisor received the Nobel Prize in 1974 for Burnell's dissertation research on pulsars in the late 1960s.[8]

Many other lesser-known women have also made significant contributions to scientific knowledge. The STEM community has recently been (re)discovering its women's history, including through projects such as International Ada Lovelace Day, founded in 2009, to raise awareness of women in STEM. Wikipedia "edit-a-thons" create pages for notable women in STEM,[9] many of whom had no visibility before in this online, community-driven encyclopedia. And while it is true that feminist historians of science have been documenting the missing women in STEM for decades,[10] new and "disruptive" technologies like Twitter have been used in recent years to call out ongoing sexist and exclusionary practices in science faculties generally, including biology and ecology programs, and also to highlight historical contributions by women to STEM. Popular social media "hashtags" like Kate Clancy's *#GirlsWithToys* on Twitter or Instagram's *#WomenInSTEM* or Biodiversity Heritage Library's open access virtual collection curated under *#WomenInBHLib* serve to disrupt conventional circuits of knowledge about whose scientific and technological activities are important, and they create opportunities for bringing women's STEM contributions into view.[11]

Thinking about disruptive technologies in this way also offers a new lens for seeing the work of women and science in the past. If social media such as websites, apps, and blogs have served to disrupt elite or professional control of scientific knowledge today, a different kind of disruptive technology had a similar effect in the late nineteenth and early twentieth centuries. This chapter argues that public education, mass literacy, and the evolution of cheap, mass publications combined to become the disruptive technologies of the 1870–1920 era, which enabled many "ordinary" women and men to produce and disseminate scientific knowledge. This chapter focuses on women involved in botanical study and work in Ontario and argues that, although Canadian women were excluded from "professional" or "institutionally credentialed" botanical knowledge until World War I and after, they actively engaged in "public" botanical activities as authors, illustrators, members of horticultural societies, and owners of commercial enterprises.

HOW FLORA'S DAUGHTERS WERE FROZEN OUT OF PROFESSIONAL BOTANY: 1760–1860

How did botanical science become a part of what so many women in STEM regard as ongoing gendered science?[12] In her path-breaking book *Cultivating Women, Cultivating Science*, Ann Shteir explored how the "polite botanical culture" in which women participated in late eighteenth and early nineteenth-century England became "botanical science" by the mid-nineteenth century, with its own scientific institutions and a concomitant gendering of the field as masculine. By 1850 women's botanical knowledge was concentrated in domestic, familial, personal, and artistic expression.[13] Historians such as Suzanne Le-May Sheffield have built on Shteir's work to examine how American and British women scientists continued to be marginalized from scientific institutions.[14]

Canadian women, too, were excluded from the nineteenth century's formal, institutionalized structures of the scientific study of plants – that is, "professional" botany. The experiences of Ontario's Catharine Parr Traill are often used to illustrate this exclusion. Michael Peterman and others have arguably made Catharine Parr Traill the most famous Ontario botanist, amateur or otherwise, of the nineteenth century. Traill, a prolific author, studied and wrote about plants from a young age and later supported her family through her published writing; yet in her time she gained little access to the ranks of professional science.[15] Other scholars have pointed to the career of Carrie Derick who completed her PhD at the University of Bonn in 1901 and worked for fourteen years at McGill's Botany Department for no or low pay until finally being granted a university professorship in 1912. Derick's 1912 appointment is often cited as the turning point where women successfully reversed their exclusion from botanical study in Canada.[16]

The struggles and successes of these high-profile individuals, though, is only one part of the story. Over the same decades when Derick was challenging the male-dominated ranks of university botany, other women were pursuing the production and dissemination of botanical knowledge through other means. Art historian Kristina Huneault's analysis of women's botanical drawing in late nineteenth-century Canada focuses on the subjectivities women developed and expressed through their drawing.[17] This article takes a different tack, exploring not the self-fulfillment female botanists may have enjoyed but rather their intrusion into the public domains of botanical study and commerce. Women's engagement in local horticultural societies, in commercial gardening and landscaping ventures, and in the disruptive work of authoring and illustrating botanical books all point to an important, if hidden, history of women in Canadian science.

WOMEN IN ONTARIO'S HORTICULTURAL SOCIETIES AND COMMERCIAL VENTURES: 1860S–1920S

Throughout the late nineteenth and early twentieth centuries, local horticultural and gardening organizations constituted significant sources of plant knowledge in Ontario. By 1806, agricultural societies in communities like York (Toronto) had been formed "to enable farmers to excel in various branches of agriculture and rural economy," and in 1830, these societies became eligible for government grants, a privilege codified in the 1854 Act for "the encouragement of agricultural societies and agriculture in Upper Canada."[18] In contrast with agricultural societies, which focused their energies on supporting local practices of raising grazing animals and grass crops, horticultural societies were dedicated to developing and disseminating knowledge of non-field crops, especially fruits and flowers. One of the first such groups was the Hamilton Horticultural Society formed in 1850, though not much is known about its members and activities.[19] The 1857 provincial Agricultural Statute made a provision for local horticultural societies of more than twenty-five members to organize and thus be entitled to apply for the provincial funds administered through the agricultural societies. This already-crowded organizational landscape was complicated further in 1859 when the Ontario Fruit Growers' Association was formed. For the next forty years, tussles over provincial funding percolated as horticulturalists, fruit growers, and agronomists sought grants to support their educational and promotional work.[20] In 1878, the Fruit Growers' Association gained a distinct advantage when it founded its monthly publication *The Canadian Horticulturist*. Dedicated to "Fruits, Flowers, Forestry," the journal cost $1.00 per year, a fee which also garnered subscribers a membership in the province-wide Fruit Growers' Association.[21]

In the late 1890s, the Ontario government proposed that the local horticultural societies unite into a province-wide organization which, when finally established in 1906 as the Ontario Horticultural Association (OHA), gave horticultural groups autonomy from the agricultural societies, including permitting horticulturalists to access provincial grant money directly.[22] This reorganization exacerbated cleavages in the associational politics of Ontario growers. Horticulturalists complained that the Fruit Growers' prioritized the interests of commercial growers, as evidenced by a 1898 lecture to the Hamilton Horticultural Society wherein L. Woolverton of the Ontario Fruit Growers'

12.1
The Canadian Horticulturist 4, no. 2 (February 1881): i and 26–8. In the early 1880s colour plates were introduced to *The Canadian Horticulturist* magazine. In this print of Autumn Berries, the anonymous artist has included berries known as Climbing Bitter-sweet, Dogwood, European Barberry, Whorled Winterberry (Black Alder), Virginia Creeper, European Spindle Tree, Snowberry, and American Holly, each of which was explained to readers in an accompanying article.

Association spoke against growing apples in the "City Fruit Garden" and thus exposed a protectionist sentiment that would favour local orchards over smaller, domestic producers.[23] In 1906, H.B. Cowan, the superintendent and secretary treasurer of the newly formed OHA took aim at agricultural societies for doing so little to further their mandated government-funded activities, especially public education.[24] Meanwhile, the Fruit Growers' Association sought to retain control over its monthly publication with its profitable membership base, fending off competition not only from the newly formed OHA but also from American publications. *The Canadian Horticulturist*'s nationalist ambitions were realized in 1905 when the journal became the official voice of not only Ontario fruit and vegetable growers but also those from British Columbia, Quebec, and Prince Edward Island. In 1906 correspondents from other provinces were included, all with the goal of publishing a paper "of which Canadian fruit, vegetable and flower growers may have reason to feel proud."[25]

Like the provincial agricultural organizations and the Fruit Growers' Association, the 1906 Ontario Horticultural Association's inaugural roster of officers was all male, but the gendered composition of societal leadership did not mean that women were not playing important roles in the province's societies. The presence and activity of women in the agricultural and horticultural realm during the nineteenth century is evident from the prize given in the category "Gardening for Ladies" at the 1846 Provincial Fair in Toronto, precursor to the Canadian National Exhibition. Female participation was a regular part of local and regional horticultural and agricultural competitions and displays throughout the nineteenth and twentieth centuries.[26] While comprehensive membership lists for the Ontario Fruit Growers' Association are not available, its *Canadian Horticulturist* did publish names of new subscribers. For example, the February 1886 issue listed fifty new subscriptions received in December 1885, of which seven named were women – five listed as "Mrs" and two as "Miss."[27] Other women may also have been organizing subscriptions for their household, under their husband's or father's or brothers' names.[28] Meanwhile, the number of women participating in the rival Ontario Horticulture Association grew from one in the 1906 inaugural meeting to "many" in 1907.[29] Women were reported to have been 30 per cent of the audience at a 1913 OHA meeting, at which Ada (aka "Mrs. R.B.") Potts, secretary of the Hamilton society, advocated introducing gardening into the school curriculum as "better than calisthenics."[30] By 1921 the OHA bragged that "unlike most provincial organizations … its membership is composed almost equally of women and men." That year a Miss Yates was elected president.[31]

Women's involvement in horticulture was often legitimized through claims to ideals of self-improvement and personal fulfillment. For example, in its inaugural year, *The Canadian Horticulturist* published an article entitled "Woman's Work in Horticulture" by Mrs H.M. Lewis of Madison, Wisconsin, reprinted from the transactions of the Wisconsin State Horticultural Society. Lewis began the article by clarifying that she was "not an advocate of woman's rights in the full sense of the word" but nonetheless exhorted women "who feel in their hearts that they ought to have some work to do … [to] … take up the work of horticulture." Lewis proclaimed, "Horticulture promotes health, furnishes appetizing and invigorating food, is a most delightful means of recreation, cultivates a refined taste, induces a spirit of cheerfulness, and awakens a sympathy with nature and a love for all the Creator's works."[32]

Even as Lewis justified women's interest in developing knowledge of plants in personal, cultural, and religious terms, she also admitted that economic and professional ambition motivated many female members. Lewis recognized

12.2
"Members of the Canadian Horticultural Association Out for a Good Time," *The Canadian Horticulturist* 28, no. 9 (September 1905): 332. This 1905 photograph of some delegates to the annual convention of the Canadian Horticultural Association in Montreal includes ten or eleven women among the membership of "professional florists and gardeners."

that "some women have a natural talent for fruit growing and market gardening" and claimed she could "see no reason why [women] should not succeed as professional florists," earning as much as their male peers.[33] Many did. A 1905 photo titled "Members of the Canadian Horticultural Association Out for a Good Time" included eleven women who were among the "professional florists and gardeners" attending the annual conference.[34]

The career of Mary Eliza Blacklock (1860–1956) illustrates how some women pursued the commercial possibilities of horticulture. An active member of the Ontario Horticultural Society, in 1910 "Miss Blacklock" was the first "lady on the program," speaking about "Some Gardens and Gardeners in the Old Land" (UK).[35] In 1914, Blacklock established a commercial garden centre, called Rowancroft Gardens, on five acres of land on the banks of the Credit River, near Meadowvale, Mississauga. With the assistance of Minerva Swan Castle (who would later serve as Meadowvale's inaugural librarian), Blacklock introduced many varieties of peonies, lilacs, irises, and lilies and was also a

12.3
Mary Eliza Blacklock (1860–1956) had a lengthy career in horticulture. She founded the nursery and garden centre Rowencroft Gardens in Meadowvale, Ontario.

renowned plant breeder, winning medals not only in the Canadian horticultural society competitions but also in England.[36] Interestingly, as a young girl, Mary developed her business instincts in her parents' garden on Dovercourt Road in Toronto, raising money for the building fund of St Anne's Anglican Church. She was an avid painter of flowers, and copies of her corpus of work have recently been acquired by the Royal Botanical Gardens (Canada).[37] Blacklock's career exemplifies the ways in which the pursuits of some women horticulturalists spanned art, volunteer associations, and commercial success.

Expertise with plants and training in horticulture as a science-based field also animated the careers of women like Lorrie Dunington (1877–1945). Dunington spent her childhood moving around the British Empire before returning to her native England to attend Swanley Horticultural College in Kent.[38] Upon graduation, Dunington became involved in the English garden design movement and then sought the training needed to move into landscape architecture. Her marriage to landscape architect Howard Grubb provoked the pair to change their names (now the Dunington-Grubbs) and location. Migrating to North America, the Dunington-Grubbs landed in Toronto where they established a landscape architectural firm and then in 1913 Sheridan Nurseries.[39] Her contemporary observers did not always appreciate Lorrie Dunington-Grubb's contributions; a later president of Sheridan Nurseries acknowledged that while "Mrs. Grubb was every bit as smart and an accomplished garden architect" she "perhaps did not get so much of the limelight as her famous husband."[40]

A contemporary of Dunington's, Isabella Preston (1881–1965) also parlayed her interest in horticultural into an impressive career. Born in England in 1881, Preston's parents were avid gardeners and Preston enrolled in Swanley Horticultural College, completing a "short course" which, according to her biographer Edwinna von Baeyer, constituted "nearly her entire formal education in horticulture."[41] After the death of her parents, Preston followed her sister to Guelph, Ontario, where her sister had secured a position as a music teacher in a girls' school. Preston worked picking fruit before enrolling in 1912 at the Ontario Agricultural College (OAC). OAC did not offer degrees to women and, according to von Baeyer, Preston was not "taken seriously by many of her fellow students." Preston's enrolment seems to have lapsed within the year, but she remained on as a technician of sorts, overseeing an experiment that Prof. J.W. Crow was running. Preston learned grafting and hybrid techniques from Crow, and by 1917 her work producing a new hybrid lily had earned her an international reputation. In 1920 Preston secured a position in the Division of Horticulture within the federal government's Department of Agriculture, but it took two years for her position to be made permanent.[42]

THE DISRUPTIVE WORK OF WRITING

For both Dunington and Preston, early-life interest in gardening and their training at a horticultural college led to important professional careers. Both women also used writing to cement their expert status. Preston's extensive publication record was perhaps not surprising given her professional appointment with the federal government; throughout the 1920s and 1930s Preston penned numerous technical notes about specific hybrids and her hybridization research. Dunington too was a prolific writer and speaker. She co-authored at least one technical report with her husband and business partner Howard and sole-authored an early contribution to the *Journal of the Society of Architects*.[43]

In addition to these contributions to professional and scientific publications, both women also wrote for popular and domestic audiences – Preston contributed to *The Canadian Horticulturist* and Dunington became well known through her many submissions about botanical gardens and gardening to journals like *Maclean's* as well as those aimed at female consumers such as *Women's Century* and *Canadian Homes and Gardens.*[44]

Preston and Dunington were not alone in using the written word to assert their botanical and horticultural expertise. From at least the 1870s, when *The Canadian Horticulturist* was launched, female horticultural enthusiasts used magazines to share their knowledge. Sometimes, it is true, women writers appeared to be writing for audiences of other women – titles such as "Woman's Work in Horticulture" or "Home Gardens and the Homemakers" in the annual report of the Horticultural Societies of Ontario for 1914 seemed to reinforce a "separate sphere" of horticultural activity for women.[45] Yet, other articles penned by women offered technical knowledge about cultivating pansies or growing grapes and were aimed at all readers. Regardless of the intended audience, these women writers inserted themselves into public discussions of scientific knowledge by capitalizing on the new technologies of print culture. In the late nineteenth century, a mass market for affordable books and magazines developed throughout the industrial world. Cheap newsprint paper, steam-powered presses, and telegraph communication combined to make inexpensive publications available at a time when literacy rates were rising with the expansion of mandatory, public education, such as the 1871 Ontario Comprehensive School Act which mandated school for all children.[46] Together, publications and their readers constituted the disruptive technology of their day, and women writing about plants from botanical and horticultural perspectives – like scientists in many other fields – used this disruptive technology to make their knowledge public.[47]

Gaining access to these technologies was not always easy, particularly in book publishing. For example, Catharine Parr Traill and her niece Agnes Fitzgibbon initially struggled to get their large *Canadian Wild Flowers* – which because of its size would, today, be described as a coffee table book – published in 1868. Fitzgibbon had to complete the lithographic drawings herself and essentially crowd-fund the publication of the book because the Montreal printer would not commit to 500 copies unless they were pre-ordered.[48] Demand for the book clearly grew. By 1895 the Toronto publisher William Briggs had printed a fourth edition of the text.[49]

Briggs himself was a leader in transforming the Canadian publishing industry. A Methodist and a pastor of the Toronto Metropolitan Church, William

12.4
Frontispiece from the fourth edition of *Canadian Wild Flowers* by Agnes Fitzgibbon (née Moodie and later Chamberlin) and Catharine Parr Traill, published by William Briggs in Toronto in 1895. The first edition was published by subscription in 1868. John Lovell published the second and third editions in Montreal.

Briggs was a steward of the Methodist Book and Publishing House, which eventually became Ryerson Press. As historian Janet Friskney's important work shows, in the 1890s, changes in North American copyright law combined with the rise of the social gospel and growing Canadian cultural nationalism to foster profound changes in Canada's print culture. Briggs transformed the Methodist Book and Publishing House from one dedicated to religious texts to an enterprise in which a range of nonreligious works – poetry, fiction, and history – were published under his own imprint. Not only did Briggs champion women botanical authors and illustrators, he also published the early works of Canada's foremost feminist author, Nellie McClung.[50]

Briggs was evidently one of several Canadian publishers responding to the demand for affordable botanical trade books. In 1895, the *Montreal Star* published a weekly portfolio of *Wild Flowers of Canada* "by special artists and botanists and Endorsed by University Botanists of both Continents" that was so popular that the newspaper was forced to publish a special notice: "The

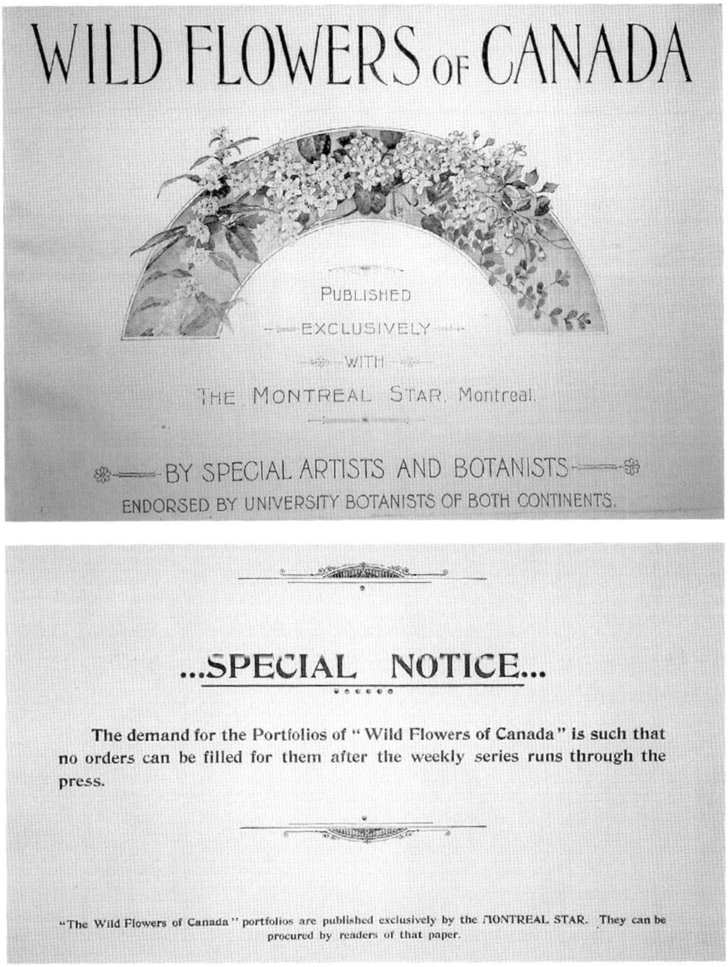
WILD FLOWERS OF CANADA

PUBLISHED
EXCLUSIVELY
WITH
THE MONTREAL STAR, Montreal.

BY SPECIAL ARTISTS AND BOTANISTS
ENDORSED BY UNIVERSITY BOTANISTS OF BOTH CONTINENTS.

...SPECIAL NOTICE...

The demand for the Portfolios of "Wild Flowers of Canada" is such that no orders can be filled for them after the weekly series runs through the press.

"The Wild Flowers of Canada" portfolios are published exclusively by the MONTREAL STAR. They can be procured by readers of that paper.

12.5
"The Wild Flowers of Canada," *Montreal Star*, 1895. The *Montreal Star* enlisted "special artists and botanists" to illustrate its insert, "The Wild Flowers of Canada." The insert was so popular that the *Star* had to tell readers that extra copies were not available.

demand for Portfolios … is such that no orders can be fulfilled for them after the weekly series runs through the press."[51]

We do not know how many women may have been involved in this undertaking, one that was very likely inspired by Fitzgibbon and Traill's *Canadian Wild Flowers*; Kimberlie Robert's fine thesis documents the existence of accomplished female botanical illustrators working in Ontario and Quebec in the late nineteenth century.[52]

Canadian Wild Flowers and the *Montreal Star* series were designed for domestic consumption – that is, they would have been enjoyed in readers' homes. Their success reflected a wider market for household books about plants, both wild and domestic. Canadian readers could enjoy botanical publications reprinted from American and British publishers, especially when content was appropriate for local markets. For example, American botanist and author Alice Lounsberry's *A Guide to the Trees*, published in 1900 by Frederic Stokes in the USA, was soon after acquired by William Briggs. This book, illustrated by accomplished Australian botanical artist Ellis Rowan, covered species with ranges from the Carolinas northwards to southwestern Ontario.[53]

In addition to these relatively large, often technically oriented books, Canadian readers also bought smaller, more portable publications aimed at travellers and naturalists going "to the field." This latter category was the precursor to the modern "field guide."[54] Though less-comprehensive than today's versions, the earlier iterations offered accessible ways of identifying plant species in the field and in plain language without need for an herbarium, microscope, and other technical identification tools. They were published as portable "pocket books," anywhere from six to ten inches tall and six to eight inches wide. Women authors took up this genre. American Maude Gridley Peterson's *How to Know Wild Fruits* (1905) is considered to be a classic work in this sub-genre.[55] Canadian readers often consulted these American field guides, especially those that covered the shared biome. Alice Lounsberry and Ellis Rowan wrote *A Guide to the Wild Flowers* (1899), for which the fourth edition, acquired by Toronto publisher McLelland and Goodchild Ltd, included many Ontario plants. Maud Going wrote *With the Wild Flowers*, published by the Baker and Taylor Company in New York in 1894. A revised edition was published in 1901, and the Royal Botanical Gardens (Canada) holds the copy that was owned and inscribed in 1907 by Milton Musgrove of Prescott, Ontario, near Ottawa.[56]

Canadian authors also wrote for this market. *The Canadian Garden: A Pocket Help for the Amateur* (1903) by Annie Jack went on to a second edition in 1910 and could be ordered as a subscription gift from *The Canadian Horticulturist* for $1.00 as late at 1914.[57] The British-born and American-educated Jack lived her adult life in Quebec, where she and her husband established a successful market garden. Jack's writing career began in the late 1870s, and until her death in 1912 she contributed articles on horticulture to numerous journals and magazines in Quebec, Ontario, and Massachusetts, in addition to penning her hugely successful *The Canadian Garden*.[58]

12.6
Maud Going, *With the Wild Flowers*, 1907. Maud Going first published *With the Wild Flowers* in 1894. This is the cover of the reprinted 1907 edition.

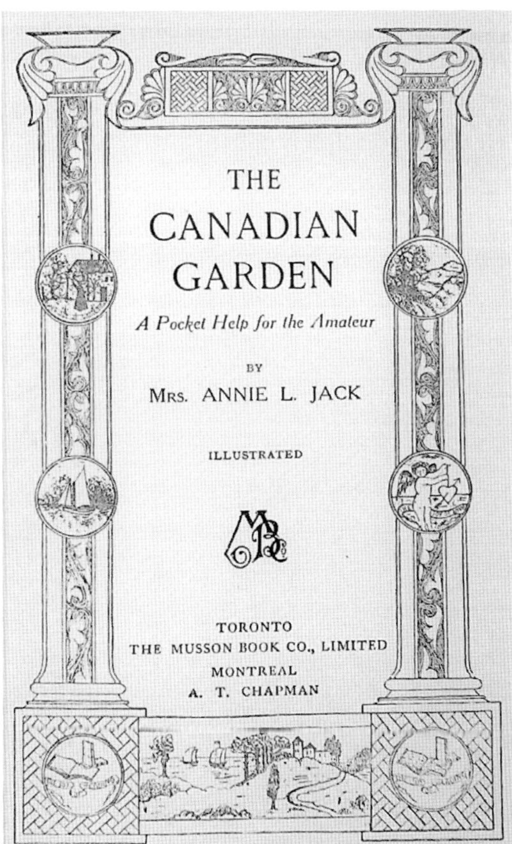

12.7
Front cover of Annie Jack's *The Canadian Garden*, 1910. Mrs Annie Jack published *The Canadian Garden: A Pocket Help for the Amateur* with the Musson Book Company of Toronto as well as A.T. Chapman of Montreal.

Canadian horticultural enthusiasts thus enjoyed a range of publications written by women; the significance of these works was substantiated by the powerful allies Canadian authors enrolled to endorse their work. Philosopher and sociologist of science Bruno Latour has argued that in order to establish their knowledge claims, scientists and engineers create a web of relations, or network, by connecting or enrolling human allies and also material objects. In the late nineteenth and early twentieth centuries, scientists, their publications, and powerful members of society constituted a network that women writers could mobilize to establish their credentials.[59] For instance, Jack and her *The Canadian Garden* enjoyed the patronage of *The Canadian Horticulturist* magazine. The *Montreal Star* invoked testimonials from Lady Aberdeen, wife of the governor general, the Supreme Court justice H.E. Taschereau, famed engineer Sandford Fleming, and even Wilfrid Laurier – who in three years would be elected Canada's prime minister. Alice Lounsberry's work was introduced by Dr N.L. Britton, emeritus professor of botany from Columbia University and also director-in-chief of the New York Botanical Garden. Julia W. Henshaw's 1906 text *Mountain Wildflowers of Canada*, published by William Briggs, included an endorsement from both James Fletcher, Dominion entomologist and botanist, and John Macoun, famed naturalist with the Geological Survey of Canada.

Macoun pronounced that Henshaw's book would "serve a splendid purpose in attracting attention to our grand Canadian mountains" and that "until the tourist had in his hands some such book that would enable him to identify the many flowers that grow there in profusion he must feel lost among the unnamed beauties which would surround him."[60] Building on popular enthusiasm for the natural beauty of the Canadian west, Stewardson Brown secured the patronage of the Canadian Pacific Railway to produce his 1907 *Alpine Flora of the Canadian Rocky Mountains*, for which "Mrs. Charles Schäffer" provided illustrations.

Like Traill and Fitzgibbon, the women who served as authors and illustrators and earned a living from their botanical knowledge had their expertise validated by powerful economic and intellectual allies – allies invested in what historian Suzanne Zeller calls the "conceptual relationship between science and nation-building" in nineteenth-century Canada.[61] This relationship was evident in John Macoun's appraisal of Julia Henshaw's work: "Your choice of English names when such had not before been given to our alpine flowers is excellent … I am glad to note, too, that the scientific names you have used are strictly in accordance with our Canadian nomenclature as indorsed [*sic*] by the Canadian Dominion Government Botanists."[62] Macoun's endorsement

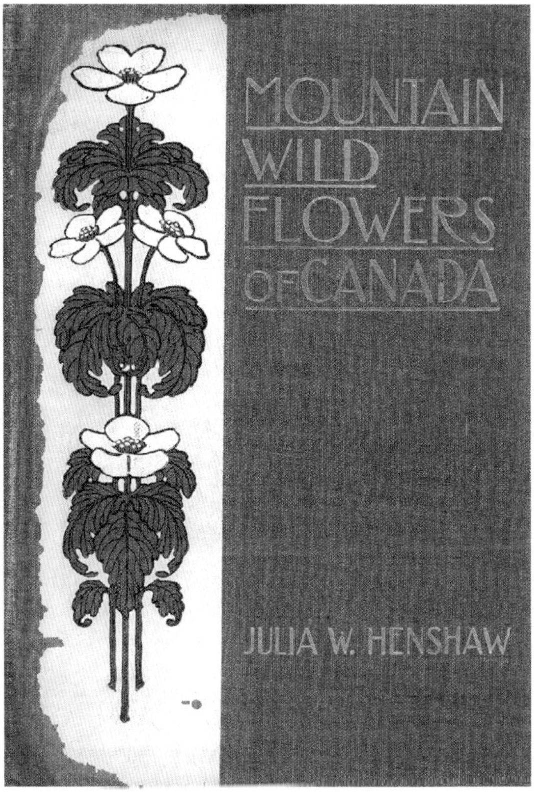

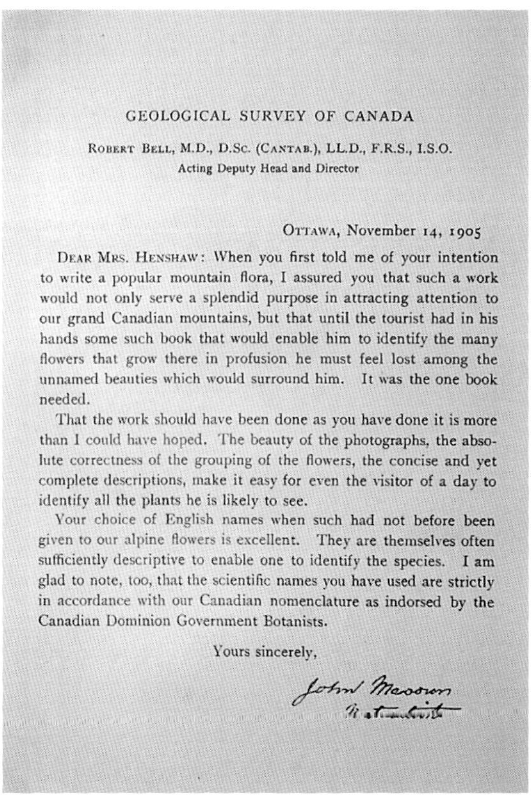

GEOLOGICAL SURVEY OF CANADA

ROBERT BELL, M.D., D.Sc. (CANTAB.), LL.D., F.R.S., I.S.O.
Acting Deputy Head and Director

OTTAWA, November 14, 1905

DEAR MRS. HENSHAW: When you first told me of your intention to write a popular mountain flora, I assured you that such a work would not only serve a splendid purpose in attracting attention to our grand Canadian mountains, but that until the tourist had in his hands some such book that would enable him to identify the many flowers that grow there in profusion he must feel lost among the unnamed beauties which would surround him. It was the one book needed.

That the work should have been done as you have done it is more than I could have hoped. The beauty of the photographs, the absolute correctness of the grouping of the flowers, the concise and yet complete descriptions, make it easy for even the visitor of a day to identify all the plants he is likely to see.

Your choice of English names when such had not before been given to our alpine flowers is excellent. They are themselves often sufficiently descriptive to enable one to identify the species. I am glad to note, too, that the scientific names you have used are strictly in accordance with our Canadian nomenclature as indorsed by the Canadian Dominion Government Botanists.

Yours sincerely,

John Macoun
Naturalist

12.8
Julia W. Henshaw, *Mountain Wild Flowers of Canada*, 1905. Book cover and xv. This book is one of the early flower guides authored by a woman. It was endorsed by John Macoun, the self-educated naturalist who contributed to Sandford Fleming's railway survey of western Canada. Macoun's endorsement came from the desk of Robert Bell, acting deputy head and director of the Geological Survey of Canada.

reveals how the particular brand of late nineteenth and early twentieth-century Anglo-Canadian imperialism – including, of course, territorial expansion and colonization of Indigenous Peoples and their land – helped create cultural space for women botanists to intervene in the production and dissemination of scientific knowledge.[63]

CONCLUSIONS

Of the individuals discussed in this article, only Catharine Parr Traill is widely known in Canada, as much for her autobiographical accounts of early settler-colonial life in Ontario as for her botanical expertise.[64] Yet as the research presented here reveals, many other female writers, illustrators, landscape architects, business owners, flower breeders, and members of voluntary organizations engaged in the production, dissemination, and consumption of knowledge about plants in late nineteenth and early twentieth-century Ontario. These women constituted the citizen scientists of their day, blurring the lines between "amateur" and "professional" at that time and defying the gender limits imposed on them to assert their expertise in public fora.

More can be done to document this remarkable array of female botanical experts. The women presented here represent only a narrow swath of Ontario women who in the years 1870–1920 expressed their interest in botanical and horticultural pursuits. The sources consulted for this paper did not reflect the botanical knowledge of Indigenous women, Franco-Ontarian women, African Canadian women, or the many European immigrants who were likely marginalized from white-dominated and/or English-speaking organizations or businesses.

Knowing more about the women who were, and were not, welcomed into horticultural organizations, business ventures, and the publishing worlds of the 1870–1920 era will support current STEM campaigns to foreground women's contributions to scientific communities. Such research exposes patterns of gender (as well as class, race, ethnicity, and Indigeniety) inclusion *and* exclusion in STEM. History, then, adds depth and complexity to twenty-first century campaigns to democratize science. This analytic influence cuts both ways. Recognizing that digital technologies today – whether the Bioblitz[65] or Wikipedia edit-a-thons, Twitter or plant-apps – disrupt established power relations in STEM and foster communities of citizen scientists prompts historians to ask new questions about the disruptive technologies of the past. As this chapter argues, the disruptive technologies of mass education and the growth of print culture helped create citizen scientists in the late nineteenth and early twentieth centuries. Affordable, mass-produced books and magazines dedicated to describing and depicting plant species and habitats offered a vehicle through which women with botanical interests and knowledge could insert themselves into public discussions of science. Support from powerful male allies also played a role in botanical print culture, as it often does in scientific careers today. Through such research, the value of feminist interdisciplinary research is made evident and suggests the need for greater scholarly collaboration across the STEM/liberal arts divide.[66]

ACKNOWLEDGMENTS

The authors acknowledge the support of the staff of the Library of Royal Botanical Gardens (Canada), especially Erin Ault, Stephanie Bellissimo, and Dr Jim Pringle. We also thank Dr Janet Friskney, York University, Toronto, Ontario, for her guidance in Canadian book history.

NOTES

1 The acronym STEM was coined by the USA National Science Foundation in 2001 to reflect a wide range of related scientific/mathematical/applied disciplines.

2 Science literacy or scientific literacy is generally understood to mean how well a person understands the process of doing science. In 1993, the American Association for the Advancement of Science published the monograph *Benchmarks for Science Literacy* which defines the attributes of the scientifically literate American. American Association for the Advancement of Science, *Benchmarks for Science Literacy* (New York: Oxford University Press, 1993).

3 Mooney and Kirschenbaum, *Unscientific America*; Oreskes and Conway, *Merchants of Doubt*; Otto, *Fool Me Twice*.

4 Baron, *Escape from the Ivory Tower*. Public science is a term that has several different meanings and has been taken up by science and technology studies scholars; see https://www.yorku.ca/gradstudies/sts/.

5 "Citizen Science," *The English Oxford Living Dictionaries*, accessed 18 October 2017. The project "Let's Talk Science" received a grant of more than a million dollars from NSERC to promote citizen science and science literacy in Canada in 2014 (https://www.nserc-crsng.gc.ca/NSERC-CRSNG/Reports-Rapports/DPR RMR/2013-2014/index_eng.asp); see also, Trouille, "Build Your Own Citizen Science

Project with Zooniverse"; and Dickinson, Zuckerberg, and Bonter, "Citizen Science as an Ecological Research Tool," 149–72.

6 Bazely, Perkins, Duailibi, and Klenk, "Strengthening Resilience," 119–32; see also Van Noorden, "Online Collaboration," 126–9.

7 Evans, "The Genius of Twitter"; species-identification platforms include apps like *iNaturalist* or *picturethis*.

8 Both women are often featured in current mainstream media; Foster, "Female Physicist Overlooked for Nobel Prize Finally Receives Recognition as Woman of the Year"; see also Cobb, "Sexism in Science: Did Watson and Crick Really Steal Rosalind Franklin's Data?"

9 International Ada Lovelace Day: https://findingada.com/; "How to run an edit-a-thon": https://en.wikipedia.org/wiki/Wikipedia:How_to_run_an_edit-a-thon.

10 Regarding Canada, see Ainley, *Despite the Odds* and Heap, "Women and Gender in Canadian Science." Sarah Carter's celebrated study *Imperial Plots* explores settler-colonial women's attempts to establish themselves as farmers in the Canadian west, in the face of federal policy which excluded most women from claiming "free" homesteads. Carter also reminds readers of the long tradition of Indigenous women's farming and gardening.

11 Costantino, "Writing Women Back Into the History of STEM."

12 Pollack, *The Only Woman in the Room.*

13 Shteir, *Cultivating Women, Cultivating Science*, esp. ch. 6.

14 Sheffield, *Women and Science.*

15 Peterman, "'Splendid Anachronism.'" See also the chapter by Michael Peterman in this volume.

16 Gillett, "Carrie Derick," 74–87.

17 Huneault, *I am Not Myself at All*, 153. See also Kristina Huneault's chapter in this volume.

18 Dodds and Markle, *The Story of Ontario Horticultural Societies*, 27.

19 Hamilton Centennial Society Editorial Committee, *Centennial Year Book and Garden Guide, 1850–1950.* Hamilton, ON: Hamilton Horticultural Society, 1950, 6.

20 Dodds and Markle, *The Story of Ontario Horticultural Societies*, 27.

21 *The Canadian Horticulturist* 1, no 1, (1878); Dodds and Markle, *The Story of Ontario Horticultural Societies*, 28.

22 Ibid., 28.

23 Hamilton Centennial Society Editorial Committee, *Centennial Year Book and Garden Guide*, 7.

24 Dodds and Markle, *The Story of Ontario Horticultural Societies*, 36–7.

25 "May Success Attend You," *The Canadian Horticulturist* 29, no. 1 (January 1906): 14.

26 Dodds and Markle, *The Story of Ontario Horticultural Societies*, 29.

27 *The Canadian Horticulturist* 9, no. 2 (February 1886): 48.

28 *The Canadian Horticulturist* 2, no. 12 (December 1879): 185.

29 Dodds and Markle, *The Story of Ontario Horticultural Societies*, 36–45.

30 "Ontario Horticulturalists Meet and Confer," *The Canadian Horticulturist* 36, no. 12 (December 1913): 297; Dodds and Markle, *The Story of Ontario Horticultural Societies*, 53. "Lady Woke Delegates Up – The education of children to a proper appreciation of flowers was advocated in a talk by Mrs. R.B. Potts, secretary, Hamilton."

31 "Ontario Horticulturists Plan Forward Moves," *The Canadian Horticulturist* 42, no. 3 (March 1921): 32. Women were also important constituents in nineteenth-century American botanical organizations. Rudolph's analysis of botanical journals and club memberships lists found 1,185 women with an active interest in botany (Rudolf, "Women in Nineteenth Century American Botany," 1346).

32 Lewis, "Woman's Work in Horticulture," 84–7.

33 Ibid. Lewis wrote, "if a woman succeeds as a florist, she can do what she cannot in many other fields; she can command the same price for her produce that a man can, and that is most encouraging," 86.

34 "Members of the Canadian Horticultural Association Out For a Good Time," *The Canadian Horticulturist* 28, no. 9 (September 1905): 332.

35 Dodds and Markle, *The Story of Ontario Horticultural Societies*, 48.

36 "Rowancroft Garden History," http://canadianpeonysociety.blogspot.com/2010/04/rownacroft-garden-history.html.

37 Stensson, "Great Canadians In Horticulture," 24; see also, "Library Shelf," *Pappus* 8, no. 4 (Spring 1989): 6–7; and Hicks, *Meadowvale*, 113.

38 On the history of how Swanley Horticultural College became a woman-only institution, see Opitz, "A Triumph of Brains over Brute."

39 Milovsoroff, "For the Love of Gardens," 101–133.

40 Stensson, "Foreword," 99.

41 von Baeyer, *Rhetoric and Roses*, 126; see also, "Isabella Preston," *Ontario Agricultural College History*, University of Guelph, https://www.uoguelph.ca/oac/140faces/isabella-preston.

42 von Baeyer, *Rhetoric and Roses*. On the integration of women into the federal agricultural services see Lloydlangston, "Women in Botany."

43 Crawford, "Preliminary Bibliography," 155–60, documents that Dunington wrote under both her name at birth and her hyphenated married name.

44 Ibid.

45 Lewis, "Woman's Work in Horticulture"; Potts, "Home Gardens and the Homemaker."

46 On the growth of print technology and the periodical press, see, for example, Hopkin, "Technology and the Periodical Press," 184–97; on the growth of public education see Axelrod, *The Promise of Schooling.*

47 On science and periodical culture, see Cantor and Shuttleworth, *Science Serialized.*

48 Traill and Fitzgibbon, *Canadian Wild Flowers.*

49 Ibid.

50 Friskney, "Beyond the Shadow of William Briggs."

51 *Wild Flowers of Canada, Montreal Star*, 1.

52 Robert, "Women's Botanical Illustration."

53 Lounsberry, *A Guide to the Trees.* Marian Ellis Rowan, https://www.nla.gov.au/selected-library-collections/rowan-collection. See reference to Ellis Rowan in the chapter by Sara Maroske in this volume.

54 According to Dr Jim Pringle, the first truly comprehensive, professional plant field guide for North America was the 1968 Peterson's series *A Field Guide to Wildflowers of Northeastern and North Central North America*, by Roger Tory Peterson and Margaret McKenny. James Pringle, pers. comm.

55 Peterson, *How to Know Wild Fruits*; on Peterson's status in the field see, Palter, "Some Thoughts on Wild Fruit," 254.

56 Going, *With the Wild Flowers.* Going penned other books including *Field, Forest and Wayside Flowers: With Chapters on Grasses, Sedges and Ferns* in 1899 and *Field and Forest Trees* in 1916.

57 Jack, *The Canadian Garden.*

58 Chiasson, "Hayr, Annie Linda."

59 Latour, *Science in Action*, 131.

60 Henshaw, *Mountain Wildflowers of Canada,* xv.

61 Zeller, *Inventing Canada,* 6.

62 Henshaw, *Mountain Wildflowers of Canada*, xv

63 Mary Louise Pratt's path-breaking work *Imperial Eyes* (building on Spivak) argues that travel books by nineteenth-century Europeans created "the 'domestic subject' of Euroimperialism," 4. On women, botany, and imperialism, see for example, Alison Blunt's analysis of Mary Kingsley in West Africa, *Travel, Gender and Imperialism.* On how colonialism informed botanical drawing in nineteenth-century Canada, see Robert, "Women's Botanical Illustration."

64 Traill, *The Backwoods of Canada.*

65 The *Bioblitz* is a popular public science activity, in which a group of taxonomic experts and amateur naturalists, or those interested in learning to identify species, come together to identify as many taxa as possible in a defined area over a specific period of time. See http://bioblitzcanada.ca.

66 Bazely, "What can a biologist learn?"

BIBLIOGRAPHY

Ainley, Marianne Gosztonyi, ed. *Despite the Odds: Essays on Canadian Women and Science.* Montreal: Véhicule Press, 1990.

Axelrod, Paul. *The Promise of Schooling: Education in Canada, 1800–1914.* Toronto: University of Toronto Press, 1997.

Baron, N. *Escape from the Ivory Tower: A Guide to Making Your Science Matter.* Washington: Island Press, 2010.

Bazely, Dawn R. "What can a biologist learn about the science–policy–politics spectrum from working with social scientists?" *CanadaWatch: Practical and Authorative analysis of key national issues.* Toronto, ON: Robarts Centre for Canadian Studies., Fall Issue (2015): 7–9. http://robarts.info.yorku.ca/files/2015/09/CW_Fall2015_FINAL.pdf.

Bazely, Dawn R., P.E. Perkins, M. Duailibi, and N. Klenk. "Strengthening Resilience by thinking of Knowledge as a Nutrient connecting the local person to global thinking: The case of Social Technology/Tecnologia Social." In *Planetary Praxis & Pedagogy: Transdisciplinary Approaches to Environmental Sustainability*, edited by R.C. Mitchell, and S.A. Moore, 119–32. Rotterdam: Sense Publishers, 2015.

Blunt, Alison. *Travel, Gender, and Imperialism: Mary Kingsley and West Africa.* New York: Guilford Press, 1996.

Cantor, Geoffrey, and Sally Shuttleworth, eds. *Science Serialized: Representations of the Sciences in Nineteenth-Century Periodicals.* Cambridge Massachusetts: MIT Press, 2004.

Carter, Sarah. *Imperial Plots: Women, Land, and the Spadework of British Colonialism on the Canadian Prairies.* Winnipeg: University of Manitoba Press, 2016.

Chiasson, Paulette M. "Hayr, Annie Linda." In *Dictionary of Canadian Biography*, vol. 14. Toronto: University of Toronto/Université Laval, 2003.

Cobb, Matthew. "Sexism in Science: Did Watson and Crick Really Steal Rosalind Franklin's Data?" *The Guardian*, 23 June 2015.

Costantino, Grace. "'Writing Women Back into the History of STEM': BHL Supports Research on Women in Science." *BHL: Biodiversity Heritage Library.* https://blog.biodiversitylibrary.org/2020/03/bhl-supports-research-on-women-in-science.html.

Crawford, Pleasance. "Preliminary Bibliography of the Published Writings of Lorrie Alfreda Dunington (1877–1944) and Howard Burlington Grubb (1881–1965)." *Canadian Horticultural History: An Interdisciplinary Journal* 2, no. 3 (1990): 155–60.

Dickinson, J.L., B. Zuckerberg, and D.N. Bonter. "Citizen Science as an Ecological Research Tool: Challenges and Benefits." *Annual Review of Ecology, Evolution and Systematics* 41 (2010): 149–72.

Dodds, P.F., and H.E. Markle. *The Story of Ontario Horticultural Societies, 1854–1973: And Their Contribution to Making the Province a More Beautiful and Better Place in Which to Live.* Hamilton, ON: Ontario Horticultural Association, 1973.

Evans, J. "The Genius of Twitter: A Paean." *TechCrunch*, 2013. https://techcrunch.com/2013/09/28/the-genius-of-twitter-a-paean/.

Foster, Patrick. "Female Physicist Overlooked for Nobel Prize Finally Receives Recognition as Woman of the Year." *The Daily Telegraph*, 19 October 2015. https://www.telegraph.co.uk/news/science/11941453/Female-physicist-overlooked-for-Nobel-Prize-finally-receives-recognition-as-Woman-of-the-Year.html.

Friskney, Janet B. "Beyond the Shadow of William Briggs Pt 2: Canadian-Authored Titles and the Commitment to Canadian Writing." *Papers of the Bibliographical Society of Canada* 35, no. 2 (1997): 161–207.

Gillett, Margaret. "Carrie Derick (1862–1941) and the Chair of Botany at McGill." In *Despite the Odds: Essays on Canadian Women and Science*, edited by Marianne Gosztonyi Ainley, 73–87. Montreal: Véhicule Press, 1990.

Going, Maud. *Field, Forest and Wayside Flowers: With Chapters on Grasses, Sedges and Ferns.* New York: The Baker and Taylor Company, 1899.

– *Our Field and Forest Trees.* Chicago: A.C. McClurg and Company, 1916.

– *With the Wild Flowers.* New York: The Baker and Taylor Company, 1894.

Hamelin, Danielle. "William Briggs." In *Dictionary of Canadian Biography*, vol. 15. Toronto: University of Toronto/Université Laval, 2005–2019.

Heap, Ruby, ed. "Women and Gender in Canadian Science, Engineering and Medicine." *Scientia Canadensis*, special issue 29, no. 2 (2006).

Henshaw, Julia. *Mountain Wildflowers of Canada.* Toronto: William Briggs, 1906.

Hicks, Kathleen A. *Meadowvale: Mills to Millennium.* Mississauga: Friends of the Mississauga Library System, 2004.

Hopkin, Deian. "Technology and the Periodical Press." In *Investigating Victorian Journalism*, edited by Laurel Brake, Aled Jones, and Lionel Madden, 184–97. Basingstoke: Macmillan, 1990.

Huneault, Kristina. *I'm Not Myself at All: Women, Art, and Subjectivity in Canada.* Montreal and Kingston: McGill-Queen's University Press, 2018.

Jack, Mrs Annie L. *The Canadian Garden: A Pocket Help for the Amateur.* Toronto: The Musson Book Co., Limited, 1910.

Latour, Bruno. *Science in Action: How to Follow Scientists and Engineers through Society.* Cambridge, MA: Harvard University Press, 1987.

Lewis, H.M. "Woman's Work in Horticulture." *The Canadian Horticulturist* 1, no. 6 (June 1878): 84–7.

Lloydlangston, Amber. "Women in Botany and the Canadian Federal Department of Agriculture, 1887–1919." *Scientia Canadensis: Canadian Journal of the History of Science, Technology and Medicine/Revue canadienne d'histoire des sciences, des techniques et de la medecine* 29, 2 (2006). DOI: https://doi.org/10.7202/800521ar.

Lounsberry, Alice. *A Guide to the Trees with Sixty-Four Coloured and One Hundred and*

Sixty-Four Black-and White Plates and Fifty-Five Diagrams by Mrs. Ellis Rowan, with an Introduction by Dr. N.L. Britton. Toronto: William Briggs, 1900.

Macoun, John. "Preface, November 14, 1905." In *Mountain Wildflowers of Canada*, by Julia Henshaw, xv. Toronto: William Briggs, 1906.

Milovsoroff, Ann. "For the Love of Gardens: A Biography of H.B. and L.A. Dunington-Grubb." *Canadian Horticultural History: An Interdisciplinary Journal* 2, no. 3 (1990): 101–33.

Mooney, C., and S. Kirschenbaum. *Unscientific America: How Scientific Illiteracy Threatens Our Future*. Philadelphia, PA: Basic Books, 2009.

Opitz, Donald L. "'A Triumph of Brains over Brute': Women and Science at the Horticultural College, Swanley, 1890–1910." *Isis* 104, 1 (2013): 30–62.

Oreskes, N., and E.M. Conway. *Merchants of Doubt: How a Handful of Scientists Obscured the Truth on Issues from Tobacco Smoke to Global Warming*. New York: Bloomsbury Press, 2010.

Otto, Shawn Lawrence. *Fool Me Twice: Fighting the Assault on Science in America*. Emmaus, PA: Rodale, 2011.

Palter, Robert. "Some Thoughts on Wild Fruits." In *Wild Food: Proceedings of the Oxford Symposium on Food and Cookery, 2004*, edited by Richard Hosking, 246–55. Totnes, Devon: Prospect Books, 2006.

Peterson, Roger Tory, and Margaret McKenny. *A Field Guide to Wildflowers of Northeastern and North-Central North America*. Boston: Houghton Mifflin, 1968.

Peterman, Michael. "'Splendid Anachronism': The Record of Catharine Parr Traill's Struggles as an Amateur Botanist in Nineteenth-Century Canada." In *Re(Dis)covering Our Foremothers: Nineteenth-Century Canadian Women Writers*, edited by L. McMullen, 173–86. Ottawa: University of Ottawa Press, 1990.

Peterson, Maude Gridley. *How to Know Wild Fruits: A Guide to Plants When Not in Flower by Means of Fruit and Leaf*. Illustrated by Mary Elisabeth Herbert. New York: The Macmillan Company, 1905.

Pollack, E. *The Only Woman in the Room: Why Science Is Still a Boy's Club*. Boston: Beacon Press, 2015.

Potts, R. B. "Home Gardens and the Homemaker." *Ninth Annual Report of the Horticultural Societies of Ontario for the Year 1914*. Toronto: Ontario Department of Agriculture, 1915.

Powell, Clive. "Canadian Wild Flowers." *The Ryerson Archive*. Accessed 13 June 2017. https://mychangingtimes.wordpress.com/2017/06/13/canadian-wild-flowers/.

Pratt, Mary Louise. *Imperial Eyes: Travel Writing and Transculturation*. London: Routledge, 1992.

Robert, Kimberlie M. "Women's Botanical Illustration in Canada: Its Gendered, Colonial and Garden Histories, 1830–1930." MA thesis, Concordia University, 2008.

Rudolph, Emanuel. "Women in Nineteenth Century American Botany; A Generally Unrecognized Constituency." *American Journal of Botany*, 69 (1982): 1346–55.
Sheffield, Suzanne Le-May. *Women and Science: Social Impact and Interaction*. New Brunswick, NJ: Rutgers University Press, 2006.
Shteir, Ann B. *Cultivating Women, Cultivating Science: Flora's Daughters and Botany in England 1760 to 1860*. Baltimore and London: The John Hopkins University Press, 1996.
Stensson, Howard H. "Foreword." *Canadian Horticultural History: An Interdisciplinary Journal* 2, no. 3 (1990): 99.
Stensson, J. Vilhelm. "Great Canadians in Horticulture." *Canadian Nurseryman Centennial Yearbook*, 1967. Ottawa: Canadian Nursery Landscape Association.
Traill, Catharine Parr. *The Backwoods of Canada: Being Letters from the Wife of an Emigrant Officer*. London: Charles Knight, 1836.
Traill, Catharine Parr, and Agnes Fitzgibbon. *Canadian Wild Flowers*. Montreal: John Lovell, 1868; 4th edition by Ryerson Press, 1895.
Trouille, Laura. "Build Your Own Citizen Science Project with Zooniverse." Accessed 18 October 2017. https://aas.org/posts/news/2017/06/build-your-own-citizen-science-projectzooniverse.
Van Noorden, R. "Online collaboration: Scientists and the Social Network." *Nature* 512 (2014): 126–9.
von Baeyer, Edwinna. "The Horticultural Odyssey of Isabella Preston." *Canadian Horticultural History: An Interdisciplinary Journal* 1, no. 3 (1987): 125–75.
– *Rhetoric and Roses*. Markham, ON: Fitzhenry & Whiteside, 1984.
Zeller, Suzanne. *Inventing Canada: Early Victorian Science and the Idea of a Transcontinental Nation*. Kingston and Montreal: McGill-Queen's University Press, 1987.

Afterword

Finding Meaning in the Understory

SUZANNE ZELLER

> The tree is saying things, in words before words … *If your mind were only a slightly greener thing, we'd drown you in meaning.*
>
> – Richard Powers, *The Overstory* (2018)

With thanks to Ann (Rusty) Shteir for her kind invitation – first to participate in the workshop that inspired this welcome collection of scholarly essays and then to offer some afterthoughts – I dedicate these comments to the memory of Professor Marianne Ainley (1937–2008). Marika's pioneering enthusiasm for the history of women in science, including the theme of British North American women and plants as they intertwined in empire,[1] opened doors for many others to situate their scholarly pursuits at the intersections of science and gender, not only in Canada, but also farther afield. She would undoubtedly have joined me in appreciating the range of both scholarly expertise and creative research so clearly evident in these essays, since trans-/multi-/interdisciplinary and collaborative approaches now seem indispensable to any well-rounded elucidation of science in culture. She would also have approved the gentle language of sympathy that graces these noise-cancelling efforts to recover the historical voices of botany's gendered – and all too often anonymous – understory.

The special interest of these authors in understanding "Who shaped access to knowledge of plants in nineteenth-century Canada" rightly emphasizes its crucial underpinnings of class as well as gender. These key social categories, of course, represent analytical variables more than constants, their historical roles and values modulating as science too shifted its assumptions, priorities, approaches, and methods over time. Ongoing quests to identify "a collectivity that has not been sketched thus far," as Professor Shteir eloquently explains, of "women engaged panoramically with plants across provinces, colonies, and territories, across decades and across boundaries of language and culture" may ultimately also reveal multiple timebound collectivities, separated (and perhaps even differentiated) by changing cultural contexts and physical environments, their members in most cases mirroring the larger social structures that inexorably shaped even their own expectations. Such historical strictures make exceptions to the rule all the more noteworthy and call for explanations.

Consider, for example, the plant work performed by the women in these essays when we reorder them chronologically, in terms of the so-called *Ways of Knowing* that constitute John V. Pickstone's insightful *New History of Science, Technology, and Medicine*. Pickstone's natural historical, analytical, and experimentalist ways of knowing succeeded one another always in that order, each in this historical sequence *dis*placing – but explicitly not *re*placing – its predecessor to represent a dominant set of outlooks and practices that determined, for a time, at least, who participated – and how and why they did so – in the scientific projects that directed and lent meaning to their activities.[2]

I. NATURAL HISTORICAL

Pickstone defined as natural historical ways of knowing practices that dominated during much of the extended period from the Renaissance through the Enlightenment (1500–1800), as Europe's New World intrusions prompted the encounter and possession of nature by collecting, describing, classifying, and displaying specimens that soon shattered the long-standing confines of classical natural knowledge. As botany and botanical gardens claimed their place in "colonial machines" constructed to process overwhelming amounts of incoming information, so too were colonists and various visitors engaged to fill gaps in Europe's rapidly expanding knowledge system. While the British colonial apparatus in what is now Canada appears at first glance more loosely organized after 1763 than that imposed by the French bureaucracy from 1663,[3] it nevertheless found a powerful centripetal force in the Baconian philosophical revival that persisted in the culture of British science through much of the nineteenth century. Baconianism promised to improve the quality of life for all, emphasizing inductive reasoning based on direct observations of the natural world in a hierarchically organized humanist project to which all, in turn, could contribute. This Baconian philosophy gendered even the project itself, promoting a dynamic male science to interrogate a passive female "nature and all her children" in humanity's service. More broadly, a British culture of natural theology, fortified by a deeply rooted popular amateur naturalist tradition, lent Baconian projects firm social foundations.[4]

Seen in this light, the botanical collectors Lady Dalhousie, Anne Mary Perceval, Harriet Sheppard, Catharine Parr Traill, and Mary Brenton during the 1820s and '30s may have had more in common with Catherine Jérémie in the 1720s – and even with Marie de l'Incarnation in the 1660s – in New France than they did with Alice Hollingworth and her contemporaries in the later nineteenth century. Whether as settlers or as sojourners, the earlier generations anchored their natural historical ways of knowing in the formal military, political, and/or religious institutions that structured their colonial lives and, for that matter, furnished them with their botanical vantage points on the New World scene.[5] Herself a botanizing satellite orbiting Queen Charlotte in Britain, Dalhousie soon attracted kindred spirits to her own colonial social circles, first in Halifax, then at Quebec, and ultimately in India. As the three papers by Deborah Reid, Virginia Vandenberg, and Jacques Cayouette and Faye-Yin Khoo confirm, Dalhousie's personal botanical filter, an aesthetic focus on North American ornamentals (including orchids, rhododendrons, lilies, and ferns), was intended first and foremost to enrich her extensive gardens at home in Scotland. Her reading of John Milton furthermore suggests a Calvinist tendency to imbue "natural" landscapes with ideal Edenic qualities deemed closer to heaven than to human-inflicted corruption.[6] These personal passions then ignited in Lady Dalhousie a collector's ambition fuelled by (Sir) William Jackson Hooker (1785–1865) at Glasgow from his elevated perch on the Baconian inductive ladder. Deferential appeals (and flattering responses) accordingly characterized their exchanges: even Charles Darwin employed these strategies to ground *The Origin of Species* in his critics' Baconian expectations.[7]

Hooker had served as regius professor of botany at Glasgow University since 1820, his geographical focus on plants especially in northern latitudes sharpened after a shipboard fire destroyed all of the specimens laboriously assembled during his botanical survey of Iceland in 1809. In Paris in 1814 he had visited the Prussian scientific traveller, philosopher, and global networker *par excellence* Alexander von Humboldt (1769–1859), who entrusted to Hooker the task of working up the non-flowering collections from Humboldt's famous voyage to South America a decade earlier. Like the influential Humboldt, Hooker proceeded on Baconian principles to cultivate a worldwide web of geographically dispersed correspondents whose very location – like that of the specimens they supplied from the world's plant diaspora – placed their contributions at a premium: Hooker hoped to mitigate his collectors' lack of specialized scientific training and experience by sending them books and instructions.[8]

In his case, the British Empire's global reaches defined a botanical playground as wide as the world itself, including the *terra incognita* (to Europeans, at least) of the Hudson's Bay Company's vast northwestern territories, where fur traders, for their part, valued opportunities to connect usefully and meaningfully to their distant homeland.[9] In the formal colonies, it made sense for Hooker to appeal for collaborators through government officers, whose interested contacts included women in administrators' families or close social circles who in turn recruited others at greater social distances. If overcoming some of the considerable research challenges involved has allowed these essays finally to recover important aspects of such collectors' contributions, there may be hope yet of learning still more about those subcontracted by Mary Brenton at Hooker's expense, or even about Traill's Indigenous informants. Yet even in the inner circles considered here, questions persist: what, for example, did it mean for Lady Dalhousie to demur when Hooker offered to publish her findings in his *Journal of Botany*?

II. ANALYTICAL

Deborah Reid's and David Galbraith's respective accounts of the Dalhousies' ill-fated proposal during the 1820s to establish a King's Garden on Île Ste-Hélène, with its noteworthy attention to the value of Indigenous knowledge, speaks volumes about the time it took for Pickstone's analytical ways of knowing to displace the natural historical in British North America. They began to do so, formally, at least, after the end of the Napoleonic Wars in 1815 pushed military officers – highly trained in science – to ward off the spectre of superannuation through peacetime deployments of their analytical skills in subduing nature by coordinating, synchronizing, graphing, charting, and mapping through isolines newly standardized observations intended to reveal patterns of geographical distribution and the natural laws behind them.

By far the most influential analytical model, the work of Alexander von Humboldt, inspired, among other Humboldtian projects, the British "Magnetic Crusade" that established Toronto's Imperial Magnetic and Meteorological Observatory (1839) as a link in a worldwide chain and the Toronto Observatory's grammar school meteorological stations (1850s) as a proud provincial undertaking.[10] Yet the respective accounts by Virginia Vandenberg,

Ann Shteir, Michael Peterman, David Galbraith, and Karen Stanworth illuminate other inroads: in the geographical data solicited by Hooker from his Canadian collectors during the 1820s; in Mary Brenton's growing interest during the 1830s in problems of plant geographical distribution in Newfoundland, including regional differences and climatic effects; in Catharine Parr Traill's decades-long concern from the 1830s to assess in her adopted backwoods home the environmental impact of "the onward march of civilization";[11] in acclimatization studies presented in Kingston to the short-lived Botanical Society of Canada by Lucy Lawson and Miss Gildersleeve in 1861; and in Isabella McIntosh's application of a modernized Presbyterian-inspired environmental outlook in her hands-on educational approaches to nature study from the 1860s.[12]

When Hooker complained in 1865 that only Canada had not responded to his formal appeals for help with his series of colonial floras (including the multi-volume *Flora Boreali-Americana* published in 1829 and 1840, when he moved to London to direct the Royal Botanic Gardens, Kew), he had already arranged for his son, the botanist and explorer (Sir) Joseph Dalton Hooker (1817–1911), to succeed him. A sympathetic confidant of Charles Darwin's while Darwin was constructing his theory *On the Origin of Species by Means of Natural Selection* (1859), with geographical distribution one of its key building blocks, J.D. Hooker had drawn upon Kew's global collections to reinforce the theory with his own classic "Outlines of the Distribution of Arctic Plants" (1862), culminating in his definitive address to the British Association for the Advancement of Science "On Geographical Distribution" (1881). These Humboldtian – and ultimately post-Darwinian – efforts to "shed light on the natural laws which shaped the plant kingdom" included those of the German botanist Ferdinand von Mueller (1825–1896), whose indispensable firsthand contributions to the *Flora Australiensis* published at Kew (1880–84) in turn comprised, as Sara Maroske shows, a remarkable Australian nationalizing project in the eyes of some 1,800 colonial collectors. Von Mueller and his Australian collaborators thus resisted for their own purposes Hooker's tactics of "encourag[ing] deference" not only from colonial women collectors but also from the men to whom they turned for advice – and whose speculations, Hooker feared, undermined his efforts to conform botanical practice to the analytical ways of knowing that defined scientific expectations in his day.[13]

James Pringle's fascinating essay situates Alice Hollingworth near the zenith of botany's analytical ways of knowing in Canada. After undertaking her own regional flora of Ontario's Muskoka region, assembling a broader range of species during the 1880s than earlier efforts had achieved, Hollingworth joined the Biological Subsection of Toronto's Royal Canadian Institute as an associate member, duly reporting her results in the group's *Biological Review of Ontario*. From there she went on to serve the Royal Society of Canada's Botanical Club (established in 1891) in a quintessentially Humboldtian project to organize a clearing-house for a national "army of explorers" to collect seasonal data in coordinated, increasingly complex phenological observations. The Botanical Club's founder, the same George Lawson (1827–1895) who had launched the Botanical Society of Canada in Kingston some thirty years before, now lamented in the rise of new anatomical and physiological approaches to plant study a critical decline in the numbers of field workers emerging from universities. He sought hope in a new set of potential collaborators, growing numbers of nature studies teachers in schools, many of whom were women, who could – along with their students – be recruited to form the "dedicated band of gleaners" that analytical work continued to require. Lawson's successor at the helm of the Botanical Club of Canada, the Nova Scotian educator A.H. MacKay (1848–1929), steadfastly expanded the project to a national scale until his retirement in 1924, pioneering what he termed "phenochrons" – graphs distilled "inductively" from thousands of standardized questionnaires that ultimately encoded some 200 categories of seasonal floral, faunal, and meteorological markers. The voluminous records of MacKay's extraordinary undertaking, which inspired similar efforts abroad, are preserved in the Nova Scotia Museum of Natural History as a crucial baseline for contemporary studies of climate change.[14]

III. EXPERIMENTALIST

History had placed Hollingworth, like Lawson, on the edge of the biological turn in botany that so repelled Catharine Parr Traill (1802–1899), whose long life destined her to witness all three ways of knowing in succession, as experimentalism now acceded to the cutting edge of scientific practice. The privileging of laboratory research in science from the 1890s, as Marianne Ainley has shown, actively devalued fieldwork until well into the twentieth century.[15] While laboratories soon required women assistants and demonstrators – and the botanist Carrie Derick (1862–1941) at McGill University took this route to become Canada's first female professor[16] – fieldwork in the early twentieth century found itself notoriously relegated, by the likes of the experimental physicist and Nobel Prize laureate Sir Ernest Rutherford, also at McGill, to mere "stamp collecting." In contrast, the influential British cultural critic John

Ruskin (1819–1900) had rejected such signs of "progress" in modern science, resigning his position at Oxford University in 1884 to protest the concomitant practice of vivisection. Ruskin scholars have recognized in his later writings the clear distinction of – and an outspoken preference for – an alternative "feminine" science of looking and listening, over the (quite literal) cut and thrust of "male" scientific practices.[17]

It was, for Ruskin, at least, a losing battle: not until well after 1945 did the faint stirrings of an ecological outlook begin to restore widespread appreciation of both current as well as historical fieldwork – the latter recorded not only in the plant collections that remain for re-analysis but also in the botanical quilts studied here by Vanessa Nicholas – as indispensable to our understanding of nature's complex interconnections over time. Meanwhile, however, as the three essays by Sara Maroske, Kristina Huneault, and Dawn Bazely and Kathryn McPherson point out, from the late nineteenth century women with botanical inclinations found viable outlets in other ways, drawing upon their artistic training – often based in a more traditionally egalitarian agricultural context in Canada – to transform Sophie Pemberton's floral albums into illustrated botanical books and from there into some of the first botanical field guides.

CONCLUSION

Looking back from this side of the experimentalist divide, recent scholarship supports increasingly vigorously the need to continue looking and listening for alternative sources of understanding in science's rich and intriguing understory. The historian Helen Tilley in *Africa as a Living Laboratory* rejects the very idea of "colonial science" as misleading. What really matters, in Tilley's view, is "the dynamic interplay between scientific fieldwork and research across metropolitan and colonial contexts." A decentred imperial science understood as neither a one-way nor a monolithic process of appropriation, she explains, exposes Europe's need for site-specific knowledge as corrosive to the traditional assumptions and practices that shaped its self-image.[18]

Tilley's caveat cuts both ways in these essays, validating Kristina Huneault's theoretical discussion of identity in which a persistent tradition of philosophical idealism privileged the generalized over the individual "other." For while J.D. Hooker's insistence on his imperial prerogative as a "lumper"[19] applied to his women collectors every bit as much as it did to the plant species that he reclassified over the efforts of colonial "splitters," his collectors too

A.1
Sir Joseph Dalton Hooker in the Rhododendron area of the Himalaya, mezzotint by W. Walker, after F. Stone, 1854.

became lumpers vis-à-vis their own anonymous collaborators, in downward-spiralling cycles of what Virginia Vandenberg here cites as nothing short of "epistemic violence." Scholarly efforts, furthermore, to historicize concepts of plant indigeneity remind us, in turn, of our long-standing responsibility – especially in a place like Canada – to recognize and to recover the Indigenous roots of our plant knowledge, where a dynamic cultural exchange also saw Indigenous art patterns adopt the thistle after contact with Scottish fur traders. That race must indeed join gender and class in this discussion is effectively attested by the classic portrait *Sir Joseph Dalton Hooker in the Rhododendron Area of the Himalaya* (1854), reprinted here. [20]

As scientific considerations too return to the place of plants on the human–nonhuman continuum as more than inanimate entities and Pulitzer Prize winning novels showcase individual trees and their communities as literary characters,[21] one cannot help but appreciate the timeliness of this hard-won collection of essays. What makes it worth the risk, as these intrepid researchers have shown, of being ground down among the many moving parts involved in their pursuit of the botanical understory remains – at least on a good day – the tantalizing promise of a movable feast of new insights at the end of the struggle to manage them.

NOTES

1 See, for example, her "Last in the Field?," "Science in Canada's Backwoods," and *Creating Complicated Lives.*

2 Pickstone, *Ways of Knowing*. See esp. ch. 3–6.

3 McClellan and Regourd, "The Colonial Machine." On early modern Europe's stagewise efforts to cope with "knowledge overload," see Heilbron, "History of Science." On the Canadian context, see Suzanne Zeller, "Canada."

4 See Weltman, *Performing the Victorian*, ch. 2. Among the vast literature on Baconianism, see Quinton, *Francis Bacon*, and Yeo, "An Idol of the Market-Place."

5 See Fortin-Morisset, "Jérémie, Lamontagne, Catherine," and Chabot, "Guyart, Marie." Christopher M. Parsons focuses insightfully on plants in both "Wildness without Wilderness" and *A Not-So-New World.*

6 On the Calvinist connection, see Stoll, *Inherit the Holy Mountain*, esp. ch. 1.

7 Darwin supplemented his gentlemanly approach in *The Origin* with a reiteration, quoted as late as the 1880s, that "I worked on true Baconian principles" (*Autobiography*, 40).

8 Fitzgerald, "Hooker, Sir William Jackson."

9 Zeller, "The Spirit of Bacon."

10 On Humboldt's influence in Canada, see Zeller, "Recalibrating Empire"; "Humboldt and the Habitability of Canada's Great Northwest"; and *Inventing Canada*, part 2.

11 On nature as an environmental laboratory, see Zeller, "'Merchants of Light.'"
12 Stoll, ch. 5; see also Armitage, *The Nature Study Movement*, and Allen, *The Victorian Fern Craze.*
13 Cf. Endersby, *Imperial Nature*, 15–19.
14 Zeller, "Reflections on Time and Place."
15 Ainley, "Rowan vs. Tory." See also Zeller, "Reflections on Time and Place."
16 Margaret Gillett, "Carrie Derick."
17 Weltman, *Performing the Victorian*, ch. 2.
18 Tilley, *Africa as a Living Laboratory*, 12–13, ch. 7.
19 Endersby, *Imperial Nature*, 14.
20 See, for example, Cooper, *Inventing the Indigenous*, and Coates, *American Perceptions of Immigrant and Invasive Species. Sir Joseph Dalton Hooker in the Rhododendron Area of the Himalaya*, mezzotint by W. Walker, after F. Stone, 1854. Wellcome Collection. Attribution 4.0 International (CC BY 4.0), https://wellcomecollection.org/works/wt2csv95 (accessed 21 August 2020).
21 Wohlleben, *The Hidden Life of Trees*, and Powers, *The Overstory.*

BIBLIOGRAPHY

Ainley, Marianne Gosztonyi. *Creating Complicated Lives: Women and Science in English-Canadian Universities, 1880–1980*. Edited by Marelene Rayner-Canham and Geoffrey Rayner-Canham. Montreal and Kingston: McGill-Queen's University Press, 2012.

– "Last in the Field? Canadian Women Natural Scientists, 1815–1965." In *Despite the Odds: Essays on Canadian Women and Science*, edited by Marianne Gosztonyi Ainley, 25–62. Montreal: Véhicule Press, 1990.

– "Rowan vs. Tory: Conflicting Views of Scientific Research in Canada, 1920–1935." *Scientia Canadensis* 12, no. 1 (Spring/ Summer 1988): 3–21.

– "Science in Canada's Backwoods: Catharine Parr Traill." In *Natural Eloquence: Women Reinscribe Science*, edited by Barbara T. Gates and Ann B. Shteir, 79–87. Madison: University of Wisconsin Press, 1997.

Allen, David Elliston. *The Victorian Fern Craze.* London: Hutchison, 1969.

Armitage, Kevin C. *The Nature Study Movement.* Lawrence: University Press of Kansas, 2009.

Chabot, Marie-Emmanuel. "Guyart, Marie, named de l'Incarnation (Martin)." In *Dictionary of Canadian Biography*, vol. 1, University of Toronto/ Université Laval, 2003–. Accessed 30 October 2017, http://www.biographi.ca/en/bio/guyart_marie_1E.html.

Coates, Peter. *American Perceptions of Immigrant and Invasive Species: Strangers on the Land.* Berkeley: University of California Press, 2006.

Cooper, Alix. *Inventing the Indigenous: Local Knowledge and Natural History in Early Modern Europe*. Cambridge: Cambridge University Press, 2007.

Darwin, Charles. *Autobiography*. Edited by Francis Darwin. London: John Murray, 1887.

Endersby, Jim. *Imperial Nature: Joseph Hooker and the Practices of Victorian Science.* Chicago: The University of Chicago Press, 2008.

Fitzgerald, Sylvia. "Hooker, Sir William Jackson (1785–1865)." In *Oxford Dictionary of National Biography.* Accessed 10 September, https://doi-org.libproxy.wlu.ca/10.1093/ref:odnb/13699.

Fortin-Morisset, Catherine, "Jérémie, Lamontagne, Catherine." In *Dictionary of Canadian Biography*, vol. 3, University of Toronto/ Université Laval, 2003. Accessed 30 October 2017, http://www.biographi.ca/en/bio/jeremie_catherine_3E.html.

Gillett, Margaret."Carrie Derick (1862–1941) and the Chair of Botany at McGill." In *Despite the Odds: Essays on Canadian Women and Science*, edited by Marianne Gosztonyi Ainley, 74–87. Montreal: Véhicule Press, 1990.

Heilbron, J.L. "History of Science." In *The Oxford Companion to the History of Modern Science*, edited by J.L. Heilbron, 370–4. Oxford: Oxford University Press, 2003.

McClellan, J.E., III, and François Regourd. "The Colonial Machine: French Science and Colonization in the *Ancien Regime*." *Osiris* 15 (2001): 31–50.

Parsons, Christopher M. *A Not-So-New World: Empire and Environment in French Colonial North America*. Philadelphia: University of Pennsylvania Press, 2018.

– "Wildness without Wilderness: Biogeography and Empire in Seventeenth-Century French North America." *Environmental History* 22 (2017): 643–67.

Pickstone, John V. *Ways of Knowing: A New History of Science, Technology, and Medicine*. Chicago: University of Chicago Press, 2000.

Powers, Richard. *The Overstory*. New York: W.W. Norton, 2018.

Quinton, Anthony. *Francis Bacon*. Past Masters. Oxford: Oxford University Press, 1983.

Stoll, Mark R. *Inherit the Holy Mountain: Religion and the Rise of American Environmentalism*. New York: Oxford University Press, 2015.

Tilley, Helen. *Africa as a Living Laboratory: Empire, Development, and the Problem of Scientific Knowledge, 1870–1950*. Chicago: University of Chicago Press, 2011.

Weltmann, Sharon Aronofsky. *Performing the Victorian: John Ruskin and Identity in Theater, Science, and Education*. Columbus: Ohio State University Press, 2007.

Wohlleben, Peter. *The Hidden Life of Trees*. Translated by Jane Billinghurst. Vancouver: Greystone Books, 2016.

Yeo, Richard. "An Idol of the Market-Place: Baconianism in Nineteenth Century Britain." *History of Science* 13 (1985): 251–98.

Zeller, Suzanne. "Canada." In *The Cambridge History of Science*, vol. 8: *Modern Science in National. Transnational, and Global Context*. Edited by Hugh Richard Slotten, Ronald

L. Numbers, and David N. Livingstone, 736–51. Cambridge: Cambridge University Press, 2020.
– "Humboldt and the Habitability of Canada's Great Northwest." *Geographical Review* 96, no. 3 (July 2006): 382–98.
– *Inventing Canada: Early Victorian Science and the Idea of a Transcontinental Nation*, 2nd ed. Montreal and Kingston: McGill-Queen's University Press, 2009.
– "'Merchants of Light': The Culture of Victorian Science in Sir Daniel Wilson's Ontario, 1853–92." In *Thinking with Both Hands: Sir Daniel Wilson in the Old World and the New*, Marinell Ash and Colleagues, edited by Elizabeth Hulse, 115–38. Toronto: University of Toronto Press, 1999.
– "Recalibrating Empire: Humboldtian Climatology in the Reports of the Palliser and Hind Expeditions to British North America's Great North West, 1857–58." In *Alexander von Humboldt and the Americas*, edited by Vera M. Kutzinski, Ottmar Ette, and Laura Dassow Walls, 70–116. Berlin: Verlag Walter Frey, 2012.
– "Reflections on Time and Place: The Nova Scotian Institute of Science in Its First 150 Years." *Proceedings of the Nova Scotian Institute of Science* 48, no. 1 (2015): 5–61.
– "The Spirit of Bacon: Science and Self-Perception in the Hudson's Bay Company, 1830–1870." *Scientia Canadensis* 13, no. 2 (Fall–Winter 1989): 79–101.

Tables and Figures

TABLES

FIGURES

Contributors

DAWN R. BAZELY, professor in the Department of Biology at York University, is a plant ecologist who engages in public science and advocates for equity, diversity, and inclusion in science. Bazely co-edited *Environmental and Human Security in the Arctic* (Routledge-Earthscan, 2014).

JACQUES CAYOUETTE is a research botanist for Agriculture and Agri-food Canada, Ottawa, author of *À la Découverte du Nord* (MultiMondes, 2014), and co-author of *Curieuses histoires de plantes du Canada* (Septentrion, 5 volumes, 2014–2021).

DAVID GALBRAITH is head of science at Royal Botanical Gardens (Canada). His interests include conservation and the ecology of disturbed habitats, bridging Indigenous knowledge and Western science, and cultural heritage landscapes.

KRISTINA HUNEAULT is associate dean, faculty relations and inclusion, in the Faculty of Fine Arts, Concordia University, where she is also a professor of art history. She is a founder of the Canadian Women Artists History Initiative.

FAYE-YIN KHOO currently works in the director general's office at Crown–Indigenous Relations and Northern Affairs Canada. Her interest in the science behind everything includes flora and fauna.

SARA MAROSKE is co-editor of *Historical Records of Australian Science* and an honorary associate at Royal Botanic Gardens Victoria. Her research interests include the history of Australian science and the history of women in science, especially in the natural sciences.

KATHRYN MCPHERSON, professor in the Department of History at York University, is a historian of women, health, and work in Canada. Author of *Bedside Matters: The Transformation of Canadian Nursing, 1900–1990* (University of Toronto Press, 2003), McPherson also publishes on the history of farm women in western Canada.

VANESSA NICHOLAS is a postdoctoral fellow at Concordia University. Her research interest is nineteenth-century visual and material culture, particularly the environmental history of Canadian decorative arts and interiors.

MICHAEL PETERMAN is professor emeritus at Trent University. Beginning research on Catharine Parr Traill in the late 1970s, he persists in that work today. Other research subjects include Robertson Davies, Isabella Valancy Crawford, James McCarroll, and John Craig.

JAMES S. PRINGLE is a plant taxonomist at Royal Botanical Gardens (Canada). He has published many papers on plant taxonomy, especially on the gentian family, as well as on Canadian botanical history.

DEBORAH REID is a garden historian, lecturer, and working gardener with a particular interest in the lives and achievements of Scottish gardening women in the nineteenth and early twentieth centuries.

ANN "RUSTY" SHTEIR is professor emerita in gender, feminist, and women's studies at York University. Author of *Cultivating Women, Cultivating Science*, she writes about historical perspectives on women and nature and about botanical culture.

KAREN STANWORTH is senior scholar at York University. Her recent book, *Visibly Canadian*, examines visual culture and knowledge production in nineteenth-century Canada. She currently researches early museums in Canada.

VIRGINIA VANDENBERG is an independent scholar whose research focuses on British women's intellectual and social practices of science in the early nineteenth century, with a particular focus on botanical work conducted in colonial locations.

SUZANNE ZELLER is professor of history at Wilfrid Laurier University. Her recent publications include the essay on Canada in the *Cambridge History of Science in National, International, and Global Context.*

Index

Page numbers in *italics* denote illustrative material. Page numbers with (t) denote information in a table.